# *Electric Power Systems*

# Electric Power Systems

## Fourth Edition

**B. M. Weedy**
*University of Southampton, UK*

**B. J. Cory**
*Imperial College, London, UK*

**John Wiley & Sons**
Chichester • New York • Weinheim • Brisbane • Singapore • Toronto

*Other Wiley Editorial Offices*

John Wiley & Sons, Inc., 605 Third Avenue,
New York, NY 10158-0012, USA

WILEY-VCH Verlag GmbH, Pappelallee 3,
D-69469 Weinheim, Germany

Jacaranda Wiley Ltd, 33 Park Road, Milton,
Queensland 4064, Australia

John Wiley & Sons (Asia) Pte Ltd, 2 Clementi Loop #02-01,
Jin Xing Distripark, Singapore 129809

John Wiley & Sons (Canada) Ltd, 22 Worcester Road,
Rexdale, Ontario M9W 1L1, Canada

***Library of Congress Cataloging-in-Publication Data***
Weedy, B. M. (Birron Mathew)
    Electric power systems / B. M. Weedy — 4th ed.
        p.   cm.
    Includes bibliographical references (p.   ) and index.
    ISBN 0-471-97677-6 (alk. paper)
    1. Electric power systems.   2. Electric power transmission.
I. Title.
TK1001.W4   1998                           97-44411
621.319´1—dc21                             CIP


***British Library Cataloguing in Publication Data***

A catalogue record for this book is available from the British Library

ISBN 0 471 97677 6

Typeset in 10/12pt Times by Keyword Typesetting Services.
Printed and bound in Great Britain by Biddles Ltd, Guildford and King's Lynn.
This book is printed on acid-free paper responsibly manufactured from sustainable
forestry, for which at least two trees are planted for each one used for paper production.

# Contents

# 4  Control of Power and Frequency    163

# 5.  Control of Voltage and Reactive Power    193

# 6  Load Flows    229

# *Preface to First Edition*

In writing this book the author has been primarily concerned with the presentation of the basic essentials of power-system operation and analysis to students in the final year of first degree courses at Universities and Colleges of Technology. The emphasis is on the consideration of the system as a whole rather than on the engineering details of its constituents, and the treatment presented is aimed at practical conditions and situations rather than theoretical nicety.

In recent years the contents of many undergraduate courses in electrical engineering have become more fundamental in nature with greater emphasis on electromagnetism, network analysis, and control theory. Students with this background will be familiar with much of the work on network theory and the inductance, capacitance, and resistance of lines and cables, which has in the past occupied large parts of textbooks on power supply. In this book these matters have been largely omitted resulting in what is hoped is a concise account of the operation and analysis of electric power systems. It is the author's intention to present the power system as a system of interconnected elements which may be represented by models, either mathematically or by equivalent electrical circuits. The simplest models will be used consistent with acceptable accuracy and it is hoped that this will result in the wood being seen as well as the trees. In an introductory text such as this no apology is made for the absence of sophisticated models of plant (synchronous machines in particular) and involved mathematical treatments as these are well catered for in more advanced texts to which reference is made.

The book is divided into four main parts, as follows:

(a) Introduction, including the establishment of equivalent circuits of the components of the system, the performance of which, when interconnected, forms the main theme.

(b) Operation, the manner in which the system is operated and controlled to give secure and economic power supplies.

(c) Analysis, the calculation of voltage, power, and reactive power in the system under normal and abnormal conditions. The use of computers is emphasized when dealing with large networks.

(d)  Limitations of transmittable power owing to the stability of the synchronous machine, voltage stability of loads, and the temperature rises of plant.

It is hoped that the final chapter will form a useful introduction to direct current transmission which promises to play a more and more important role in electricity supply.

The author would like to express his thanks to colleagues and friends for their helpful criticism and advice. To Mr. J. P. Perkins for reading the complete draft, to Mr. B. A. Carre on digital methods for load flow analysis, and to Mr. A. M. Parker on direct current transmission. Finally, thanks are due to past students who over several years have freely expressed their difficulties in this subject.

**B. M. Weedy**
Southampton, 1967

# Preface to Third Edition

Since the appearance of the second edition the overall energy situation has changed considerably and this has created great interest in alternative sources of energy and energy conservation. Although this does not affect the basic theory and operation of power systems it does influence policies which have considerable impact on electric power supply. Chapter 1 has been enlarged and now includes a critical summary of new energy sources and conservation measures, and in particular their possible impact on the electricity supply industry. In addition, the influences of environmental constraints are included in the discussion of generation and transmission.

One object of the second edition was to provide a text, mainly at undergraduate level, which would cover a wide range of power-system engineering, not merely network analysis. In furthering this aim a new section on overhead line design is now combined with the previous material on underground cables.

A further major change is the bringing together of introductory network material into a new chapter called Basic Concepts. This includes a summary of three-phase theory which it is hoped will ease the transition of students into the practical world of power systems. All chapters have been revised to bring the material up to date and to improve clarity.

**B. M. Weedy**
Southampton, 1978

# Preface to Third, Revised Edition

Overall the third edition has remained up to date since its publication. However, in certain areas significant changes have occurred and to include these a new edition has been prepared. These changes are small in number and comprise decoupled load-flows, a digital method for calculating system transients, and an introduction to state estimation security analysis.

With the new material it is hoped that the book now includes a comprehensive account of power-system engineering at the senior undergraduate level.

**B. M. Weedy**
Southampton, 1987

# Preface to Fourth Edition

As a university teacher for 40 years, I have always admired the way that Dr. Birron Weedy's book has stood out from the numerous texts on the analysis and modelling of power systems, with its emphasis on practical systems rather than extensive theory or mathematics. Over the three previous editions and one revision, the text has been continually updated and honed to provide the essentials of electrical power systems sufficient not only for the final year of a first degree course, but also as a firm foundation for further study. As with all technology, progress produces new devices and understanding requiring revision and updating if a book is to be of continuing value to budding engineers. With power systems, there is another dimension in that changes in social climate and political thinking alter the way they are designed and operated, requiring consideration and understanding of new forms of infrastructure, pricing principles and service provision. Hence the need for an introduction to basic economics and market structures for electricity supply, which is given in a completely new Chapter 12.

In this edition, 10 years on from the last, a rewrite of Chapter 1 has brought in full consideration of CCGT plant, some new possibilities for energy storage, the latest thinking on electromagnetic fields and human health, and loss factor calculations. The major addition to system components and operation has been Flexible a.c. Transmission (FACT) devices using the latest semiconductor power switches and leading to better control of power and var flows. The use of optimization techniques has been brought into Chapter 6 with power-flow calculations but the increasing availability and use of commercial packages has meant that detailed code writing is no longer quite so important. For stability (Chapter 8), it has been necessary to consider voltage collapse as a separate phenomenon requiring further research into modelling of loads at voltages below 95 per cent or so of nominal. Increasingly, large systems require fast stability assessment through energy-like functions as explained in additions made to this chapter. Static-shunt variable compensators have been included in Chapter 9 with a revised look at h.v.d.c. transmission. Many d.c. schemes now exist around the world and are continually being added to so the description of an example scheme has been omitted. Chapter 11 now includes many new sections with updates on switchgear, and comprehensive introductions to digital (numerical) protection principles, monitoring and control with SCADA, state estimation, and the concept of Energy Management Systems (EMS) for system operation.

Readers who have been brought up on previous editions of this work will realize that detailed design of overhead and underground systems and components has been omitted from this edition. Fortunately, adequate textbooks on these topics are available, including an excellent book by Dr. Weedy, and reference to these texts is recommended for detailed study if the principles given in Chapter 3 herein are insufficient. Many other texts (including some 'advanced' ones) are listed in a new organization of the bibliography, together with a chapter-referencing key which I hope will enable the reader to quickly determine the appropriate texts to look up. In addition, mainly for historical purposes, a list of significant or 'milestone' papers and articles is provided for the interested student.

Finally, it has been an honour to be asked to update such a well-known book and I hope that it still retains much of the practical flavour pioneered by Dr. Weedy. I am particularly indebted to my colleagues, Dr. Donald Macdonald (for much help with a rewrite of the material about electrical generators) and Dr. Alun Coonick for his prompting regarding the inclusion of new concepts. My thanks also go to the various reviewers of the previous editions for their helpful suggestions and comments which I have tried to include in this new edition. Any errors and omissions are entirely my responsibility and I look forward to receiving feedback from students and lecturers alike.

**Brian Cory**
Imperial College, London, 1998

## Publisher's Note

Doctor B. M. Weedy sadly died in December 1997 during the production of this fourth edition.

# Symbols

Bold symbols denote phasor or complex quantities

$$
\begin{aligned}
\mathbf{A,B,C,D} &= \text{Generalized circuit constants} \\
\text{a–b–c} &= \text{Phase rotation (alternatively R–Y–B)} \\
a &= \text{Operator } 1\underline{/120^\circ} \\
C &= \text{Capacitance (farad)} \\
c_p &= \text{Specific heat at constant pressure (J/gm per } ^\circ\text{C)} \\
D &= \text{Diameter} \\
\mathbf{E} &= \text{e.m.f. generated} \\
F &= \text{Cost function (units of money per hour)} \\
f &= \text{Frequency (Hz)} \\
G &= \text{Rating of machine} \\
g &= \text{Thermal resistivity (}^\circ\text{C m/w)} \\
H &= \text{Inertia constant (seconds)} \\
h &= \text{Heat transfer coefficient (W/m}^2 \text{ per } ^\circ\text{C)} \\
\mathbf{I} &= \text{Current (A)} \\
I^* &= \text{Conjugate of } I \\
I_d &= \text{In-phase current} \\
I_q &= \text{Quadrature current} \\
j &= 1\underline{/90^\circ} \text{ operator} \\
K &= \text{Stiffness coefficient of a system (MW/Hz)} \\
k &= \text{Thermal conductivity (W/(m}^\circ\text{C))} \\
L &= \text{Inductance (H)} \\
\ln &= \text{Natural logarithm} \\
M &= \text{Angular momentum (J-s per rad or MJ-s per electrical degree)} \\
N &= \text{Rotational speed (rev/min, rev/s, rad/s)} \\
\mathbf{P} &= \text{Propagation constant } (\alpha + j\beta) \\
P &= \text{Power (W)} \\
\frac{dP}{d\delta} &= \text{Synchronizing power coefficient} \\
\text{p.f.} &= \text{Power factor} \\
p &= \text{Iteration number} \\
Q &= \text{Reactive power (VAr)} \\
q &= \text{Loss dissipated as heat (W)} \\
R &= \text{Resistance (}\Omega\text{); also thermal resistance (}^\circ\text{C/W)} \\
\text{R–Y–B} &= \text{Phase rotation (British practice)}
\end{aligned}
$$

$\mathbf{S}$ = Complex power = $P \pm jQ$
S = Siemen
s = Laplace operator
s = Slip
SCR = Short-circuit ratio
$T$ = Absolute temperature (K)
$t$ = Time
$t$ = Off-nominal transformer tap ratio
$\Delta t$ = Interval of time
$\Omega^{-1}$ = Siemen
$U$ = Velocity
$\mathbf{V}$ = Voltage; $\Delta V$ scalar voltage difference
$V$ = Voltage magnitude
$W$ = Volumetric flow of coolant ($m^3/s$)
$X'$ = Transient reactance of a synchronous machine
$X''$ = Subtransient reactance of a synchronous machine
$X_d$ = Direct axis synchronous reactance of a synchronous machine
$X_q$ = Quadrature axis reactance of a synchronous machine
$X_s$ = Synchronous reactance of a synchronous machine
$\mathbf{Y}$ = Admittance (p.u. or $\Omega$)
$\mathbf{Z}$ = Impedance (p.u. or $\Omega$)
$\mathbf{Z}_0$ = Characteristic or surge impedance ($\Omega$)
$\alpha$ = Delay angle in rectifiers and inverters—d.c. transmission
$\alpha$ = Attenuation constant of line
$\alpha$ = Reflection coefficient
$\beta$ = Phase-shift constant of line
$\beta$ = $(180 - \alpha)$—in inverters
$\beta$ = Refraction coefficient $(1 + \alpha)$
$\gamma$ = Commutation angle in converters
$\delta$ = Load angle of synchronous machine or transmission angle across a system—(electrical degrees)
$\delta_0$ = Recovery angle of semiconductor valve
$\varepsilon$ = Permittivity
$\eta$ = Viscosity (g/(cm-s))
$\theta$ = Temperature rise (°C) above reference or ambient
$\lambda$ = Lagrange multiplier
$\rho$ = Electrical resistivity ($\Omega$-m)
$\rho$ = Density ($kg/m^3$)
$\tau$ = Time constant
$\phi$ = Angle between voltage and current phasors (power factor angle)
$\omega$ = Angular frequency (rad/s)

Subscripts 1, 2, and 0 refer to positive, negative, and zero symmetrical components, respectively.

# 1
# *Introduction*

## 1.1 Historical

In 1882 Edison inaugurated the first central generating station in the U.S.A. This had a load of 400 lamps, each consuming 83 W. At about the same time the Holborn Viaduct Generating Station in London was the first in Britain to cater for consumers generally, as opposed to specialized loads. This scheme comprised a 60 kW generator driven by a horizontal steam engine; the voltage of generation was 100 V direct current.

The first major alternating-current station in Great Britain was at Deptford, where power was generated by machines of 10 000 h.p. and transmitted at 10 kV to consumers in London. During this period the battle between the advocates of alternating current and direct current was at its most intense and acrimonious level. During this same period, similar developments were taking place in the U.S.A. and elsewhere. Owing mainly to the invention of the transformer the advocates of alternating current prevailed and a steady development of local electricity generating stations commenced, each large town or load centre operating its own station.

The variation in rate of growth of electricity usage throughout the world and in the U.S.A. is shown in Figure 1.1. World growth continues at about 7 per cent per year, implying a doubling every 10 years. In the U.S.A. and in other industrialized countries there has been a tendency, since 1973, for the rate of increase to slow down, as shown by the trend in Figure 1.1. In the U.K., growth in consumption has been under 2 per cent per year for a number of years.

In 1926, in Britain, an Act of Parliament set up the Central Electricity Board with the object of interconnecting the best of the 500 stations then in operation with a high-voltage network known as the *grid*. In 1948 the British supply industry was nationalized and two organizations were set up: (1) the *area boards*, which are mainly concerned with distribution and consumer service; and (2) the *generating boards*, which are responsible for generation and the operation of the high-voltage transmission network or grid.

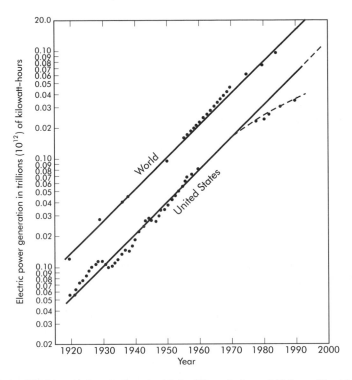

**Figure 1.1** World growth rate for electricity (*Permission of Edison Electric Institute, Historic Studies of the Electric Industry*, 1975. Subsequent updating from *Power Engineering*, Oct. 1996, p. 54)

All of this changed radically in 1989/90 when the British Electricity Supply Industry (ESI) was privatized and *separate* companies were formed in England and Wales to provide competition in power generation (with the prospect of further independent generators entering the market) and to transmit electrical energy at high voltage (H.V.) (the National Grid Company, NGC, in England and Wales), and 12 Regional Electricity Companies (RECs) were formed to distribute and supply energy to consumers. In Scotland, Scottish Power and Hydroelectric companies were allowed to continue generating, transmitting, and distributing energy as 'vertically integrated' companies who could sell power to England and Wales competitively. Because transmission and distribution are recognized as *monopolies*, a Director General of Electricity Supply (the Regulator) was established to fix the profit that the NGC and RECs could earn on their monopoly businesses.

## 1.2 Characteristics Influencing Generation and Transmission

There are three main characteristics of electricity supply that, however obvious, have a profound effect on the manner in which it is engineered. They are as follows:

1. Electricity, unlike gas and water, cannot be stored and the supplier has small control over the load at any time. The control engineers endeavour to keep the output from the generators equal to the connected load at the specified voltage and frequency; the difficulty of this task will be apparent from a study of the daily load curves in Figure 1.2. It will be seen that the load consists basically of a steady component known as the 'base load', plus peaks that depend on the time of day, popular television programmes, and other factors. The effect of an unusual television programme is shown in Figure 1.3.

2. There is a continuous increase in the demand for power, as indicated in Figure 1.1. Although in many industrialized countries the rate of increase has declined in recent years, even the modest rate entails massive additions to the existing systems. A large and continuous process of adding to the system thus exists. Networks are evolved over the years rather than planned in a clear-cut manner and then left untouched. However, with increasing emphasis on the environment, overhead-line rights of way (wayleaves) in industrialized countries are ever more difficult to obtain, thereby forcing existing systems to be used more heavily and efficiently.

3. The distribution and nature of the *fuel* available. This aspect is of great interest as coal is mined in areas that are not necessarily the main load centres; hydroelectric power is usually remote from the large load centres. The problem of station siting and transmission distances is an involved exercise in economics. The development of gas grids enables modern power stations to be sited closer to load centres.

## 1.3 Energy Conversion

### 1.3.1 Steam

The combustion of coal, gas or oil in boilers produces steam, at high temperatures and pressures, which is passed to steam turbines. Oil has economic advantages when it can be pumped from the refinery through pipelines direct to the boilers of the generating station. Gas obtained directly from extraction platforms is becoming very useful. The energy from nuclear fission can also provide energy to produce steam for turbines. The axial-flow type of turbine is in common use with several cylinders on the same shaft.

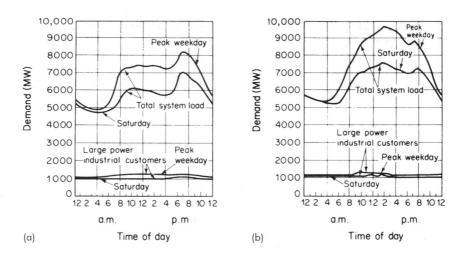

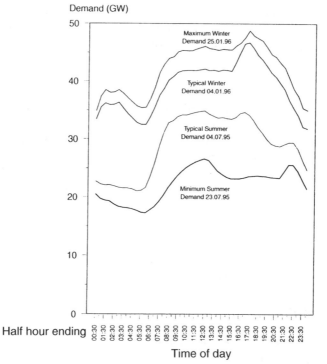

**Figure 1.2** Daily load curves showing industrial component of a U.S. utility. (a) winter, (b) summer (*Copyright © 1976 Institute of Electrical and Electronics Engineers, Inc. Reprinted by permission from I.E.E.E. Spectrum, Vol. 13, No. 9 (Sept. 1976), pp. 50–53*), (c) Actual NGC summer and winter demands for 1995/96 (not weather corrected). Typical curves for England and Wales (*Courtesy of NGC*)

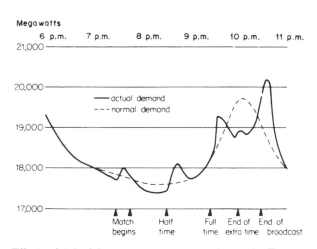

**Figure 1.3** Effect of television programmes on demand—European Cup soccer match televised in Britain on 29 May 1968. Peaks caused by connection of kettles, etc., in intervals and at end (*Permission of Central Electricity Generating Board*)

The steam power-station operates on the Rankine cycle, modified to include superheating, feed-water heating, and steam reheating. Thermal efficiency results from the use of steam at the highest possible pressure and temperature. Also, for turbines to be economically constructed, the larger the size the less the capital cost. As a result, turbogenerator sets of 500 MW and more have been used. With steam turbines of 100 MW capacity and more, the efficiency is increased by reheating the steam, using an external heater, after it has been partially expanded. The reheated steam is then returned to the turbine where it is expanded through the final stages of blading. A schematic diagram of a coal-fired station is shown in Figure 1.4. In Figure 1.5 the flow of energy in a modern steam station is shown. Despite continual advances in the design of boilers and in the development of improved materials, the nature of the steam cycle is such that efficiencies are comparatively low and vast quantities of heat are lost in the condensate and atmosphere. However, the great advances in design and materials in the last few years have increased the thermal efficiencies of coal stations to about 40 per cent.

In coal-fired stations, coal is conveyed to a mill and crushed into fine powder, i.e. pulverized. The pulverized fuel is blown into the boiler where it mixes with a supply of air for combustion. The exhaust from the L.P. turbine is cooled to form condensate by the passage through the condenser of large quantities of sea- or river-water. Where this is not possible, cooling towers are used (see section on thermal pollution).

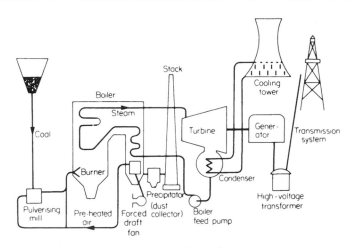

**Figure 1.4** Schematic view of coal-fired generating station

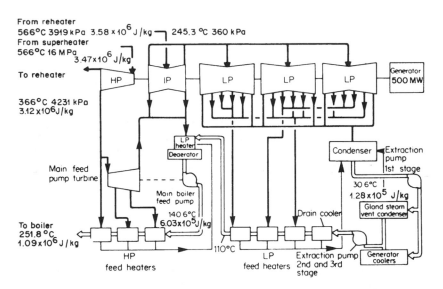

**Figure 1.5** Energy flow diagram for a 500 MW turbogenerator (*Permission of Electrical Review*)

## Fluidized-bed boilers

For typical coals, combustion gases contain 0.2–0.3 per cent sulphur dioxide by volume. If the gas flow rate through the granular bed of a grate-type boiler is increased, the gravity pull is balanced by the upward gas force and the fuel bed takes on the character of a fluid. In a travelling grate this increases the heat output and temperature. The ash formed conglomerates and sinks into the grate and is carried to the ash pit. The bed is limited to the ash-sintering temperature of 1050–1200°C. Secondary combustion occurs above the bed where CO burns to $CO_2$ and $H_2S$ to $SO_2$. This type of boiler is still undergoing extensive development and is attractive because of the lower pollutant level and better efficiency.

## 1.3.2 Energy conversion using water

Perhaps the oldest form of energy conversion is by the use of water power. In the hydroelectric station the energy is obtained free of cost. This attractive feature has always been somewhat offset by the very high capital cost of construction, especially of the civil engineering works. Today, however, the capital cost per kilowatt of small hydroelectric stations is becoming comparable with that of steam stations. Unfortunately, the geographical conditions necessary for hydro-generation are not commonly found, especially in Britain. In most highly developed countries, hydroelectric resources are used to the utmost. There still exists great hydroelectric potential in many underdeveloped countries and this will doubtless be utilized as their load grows, environmental considerations permitting.

An alternative to the conventional use of water energy, i.e. pumped storage, enables water to be used in situations which would not be amenable to conventional schemes. The utilization of the energy in tidal flows in channels has long been the subject of speculation. The technical and economic difficulties are very great and few locations exist where such a scheme would be feasible. An installation using tidal flow has been constructed on the La Rance Estuary in northern France, where the tidal height range is 9.2 m (30 ft) and the tidal flow is estimated at $18\,000\,\mathrm{m}^3/\mathrm{s}$. Proposals for a 8000 MW tidal barrage in the Severn Estuary (U.K.) are still awaiting funding.

Before discussing the types of turbine used, a brief comment on the general modes of operation of hydroelectric stations will be given. The vertical difference between the upper reservoir and the level of the turbines is known as the head. The water falling through this head gains kinetic energy which it then imparts to the turbine blades. There are three main types of installation, as follows:

1. *High head or stored*—the storage area or reservoir normally fills in longer than 400 h;
2. *Medium head or pondage*—the storage fills in 200–400 h;
3. *Run of river*—the storage (if any) fills in less than 2 h and has 3–15 m head.

A schematic diagram for type 3 is shown in Figure 1.6.

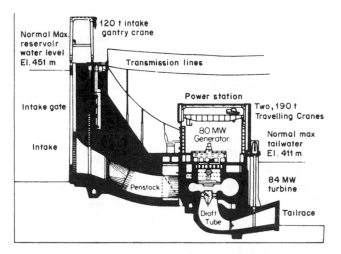

**Figure 1.6** Hydroelectric scheme—Kainji, Nigeria. Section through the intake dam and power house. The scheme comprises an initial four 80 MW Kaplan turbine sets with the later installation of eight more sets. Running speed 115.4 r.p.m. This is a large-flow scheme with penstocks of 9 m diameter (*Permission of Engineering*)

Associated with these various heights or heads of water level above the turbines are particular types of turbine. These are:

1. *Pelton*—This is used for heads of 184–1840 m (600–6000 ft) and consists of a bucket wheel rotor with adjustable flow nozzles.

2. *Francis*—This is used for heads of 37–490 m (120–1600 ft) and is of the mixed flow type.

3. *Kaplan*—This is used for run-of-river and pondage stations with heads of up to 61 m (200 ft). This type has an axial-flow rotor with variable-pitch blades.

Typical efficiency curves for each type of turbine are shown in Figure 1.7. As the efficiency depends upon the head of water, which is continually fluctuating, often water consumption in cubic meters per kilowatt-hour is used and is related to the head of water. Hydroelectric plant has the ability to start up quickly and the advantage that no losses are incurred when at a standstill. It has great advantages, therefore, for power generation to meet peak loads at minimum cost, working in conjunction with thermal stations. By using remote control of the hydro sets, the time from the instruction to start up to the actual connection to the power network can be as short as 3 min.

At certain periods when water availability is low or when generation is not required from hydro sets, it may be advantageous to run the electric machines as motors supplied from the power system. These then act as synchronous

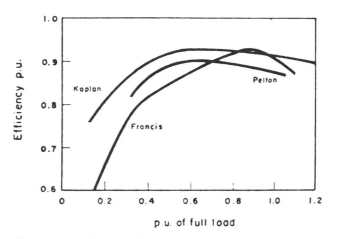

**Figure 1.7**  Typical efficiency curves of hydraulic turbines

compensators, to be discussed in Chapter 5. To reduce the amount of power required, the water is pushed below the turbine runner by compressed air. This is achieved by closing the water inlet valve and injecting compressed air which pushes the water towards the lower reservoir. The runner now rotates in air and thus requires much less motive power than in water.

The power available from a hydro scheme is given by

$$P = \rho g W H \,(\text{W})$$

where

$W$ = flow rate (m³/s) through the turbine;
$\rho$ = density (1000 kg/m³);
$g = 9.81\text{m/s}^2$;
$H$ = head, i.e. height of upper water level above the lower (m).

Substituting,

$$P = 9.81 W H \,(\text{kW})$$

## Tides

An effective method of utilizing the tides is to allow the incoming tide to flow into a basin, thus operating the turbine, and then at low tide to release the stored water, again operating a set of turbines. This gives continuous, if varying, head operation. If the tidal range from high to low water is $h$ (m) and the area of water enclosed in the basin is $A$ (m²), then the energy in the full basin

$$= \rho g A \int_0^h x \, dx$$

$$= \tfrac{1}{2} \rho g h^2 A$$

The total energy for both flows is therefore twice this value, and the average power is $\rho g A h^2 / T$, where $T$ is the period of tidal cycle, normally 12 h 44 min. The number of sites with good potential is small. Typical examples of those which have been studied are listed below, along with values of $h$, $A$, and mean power, respectively.

Passamaquoddy Bay (N. America)     5.5 (m), 262 (km$^2$), 1800 (MW)
Minas-Cohequid (N. America)         10.7, 777, 19 900
San Jose (S. America)               5.9, 750, 5870
Severn (U.K.)                       9.8, 70, 8000

### 1.3.3 Gas turbines

With the increasing availability of natural gas (methane) and its competitive price, prime movers based on the gas turbine as developed for aircraft are being increasingly used. Because of the high temperatures obtained by gas combustion, the efficiency of a gas turbine is comparable to that of a steam turbine, with the additional advantage that there is still sufficient heat in the gas-turbine exhaust to raise steam in a conventional boiler to drive a steam turbine coupled to another electricity generator. This is known as a combined-cycle gas-turbine (CCGT) plant, a schematic layout of which is shown in Figure 1.8. Combined efficiencies now being achieved are between 56 and 58 per cent.

The advantages of CCGT plant are the fast start up and shut down (2–3 min for the gas turbine, 20 min for the steam turbine), the flexibility possible for load following, the comparative speed of installation because of its modular nature and factory-supplied units, and its ability to run on oil (from local storage tanks) if the gas supply is interrupted. Modern installations are fully automated and require only a few operatives to maintain 24 h running or to supply peak load, if needed.

It is possible that with continuing further improvements demanded by aircraft in the high-temperature materials required for the gas turbine, even better efficiencies will be possible in the future. It is noteworthy that up to 10 per cent overload is available for short periods under emergency conditions to aid system recovery following stressed system operation. On the environmental side, CCGT plant with gas firing produces about 55 per cent of the carbon dioxide emission from a similarly rated coal/oil-fired plant. Many new, private operators are installing CCGT units either to produce electricity for sale

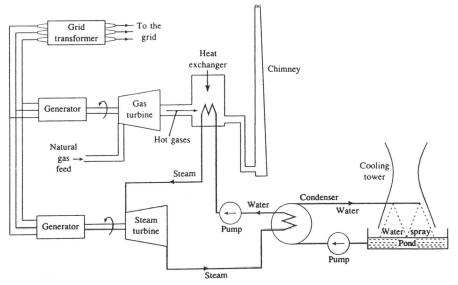

**Figure 1.8** Schematic diagram of a combined-cycle gas-turbine power station (Reproduced by permission of Butterworth/Elsevier)

through the grid or as a combined heat and power (CHP) system for industrial or commercial use.

### 1.3.4 Magnetohydrodynamic (MHD) generation

Whether the fuel used is coal, oil, or nuclear, the result is the production of steam which then drives the turbine. Attempts are being made to generate electricity without the prime mover or rotating generator. In the magnetohydrodynamic method, gases at 2500°C are passed through a chamber in which a strong magnetic field has been created (Figure 1.9). If the gas is hot enough it is electrically slightly conducting (it is seeded with potassium to improve the conductivity) and constitutes a conductor moving in the magnetic field. An electromotive force (e.m.f.) is thus induced which can be collected at suitable electrodes. Nowadays, MHD is not seen as an economically viable option compared with CCGT alternatives.

### 1.3.5 Nuclear power

#### Fission

So far, power has been successfully obtained only from the fission reaction which involves the splitting of a nucleus. Compared with chemical reactions,

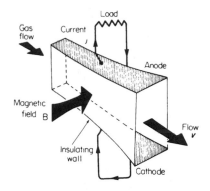

**Figure 1.9**   The principle of MHD power generation (*Permission of English Electric Co. Ltd*)

very large amounts of energy are released per atomic event with both fusion and fission, typically in the range $10–200\,\mathrm{MeV}$ $(15–300 \times 10^{-13}\,\mathrm{J})$. Metal extracted from the base ore consists mainly of two isotopes, uranium-238 (99.3 per cent by weight) and $^{233}\mathrm{U}$ (0.7 per cent). Only $^{235}\mathrm{U}$ is fissile, i.e. when struck by slow-moving neutrons its nucleus splits into two substantial fragments plus several neutrons and $3 \times 10^{-11}\,\mathrm{J}$ of kinetic energy. The fast-moving fragments hit surrounding atoms producing heat before coming to rest. The neutrons travel further, hitting atoms and producing further fissions. Hence the number of neutrons increases, causing, under the correct conditions, a chain reaction. In conventional reactors the core or moderator slows down the moving neutrons to achieve more effective splitting of the nuclei.

Fuels used in reactors have some component of $^{238}\mathrm{U}$. Natural uranium is sometimes used, and although the energy density is considerably less than for the pure isotopes it is still much better than fossil fuels. The uranium used at present comes from metal-rich ores, but these are a limited world resource (about $2 \times 10^{6}$ tons) and the requirement for a reactor which breeds fuel, the breeder reactor, is essential in the long term. The energy breakdown in the fission process is as follows (in MeV): kinetic energy of fission fragments 168, kinetic energy of neutrons 5, gamma radiation 5, beta and gammas emitted by fission products 7 and 6, respectively, and neutrinos 11.

When struck by the neutrons, certain non-fissile materials, if placed around the core of a reactor, are transformed into fissionable material. For example, uranium-238 and thorium-232 are converted into plutonium-239 and uranium-233. When more fissionable material is produced than is consumed the reactor is said to breed.

The basic reactor consists of the fuel in the form of rods or pellets situated in an environment (moderator) which will slow down the neutrons and fission products and in which the heat is evolved. The moderator can be light or heavy water or graphite. Also situated in the moderator are movable rods which absorb neutrons and hence exert control over the fission process. In some reactors the cooling fluid is pumped through channels to absorb the heat, which is then transferred to a secondary loop in which steam is produced for the turbine. In water reactors the moderator itself forms the heat-exchange fluid.

There are a number of versions of the reactor in use with different coolants and types of fissile fuel. In Britain the Magnox reactor has been used, in which natural uranium in the form of rods is enclosed in magnesium-alloy cans. The fuel cans are placed in a structure or core of pure graphite made up of bricks (called the *moderator*). This graphite core slows down the neutrons to the correct range of velocities in order to provide the maximum number of collisions. The fission process is controlled by the insertion of control rods made of neutron-absorbing material; the number and position of these rods controls the heat output of the reactor. Heat is removed from the graphite via carbon dioxide gas pumped through vertical ducts in the core. This heat is then transferred to water to form steam via a heat exchanger. Once the steam has passed through the high-pressure turbine it is returned to the heat exchanger for reheating, as in a coal- or oil-fired boiler. A schematic diagram showing the basic elements of such a reactor is shown in Figure 1.10.

A reactor similar to the Magnox is the *advanced gas-cooled reactor* (AGR). A reinforced-concrete steel-lined pressure vessel contains the reactor and heat exchanger. Enriched uranium dioxide fuel in pellet form, encased in stainless steel cans, is used; a number of cans form a cylindrical fuel element which is placed in a vertical channel in the core. Carbon dioxide gas, at a higher pressure than in the Magnox type, removes the heat. The control rods are made of boron steel. Spent fuel elements when removed from the core are stored in a special chamber for about a week and then dismantled and lowered into a pond of water where they remain until the level of radioactivity has decreased sufficiently for them to be removed from the station.

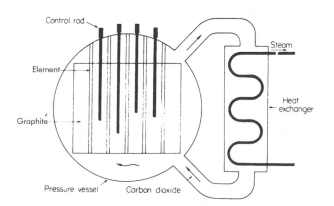

**Figure 1.10** Schematic view, nuclear reactor—British Magnox type

In the U.S.A. and many other countries pressurized-water and boiling-water reactors are used. In the *pressurized-water type* the water is pumped through the reactor and acts as a coolant and moderator, the water being heated to 315°C. The steam pressure is greater than the vapour pressure at this temperature and the water leaves the reactor at below boiling point. The fuel is in the form of pellets of uranium dioxide in bundles of stainless steel tubing. The *boiling-water reactor* was developed later than the pressurized-water type and is now used extensively. Inside the reactor, heat is transferred to boiling water at a pressure of 690 N/cm$^2$. Schematic diagrams of these reactors are shown in Figures 1.11 and 1.12.

All reactors that use uranium produce plutonium in the reaction. Most of this is not utilized in the reactor. *Fast breeder reactors* breed new fuel in considerable quantities during the reaction, as well as producing heat. In the liquid-metal fast breeder reactor, shown in Figure 1.13, liquid sodium is the coolant, which leaves the reactor at 650°C at atmospheric pressure. The heat is then transferred to a secondary sodium circuit which transfers it via a heat exchanger to produce steam at 540°C. The cost and unreliability of these reactors has meant that, to date, they have only been trialled.

Both pressurized- and boiling-water reactors use *light water*. The practical pressure limit for the pressurized-water reactor is about 167 bar (2500 p.s.i.), which limits its efficiency to about 30 per cent. However, the design is relatively straightforward and experience has shown this type of reactor to be stable and

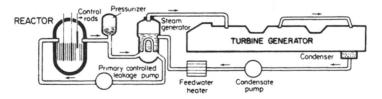

**Figure 1.11**   Schematic diagram of a pressurized-water reactor (PWR) (*Permission of Edison Electric Institute*)

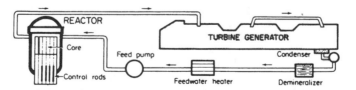

**Figure 1.12**   Schematic diagram of a boiling-water reactor (BWR) (*Permission of Edison Electric Institute*)

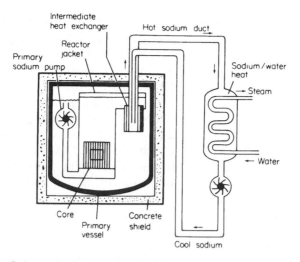

**Figure 1.13** Schematic diagram of a liquid-metal fast breeder reactor (*Permission of Edison Electric Institute*)

dependable. In the boiling-water reactor the efficiency of heat removal is improved by use of the latent heat of evaporation. The steam produced flows directly to the turbine, causing possible problems of radioactivity in the turbine. The fuel for both water reactors is enriched to 1.44 per cent $^{235}$U. These reactors are probably the cheapest to construct; however, the steam produced is saturated and requires wet-steam turbines. A further type of water reactor is the heavy-water CANDU type developed by Canada. Its operation and construction are similar to the light-water variety.

*Safety and environmental considerations*   The translation of energy states in a nucleus creates the emission of $\gamma$- and $\beta$-rays, and $\alpha$ and fission fragments. The half-lives of the substances created are: $^{3}_{1}$H$_2$ (tritium) 12.26 years, $^{90}$Sr (strontium) 28.8 years, $^{137}$Cs (caesium) 30.2 years, $^{131}$I (iodine) 8 days, $^{85}$Kr (krypton) 10.76 years, and $^{133}$Xe (xenon) 5.27 days. Generally, materials with a long half-life have a lower intensity of radioactivity than those with a short half-life. Tritium is produced in small amounts and mostly retained in the fuel. Xenon and krypton escape from fuel elements which have cladding defects and remain in free form in the coolant. Because of its long life krypton-85 constitutes the greater problem. In the water-cooled cores the fission and activation products are present in the coolant. The more active of such wastes are concentrated by evaporation, mixed with concrete, and shipped for storage. The lower level wastes are eventually released to the condenser cooling-water discharge at low concentration levels.

High-level wastes, e.g. strontium-90, are produced in processing the used fuel elements. At the moment, the wastes are concentrated in liquid form and stored in stainless steel containers. The storage of such wastes creates great controversy, the material still being active after centuries. Any mistakes made now will create serious problems for future generations. With future development such long-lived wastes will be converted to solid form (e.g. glass) and stored underground in stable geological situations such as salt domes.

Any accident involving substantial heating and rupture of the structure will involve the release of fission fragments held in the fuel rods into the atmosphere. With a breeder reactor the release of plutonium, an extremely radiotoxic material, would add to the problem. In the design and construction of reactors, great care is taken to cover every contingency. Many facilities, e.g. control systems, are at least duplicated and have alternative electrical supplies.

Over the past few years there has been considerable controversy regarding the safety of reactors. Experience is still relatively small and human error is always a possibility, such as happened at Three Mile Island in 1979 and Chernobyl in 1986. However, the health controls in the atomic power industry have, from the outset, been much more rigorous than in any other industry.

## Fusion

Energy is produced by the combination of two light nuclei to form a single heavier one. Neutron emission is not required, the reaction being sustained by the very high temperature of the reactants which maintain continual collisions. The most promising fuels are isotopes of hydrogen known as deuterium (D) (mass 2) and tritium (T) (mass 3). The product of fusion is the helium isotope (mass 3), hydrogen, neutrons, and heat. As tritium is not a naturally occurring isotope it is produced in the reactor shield by the interaction of the fusion neutrons and the lithium isotope of mass 6. The deuterium–deuterium fusion requires higher temperatures than deuterium–tritium and the latter is more likely to be used initially.

Reserves of lithium have been estimated to be roughly equal to those of fossil fuels. Deuterium, on the other hand, is contained in sea-water of a concentration of about 34 parts per million. The potential energy-resource is therefore vast.

Based on radioactive considerations, the impact of fusion reactors would be much less than with fission reactors. In fission reactors the loss of coolant accident and the 'after-heat' generated after shut-down (fission which continues after full 'shut-down' control action) may lead to vaporization and dispersal of radioactive material. With fusion there is much less power density under these conditions, possibly 1/50th of the equivalent value for a fast breeder reactor. The main source of radioactive waste from fusion reactors would be the structural material which undergoes damage due to radiation and hence requires occasional renewal. This could be recycled after a 50-year period, compared with centuries for strontium-90 and casesium-137 from fission.

Intensive international research is still proceeding to develop materials and a suitable containment method, using either magnetic fields or powerful lasers, to produce the high temperatures ($\approx 8 \times 10^7$ K) and pressures (above 1000 bar) to initiate a fusion reaction. It is unlikely that a successful fusion reactor will be available before 2020.

### 1.3.6 Generation and fuel

With new and more efficient generating sets being brought into operation, there exists a wide range of plant available for use. As previously mentioned, the load consists of a base plus a variable element, depending on the time of day and other factors. Obviously the base load should be supplied by the more efficient plant which then runs 24 h per day, with the remaining load met by the less efficient stations. In addition to the machines supplying the load, a certain proportion of available plant is held in reserve to meet sudden contingencies. A proportion of this reserve must be capable of being brought into operation immediately and hence some machines must be run at, say, 75 per cent of full load to allow for this spare generating capacity, called 'spinning reserve'.

Reserve margins are allowed in the *total* generation plant available, to cope with unavailability of plant due to faults and maintenance. It is common practice to allow a planned margin of about 20 per cent over the annual peak demand. In a power system there is a mix of plants, i.e. hydro, coal, oil, and nuclear, and gas turbine. The optimum mix gives the most economic operation, but this is highly dependent on fuel prices which can fluctuate with time and from region to region. Hence in the U.S. in regions with plentiful coal resources, in the west–north–central and also in certain eastern areas, coal plants would be expected to be predominant. In the south and other areas not endowed with coal, nuclear or gas would tend to be dominant. Of course, hydro would be exploited wherever possible for cost and environmental reasons.

Typical plant and fuel costs for the U.K. and U.S.A. are given in Table 1.1. It should be noted that fuels bought on world markets at spot prices can vary in price, dependent upon political situations, especially in the Middle East and Russia.

## 1.4   Renewable Energy Sources

There is considerable international effort into the development of alternative energy sources to supplement fossil fuels. Many of the 'novel' sources (some of

**Table 1.1**  Typical plant and fuel costs

|                   | Nuclear | Coal    | Oil     | Gas turbine | Wind |
| ----------------- | ------- | ------- | ------- | ----------- | ---- |
| Plant cost £/kW   | 1600    | 800[a]  | 800[a]  | 300         | 1000 |
| 1995 US$/kW       | 2560    | 1300    | 1300    | 480         | 1600 |
| Fuel cost p/kWh   | 0.2     | 1.6     | 3.0     | 2.0         | 0    |
| 1995 c/kWh        | 0.3     | 2.6     | 4.8     | 3.2         | 0    |

[a] Add 20 per cent if flue gas desulphurization is needed.
(Compiled from Power UK, FT Publications, London)

them, in fact, have been in use for centuries!) are manifestations of solar energy, e.g. wind, sea waves, ocean thermal gradients, and photosynthesis. The average incident solar energy received on the earth's surface is about 600 W/m$^2$, but the actual value, of course, varies considerably. In the following section the potentialities of various methods of utilizing this energy will be discussed.

### 1.4.1 Solar energy—thermal conversion

There are two distinct applications: (1) space and water heating on a domestic scale; and (2) central station, large-scale heat collection, used for steam raising to generate electricity; both of these influence power systems. The former affects the load demands and in particular the problem utilities will face in having to provide a sufficient back-up supply to customers who normally would use solar power, but in certain weather conditions would require large amounts of electricity. This involves the provision of the normal amount of utility plant but with much reduced sales of energy.

As the temperature of a solar collecting surface rises it radiates heat (infrared). The energy distributions with wavelength of solar energy and infrared radiation are shown in Figure 1.14. It is possible to design a selective cover plate over the collecting surface such that it would pass nearly all the solar radiation and reflect all the radiated infrared. Selective absorbers consist of a smooth metallic sheet covered with either a thin semiconducting surface or a finely divided metallic powder. The former reflects the infrared and provides a good thermal contact between the hot absorbing layer and cooling fluid. A diagram of a simple collecting system is shown in Figure 1.15.

The energy received by the collector per square meter (net):

$$q = I\alpha\tau - (\varepsilon_F + \varepsilon_B)\sigma(T^4 - T_0^4)$$

where

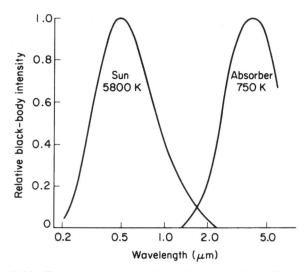

**Figure 1.14** Relative black-body intensity of radiation with wavelength

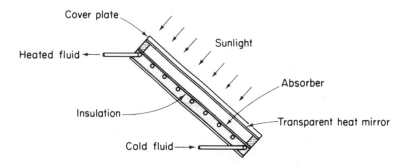

**Figure 1.15** Simple solar energy panel for water heating

$\varepsilon_F$ and $\varepsilon_B$ = front and back emissivities of absorber;

$\sigma$ = Stefan–Boltzmann constant = $5.67 \times 10^{-8} \text{W/m}^2\text{K}^4$;

$\tau$ = transmittance of cover plate (e.g. 0.93);

$T_0$ = temperature of cover plate (K);

$I$ = incident radiation normal to surface;

$T(K)$ and $\alpha$ = temperature and absorptivity of absorbing panel.

In large-scale (central station) installations the sun's rays may be concentrated by lenses or mirrors. Both require accurately curved surfaces and steering mechanisms to follow the motion of the sun. Concentrators may be designed to follow the sun's seasonal movement, or additionally to track the sun throughout the day (double-axis system). The former is less expensive and

concentration factors of 30 have been obtained. However, in the French solar furnace in the Pyrenees, two-axis mirrors are used and a concentration factor of 16 000 is achieved. A diagram of the central receiver system for major generation of electricity is shown in Figure 1.16. The reflectors concentrate the rays on to a single receiver (boiler), hence raising steam. A collector area of 1 km$^2$ for each 100 MW (e) of output has been suggested with capital costs of \$30/m$^2$ (mirrors, etc.) and thermal storage costs of \$15 per kWh of electricity. A less attractive alternative to this scheme (because of the lower temperatures) is the use of many individual absorbers tracking the sun in one direction only, the thermal energy being transferred by a fluid (water or liquid sodium) to a central boiler.

In all solar thermal schemes, storage is essential because of the fluctuating nature of the sun's energy, although it has been proposed that the schemes be used as pure fuel savers. This feature is common to all of the sources discussed, with the exception of geothermal, and constitutes a very serious drawback.

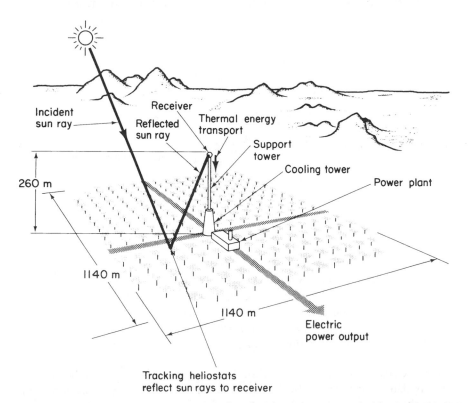

**Figure 1.16**   Central receiver scheme for electric power generated by solar energy (*Copyright © 1975 Institute of Electrical and Electronics Engineers, Inc. Reprinted by permission from I.E.E.E. Spectrum, Vol. 12, No. 12 (Dec. 1975), pp. 47–52*)

Fluctuating sources, as well as fluctuating loads, would complicate still further the process of electricity supply.

## 1.4.2 Solar energy—direct conversion to electricity

Photovoltaic conversion occurs in a thin layer of suitable material, e.g. silicon, when hole–electron pairs are created by incident solar photons and the separation of these holes and electrons at a discontinuity in electrochemical potential creates a potential difference. Whereas theoretical efficiencies are about 25 per cent, practical values are lower. Single-crystal silicon and gallium–arsenide cells have been constructed with efficiencies of 10 and 16 per cent, respectively. The cost of fabricating and interconnecting cells is high (used mainly, to date, in spacecraft). Polycrystalline silicon films having large-area grains (i.e. long continuous crystals) with efficiencies of over 16 per cent have been made by techniques amenable to mass production. Although these devices do not pollute, they will, in the large-power context, occupy large areas. It has been estimated that to produce $10^{12}$ kWh per year (about 65 per cent of the 1970 U.S. generation value) the necessary cells would occupy about 0.1 per cent of the U.S. land area (highways occupied 1.5 per cent in 1975), assuming an efficiency of 10 per cent and a daily insolation of $4\,kWh/m^2$. Automated cell production can now produce cells at around US $5 per watt.

Other forms of conversion of lesser large-scale importance come under the heading of thermoelectricity. The Seebeck effect gives a potential difference between the hot and cold ends of joins of different metals, a typical value being $150\,\mu V/K$. Solar energy can heat a cathode of a diode-type tube from which electrons will be liberated by thermionic emission. These electrons drift to the anode and return through the external circuit. It is doubtful whether these devices will make any impact on the energy situation.

## 1.4.3 Wind generators

Wind power from horizontally mounted generators on 30–50 m high towers is now becoming economically viable. Sizes between 300 and 500 kW driven by two or, more effectively, three-bladed wind turbines are an optimum but larger turbines of 2–3 MW have been built for development purposes. However, the larger towers and blades for higher outputs must be traded against the extra capital costs.

The theoretical power in the wind is given by

$$P = \tfrac{1}{2}\rho A U^3 \text{ (W)}$$

where

$\rho$ = density of air (1.201 kg/m$^3$ at NTP);
$U$ = mean air velocity (m/s);
$A$ = swept area of blades (m$^2$).

The range of operation of a wind turbine depends upon the wind speed and is depicted in Figure 1.17.

At low wind speeds, there is insufficient energy to operate the turbine coupled to the generator and no power is produced. At the 'cut-in' $U_c$ speed, between 3 and 5 m/s on the diagram, power starts to be generated until rated power $P_r$ is produced at rated wind speed U$_r$. After this point, the turbine is controlled, usually by altering the blade angle or 'pitch', to give rated output up to a maximum wind speed U$_f$, after which the blades are 'furled' and the unit is shut down to avoid excessive wind loading. Typically, wind turbines have rotors of 20 m diameter, rotate at 100–150 rpm, and are geared up to about 750 r.p.m. to drive an eight-pole induction generator excited by a 415 V three-phase (3 ph.), 50 Hz rural distribution system. If they are sited in 'windy' areas, normally found on exposed ridges, and can convert nearly half the theoretical power to electrical energy, a good site in the U.K. can produce 1800 kWh per kW of installed capacity per year.

## *Example 1.1*

Calculate the number of wind generators required to produce the equivalent of a 600 MW CCGT operating at 80 per cent load factor. Assume average wind speed is 10 km/h (2.78 m/s), blade diameter is 20 m, and conversion efficiency is 45 per cent.

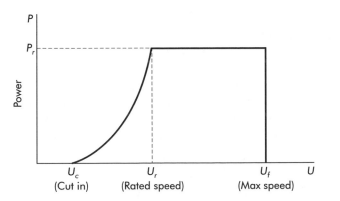

**Figure 1.17** Range of operation of wind turbine (*Courtesy of I.E.E. Power Engineering Journal, Aug. 1995*)

*Calculation*

$$P_{wind} = \tfrac{1}{2} \cdot 1.201 \cdot \pi \left(\frac{20}{2}\right)^2 \cdot 2.78^3 \times 10^{-3} = 4053 \, kW$$

$$P_{generate} = 4053 \times 0.45 = 1823 \, kW$$

$$\therefore \text{No. of wind generators for 600 MW} = \frac{600}{1.83} = 330 \text{ generators}$$

From this calculation, it is apparent that many wind generators spread over a wide area would be required. Although the ground beneath them could be used for grazing, the proliferations and the acoustic noise can be detrimental to the environment. However, the saving in $CO_2$ emissions would be of the order of 12 000 t/day provided that the wind was always blowing!

### 1.4.4 Biofuels

Biofuels are derived from decaying vegetable matter produced by agriculture or forestry operations or from waste materials collected from industry, commerce, and residential households. As an energy resource, biomass used as a source of heat by burning wood, dung, etc., in developing countries is very important and contributes about 14 per cent of the world's energy requirements. Biofuel can be used to produce electricity in two ways:

1. by burning in a furnace to produce steam to drive turbines; or
2. by allowing fermentation in landfill sites or in special anaerobic tanks, both of which produce a methane-rich gas which can fuel a spark ignition engine or gas turbine.

It is interesting to note that if crops are cultivated for combustion, either as a primary source of heat or as a by-product of some other operation, they can be considered as $CO_2$ neutral, in that their growing cycle absorbs as much $CO_2$ as is produced by their combustion. In industrialized countries, biofuels have the potential to produce up to 5 per cent of electricity requirements if all possible forms are exploited, including household and industrial waste, sewerage sludge (for digestion), agricultural waste (chicken litter, straw, sugar cane, etc.).

## 1.4.5 Geothermal energy

In most parts of the world the vast amount of heat in the earth's interior is too deep to be tapped. In some areas, however, hot springs or geysers and molten lava streams are close enough to the surface to be used. Thermal energy from hot springs has been used for many years for producing electricity, starting in 1904 in Italy. In the U.S.A. the major geothermal power plants are located in northan California on a natural steam field called the Geysers. Steam from a number of wells is passed through turbines. The present utilization is about 500 MW and the total estimated capacity is about 2000 MW. Because of the lower pressure and temperatures the efficiency is less than with fossil-fuelled plants, but the capital costs are less and, of course, the fuel is free. New Zealand and Iceland also exploit their geothermal energy resources.

The Geysers in the U.S.A. represent a dry steam field which is preferable for power generation via steam turbines. Other basic types of geothermal energy reservoirs are: hot water, hot dry rock, geopressured water, and the normal thermal gradient in the earth's crust. It is more common for wells to produce a mixture of steam and hot water, this combination being much less useful than dry steam. Electricity may be generated from hot-water wells by passing the water, under pressure, through a heat exchanger, where it causes the vaporization of a volatile liquid such as freon. The latter expands through a turbine.

The largest potential is associated with the heat in deposits of dry rock. High-pressure water is forced down a deep shaft and creates cracks in the rock at the bottom. Pressurized water is then forced through the cracks to extract heat. Hot rocks contracts as it cools, thereby creating fresh cracks and extending the catchment volume. At Los Alamos in New Mexico a 780 m shaft was drilled and then hydraulic creation of cracks was achieved at pressures of about 100 atm. All parts of the earth's surface have heat in the rock beneath, but the temperature gradients are very modest. Although the potential amount of heat is vast, the technology involved to extract this heat successfully has still to be developed and shown to be economically viable.

## 1.4.6 Other renewable resources

### Ocean temperature gradients

In 1881 D'Arsonval proposed the utilization of the temperature difference between the surface and lower layers of tropical seas. For practical purposes the layers need to be in reasonably close proximity to each other. The absorption of solar energy by the surface layers causes a thermal-syphon action, the warm surface water flowing towards the earth's poles from whence it moves back towards the tropics as cold water at a greater depth. The Gulf Stream carries, on average, about $2\,km^3/min$ of warm water, and the energy potential in this is vast. Warm and cold currents are close to each other off the Florida coast, the temperature difference range being 9–25°C. This gives a Carnot efficiency of about 3.4 per cent, and, with losses, an overall efficiency of, say, 2 per cent. Hence to produce large powers, vast quantities of water and large process plants are required. On the other hand an Arctic Ocean scheme could utilize the temperature differential between the water (at, say, +2°C) and the subzero air temperature.

A proposed plant would be situated 25 km east of Miami, where the temperature difference is 17.5°C, and this is shown in Figure 1.18. The best working fluids are ammonia ($NH_3$) or propane. A typical sea-water flow rate for a 400 MW installation is $115\,000\,m^3/min$, and the diameter of the cold-water pipe would need to be 24 m and the length 500–600 m. The plant is submerged and anchored. The warm water boils a high-pressure working fluid and the vapour expands through vapour turbines and then condenses back to a liquid in the condenser.

In 1996 the cost of producing electricity from fossil plants was 1.5–2.0 p/ kWh. Estimates for the sea thermal plant are now between 4 and 10 p/kWh. The surface temperature of the tropical oceans would be lowered by about 1°C if 60 000 GW of electricity is generated in this manner. Only if gas, oil, and coal increase in price to three times their present costs would an OTEC become worthwhile.

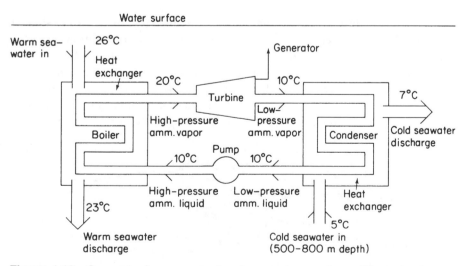

**Figure 1.18** Concept of power plant using sea-temperature differences (ocean temperature energy converter, OTEC)

## Wave power

The energy content of sea waves is very high. The Atlantic waves along the north-west coast of Britain have an average value of 80 kW/m of wave crest length. The energy is obviously very variable, ranging from greater than 1 MW/ m for 1 per cent of the year to near zero for a further 1 per cent. Hence, over several hundreds of kilometres a vast source of energy is available.

The sea motion can be converted into mechanical energy in a number of ways. One method is by using the Salter cam (or duck!), a cross-section of which is shown in Figure 1.19. The cam rotates about its axis and is shaped to minimize back-water pressure. Lengths of the cam could be moored, say, 80 km off the coast. Conversion of the float movement to electricity is difficult because of the slow oscillations. If reciprocating generators of some form are used, a device producing a 1 MW while oscillating at 0.1 Hz with an amplitude of 15° would require a torque of 1200 t-m! Owing to the randomness of the waves the system must be rated much higher than the design average output level. This implies that the cost of the installation designed to withstand the strongest winter Atlantic waves will be very high. To avoid this, shore or anchored installations using an oscillating column of water propelling air through a turbine to drive a generator are possible alternatives. However, the cost and robustness of such devices has still to be assessed.

## 1.5  Energy Storage

The tremendous difficulty in storing electricity in any large quantity has shaped the technology of power systems as they stand today. Various options exist for the large-scale storage of energy which may be converted into electricity, and, although the basic form of generation is unchanged, these methods can be used to ease operation and effect overall economies. Storage of any kind is, however, expensive and care must be taken in the economic evaluation. The options available are as follows: pumped storage, compressed air, heat, hydrogen

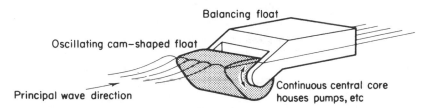

**Figure 1.19** Conversion of wave energy by floating cam (*Permission of Central Electricity Generating Board*)

gas, secondary batteries, and, of doubtful promise, flywheels, and supercon-ducting coils.

Although gas turbines are used to meet daily load peaks, very rapid changes in load (e.g. 1300 MW/min at the end of the 'Miss World' contest on British TV!) or the outage of lines or generators may occur. Gas turbines take 2 min to start up and a considerable amount of conventional steam plant must operate underloaded as a reserve. This is very expensive because there is a fixed heat loss for a steam turbogenerator regardless of output, and also the efficiency is reduced. A significant amount of storage capable of instantaneous use would be an effective method of meeting such loadings, and by far the most important method to date is that of pumped storage.

## 1.5.1 Pumped storage

A method of obtaining the advantages of hydro plant where suitable water supplies are not available is by the use of pumped storage. This consists of an upper and a lower reservoir and turbine-generators which can be used as motor-pumps. The upper reservoir has sufficient storage usually for 4–6 h of full-load generation, with a reserve of 1–2 h.

The sequence of operation is as follows. During times of peak load on the network the turbines are driven by water from the upper reservoir in the normal manner. The generators then change to synchronous-motor action and, being supplied from the general power network, drive the turbine which is now acting as a pump. During the night, when only base load stations are in operation and electricity is being produced at its cheapest, the water in the lower reservoir is pumped back into the higher one ready for the next day's peak load. Typical operating efficiencies attained are:

| | |
|---|---|
| Motor and generator | 96% |
| Pump and turbine | 77% |
| Pipeline and tunnel | 97% |
| Transmission | 95% |

giving an overall efficiency of 67 per cent. A further advantage is that the synchronous machines can be easily used as synchronous compensators if required.

Dams must be constructed if existing reservoirs are inadequate or non-exis-tent. A large scheme in Britain uses six 330 MVA pump-turbine (Francis-type reversible) generator-motor units generating at 18 kV. The flow of water and hence power output is controlled by guide vanes associated with the turbine. The maximum pumping power is 1830 MW. The machines are 92.5 per cent efficient as turbines and 91.7 per cent efficient as pumps and the speed is 500 r.p.m. Such a plant can be used to provide fine frequency control for the whole British system, when the machines will be expected to start and stop about 40 times a day.

## 1.5.2 Compressed-air storage

Air is pumped into large receptacles (e.g. underground caverns or old mines) at night and used to drive gas turbines for peak, day loads. A German utility has installed a 290 MW scheme at a capital cost in the order of £60 per kW (1975 prices). It generates 580 MWh of on-peak electricity and consumes 930 MWh of fuel plus 480 MWh of off-peak electricity. The energy stored is equal to the product of the air pressure and volume. The compressed air allows fuel to be burnt in the gas turbines at twice the normal efficiency. The general scheme is illustrated in Figure 1.20. Similar plant has been installed in the U.S.A.

One disadvantage of the above scheme is that much of the input energy to the compressed air manifests itself as heat and is wasted. Heat could be retained after compression, but there would be possible complications with the store walls due to the temperature of 450°C at 20 bar pressure. A solution would be to have a separate heat store that could comprise stacks of stones or pebbles which store heat cheaply and effectively. This would enable more air to be stored because it would now be cool. At 100 bar pressure, in the order of 30 m$^3$ of air is stored per megawatt-hour output.

## 1.5.3 Heat storage

No large-scale storage involving heat has yet evolved. Water has many advantages as a heat-storage medium, including good specific and latent heats. Liquid sodium has also been suggested and also various salts. In a steam generating plant, if the load is low, boilers may be kept at full output if the unwanted steam is bled to the feed-water heaters. During full-load output the

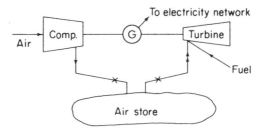

**Figure 1.20** Storage using compressed air in conjunction with gas-turbine generator G

steam produced goes wholly to generation, whilst the hot water is drawn from the store to give heated feed water. A 500 MW(e) boiler requires 405 kg/s of feed water and would require a storage of 17 000 t of water.

### 1.5.4 Secondary batteries

Large-scale battery use is unlikely and the two areas where the use of secondary batteries will have impact are in the electric car and local fluctuating energy sources such as windmills or solar power. The popular lead–acid cell, although reasonable in price, has a low energy density (15 Wh/kg). Nickel–cadmium cells are better (40 Wh/kg) but much more expensive. Still under intensive development is the sodium–sulphur battery (200 Wh/kg), which has a solid electrolyte and liquid electrodes and operates at a temperature of 300°C. Other combinations of materials are under active development in attempts to increase output and storage per unit weight. A 3 MW battery storage plant has been installed in Berlin for frequency control in emergencies.

### 1.5.5 Fuel cells

A fuel cell converts chemical energy to electrical energy by electrochemical reactions. Fuel is continuously supplied to one electrode and an oxidant (usually oxygen) to the other electrode. Figure 1.21 shows a simple hydrogen–oxygen fuel cell, in which hydrogen gas diffuses through a porous metal electrode (nickel). A catalyst in the electrode allows the absorption of $H_2$ on the electrode surface as hydrogen ions which react with the hydroxyl ions in the electrolyte to form water ($2H_2 + O_2 \rightarrow 2H_2O$). A theoretical e.m.f. of 1.2 V at 25°C is obtained. Other fuels for use with oxygen are carbon monoxide (1.33 V at 25°C), methanol (1.21 V at 25°C), and methane (1.05 V at 25°C). In practical cells, conversion efficiencies of 80 per cent have been attained. A major use of the fuel cell could be in conjunction with a hydrogen energy system.

Intensive research and development is still proceeding on various types of fuel cell—the most successful to date for power generation being the phosphoric fuel cell (Figure 1.22). This uses methane as the input fuel and operates at about 200–300°C to produce 200 kW of electrical power plus 200 kW of heat energy, with overall efficiency of around 80 per cent. Compared with other forms of energy conversion (see Figure 1.23), fuel cells have the potential of being up to 20 per cent more efficient. Much attention is now being given to the high-temperature molten carbonate cell which has the highest efficiency yet conceived.

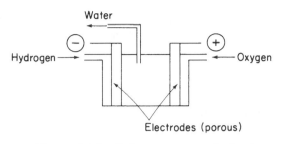

Figure 1.21 Hydrogen–oxygen fuel cell

**Figure 1.22** TEPCO's 200 kW fuel cell plant installed in the basement of the company's R&D Centre in Yokohama which went into operation in 1994. Manufactured by Toshiba, the plant is connected to TEPCO's grid and its waste heat is utilized for hot water supply and air conditioning for the R&D Centre building (*Courtesy of I.E.E. Power Engineering Journal*)

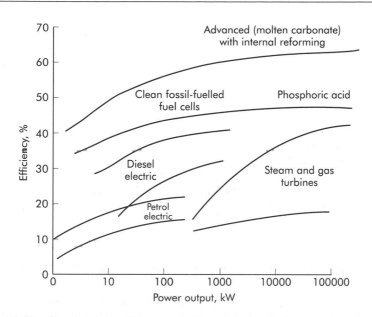

**Figure 1.23** Simple-cycle efficiency of different technologies as a function of scale (*Courtesy of I.E.E. Power Engineering Journal, June 1995*)

### 1.5.6 Hydrogen energy systems

Transmission of natural gas via a network of pipes is well established. The transmission capacity of a pipe carrying natural gas is high compared with electrical links, the installed cost being about one tenth of an equivalent capacity H.V. overhead line.

For hydrogen gas the calorific value is $12 \times 10^6 \, \mathrm{J \, m^3}$ (ATP).

$$\text{Power transmitted} = \text{volumetric flow rate} \times \text{volumetric calorific value (at working pressure)}$$

For long transmission distances the pressure drop is compensated by booster compressor stations. A typical gas system uses a pipe of internal diameter 0.914 m, and with natural gas a power transfer of 12 GW is possible at a pressure of 68 atm and a velocity of 7 m/s. A 1 m diameter pipe carrying hydrogen gas can transmit 8 GW of power, equivalent to 4–400 kV, three phase transmission lines.

The major advantage of hydrogen is, of course, that it can be stored; the major disadvantage is that it must be produced from water via electrolysis. Alternative methods of production are under laboratory development, e.g. use of heat from nuclear stations to 'crack' water and so release hydrogen; however, temperatures of 3000°C are required. Very large electrolysers can attain efficiencies of about 60 per cent. This, coupled with the efficiency of electricity production from a nuclear plant, gives an overall efficiency of hydrogen production of about 21 per cent. However, it would involve nuclear plant at off-peak periods and this, combined with the facility for storage, could be attractive.

A schematic of such a system is shown in Figure 1.24. As fossil fuel prices increase, and allowing for industry's considerable needs for hydrogen, the idea will become more attractive, even if on a limited scale. Large-scale fuel cell installations can then be used for the conversion of hydrogen energy back to electrical energy. Also, the use of hydrogen as fuel for aircraft and automobiles could give impetus to its large-scale production, storage, and distribution.

### 1.5.7 Superconducting magnetic energy stores (SMES)

Continuing development of the so-called high-temperature superconductor, where the transition temperature can be around 60–80 K (K is degrees Kelvin where 0 K is absolute zero and 273 K is 0°C) has led to the possibility to store energy in the magnetic field produced by circulating a large current (over 100 kA) in an inductance. For a coil of inductance $L$ in air, the stored energy is given by $\frac{1}{2}LI^2$ J, which can provide 100 MW for several seconds, if required to smooth out power demand peaks, with a coil diameter of the order of 20 m.

A big advantage of the high-temperature superconductor is that cooling by liquid nitrogen can be used, which is far cheaper than using helium to reach temperatures closer to absolute zero. Initially, it is expected that commercial units will be used to provide an uninterruptible supply for sensitive loads to guard against voltage sags or to provide continuity whilst emergency generators are started. Another use in transmission networks would be to provide fast

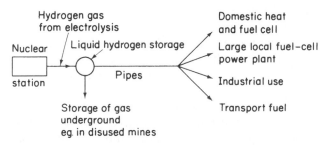

**Figure 1.24** Hydrogen storage and transmission scheme

response for enhanced transient stability and improved power quality. Figure 1.25 illustrates an early SMES unit mounted in a large trailer for countering supply interruptions at remote sites.

**Figure 1.25** Superconducting magnetic energy storage unit 750 kW (EPRI) (*Courtesy of EPRI, U.S.A.*)

## *Flywheels*

The most compact energy store known is that of utilizing high-speed flywheels. Such devices coupled to an electrical generator/motor have been employed in buses on an experimental basis and also in special industrial applications. For power systems, very large flywheels constructed of composite high-tensile-resisting materials have been proposed, but their cost and maintenance problems have so far ruled them out of economic contention compared with alternative forms of energy supply.

## *Supercapacitors*

The interface between an anode and cathode immersed in an electrolyte has a very high permittivity. This property can be exploited in a capacitor to produce a 25 V capsule with a capacitance of 0.1 F. Many units in series and parallel would have the capability of storing many MWh of energy, which can be quickly released for transient control purposes. To date, higher voltage forms of a commercially useful device have not got beyond their employment for pulse or actuator applications.

## 1.6   Environmental Aspects of Electrical Energy

Increasingly in industrialized nations, environmental pressure groups are having an impact on the development of energy resources, especially those involving electricity production, transmission, and distribution. Conversion of one form of energy to another produces unwanted side effects and, often, pollutants which need to be controlled and disposed of. In addition, safety and health are subject to increasing legislation by national and international bodies, thereby requiring all engineers to be aware of the laws and regulations governing the practice of their profession.

It must be appreciated that the extraction of fossil fuels from the earth is not only a hazardous business but also, nowadays, one controlled through licensing by governments, etc. Even hydro plants require careful study and investigation through modelling, widespread surveys, and impact statements to gain acceptance. Large installations of all kinds require often lengthy planning inquiries which are both time consuming and expensive, thereby delaying the start up of energy extraction and production. As a consequence, methods of producing electrical energy which avoid or reduce the inquiry process are to be favoured over those needing considerable consultation before receiving the go-ahead. This is likely to favour small-scale projects or the redevelopment of existing sites where industry or production facilities are already operating.

In recent years, considerable emphasis has been placed on 'sustainable development', by which is meant the use of technologies that do not harm the environment, particularly in the long term. It also implies that anything we do now to affect the environment should be recoverable by future generations. Irreversible damage, e.g. removal of the ozone layer or increase in $CO_2$ in the atmosphere, should be avoided.

## 1.6.1 Atmospheric pollution

The emissions associated with power plants are mainly sulphur oxides, particular matter, and nitrogen oxides. Of the former, sulphur dioxide accounts for about 95 per cent and is a by-product of the combustion of coal or oil. The sulphur content of coal varies from 0.3 to 5 per cent, and for generation purposes is specified by some U.S. state laws to be below a certain percentage. In the eastern U.S.A. this has led to the widespread use of coal from western states because of its lower sulphur content or the use of gas as an alternative fuel. Sulphur dioxide forms $H_2SO_4$ in the air which causes damage to buildings and vegetation. Sulphate concentrations of 9–10 $\mu g/m^3$ of air aggravate asthma and lung and heart disease. This level has been freqently exceeded in the past, a notorious episode being the London fog of 1952 (caused by domestic coal burning and not by the electricity industry). It should be noted that although sulphur does not accumulate in the air it does so in the soil.

Sulphur oxide emission can be controlled by:

- the use of fuel with, say, less than 1 per cent sulphur;
- the use of chemical reaction to remove the sulphur, in the form of sulphuric acid, from the combustion products, e.g. limestone scrubbers or fluidized-bed combustion;
- removing the sulphur from the coal by gasification or flotation processes.

In Europe (particularly Germany and the U.K.) the governments limit the amount of $SO_2$, $NO_x$, and particulate emission, as in the U.S.A. This has led to the retrofitting of 'flue gas desulphurization' (FGD) scrubbers to some coal-burning plants, thereby increasing their costs of production by up to 20 per cent. Emissions of $NO_x$ can be controlled by fitting advanced technology burners which can ensure a more complete combustion process, thereby reducing the oxides going up the stack (chimney).

Particulate matter refers to particles in the air. These, in sufficient concentrations, are injurious to the respiratory system, and by weakening resistance to infection may well affect the whole body. Apart from settling on the ground or buildings to produce dirt, a further effect is the reduction of the solar radiation entering the polluted area. Reported densities (particulate mass in $1 m^3$ of air) are $10\,\mu\mathrm{g/m}^3$ in rural areas to $2000\,\mu\mathrm{g/m}^3$ in polluted areas. The average value in U.S. cities is about $100\,\mu\mathrm{g/m}^3$.

About one-half of the oxides of nitrogen in the air in populated areas are due to power plants and originate in high-temperature combustion processes. At levels of 25–100 parts per million they can cause acute bronchitis and pneumonia. Increasingly, city pollutants are due to cars and lorries and not power plants.

A 1000 MW(e) coal plant burns approximately 9000 t of coal per day. If this has a sulphur content of 3 per cent the amount of $SO_2$ emitted per year is $2 \times 10^5$ t. Such a plant produces the following pollutants per hour (in kg): $CO_2$ $8.5 \times 10^5$, CO $0.12 \times 10^5$, sulphur oxides $0.15 \times 10^5$, nitrogen oxides $3.4 \times 10^3$, and ash. Both $SO_2$ and $NO_x$ are reduced considerably by the measures mentioned above. Gas-fired CCGT plants produce very little $NO_x$ or $SO_2$ and their $CO_2$ output is about 55 per cent of an equivalent size coal-fired generator.

The concentration of pollutants can be reduced by dispersal over a wider area by the use of high stacks. If, in the stack, a vertical wire is held at a high negative potential relative to the wall, the expelled electrons from the wire are captured by the gas molecules moving up the stack. Negative ions are formed which accelerate to the wall, collecting particles on the way. When a particle hits the wall the charge is neutralized and the particle drops down the stack and is collected. Precipitators have particle-removing (by weight) efficiencies of up to 99 per cent, but this is misleading as performance is poor for small particles of, say, less than $0.1\,\mu\mathrm{m}$ in diameter. The efficiency based on number of particles removed is therefore less. Disposal of the resulting fly-ash is expensive, but the ash can be used for industrial purposes, e.g. building blocks. Unfortunately the efficiency of precipitators is enhanced by a reasonable sulphur content in the gases. For a given collecting area the efficiency decreases from 99 per cent with 3 per cent sulphur to 83 per cent with 0.5 per cent sulphur at 150°C. This results in much larger and expensive precipitator units with the low-sulphur coal or the use of fabric filters in 'bag houses' situated before the flue gas enters the stack.

## 1.6.2 Thermal pollution

Steam from the low-pressure turbine is liquefied in the condenser at the lowest possible temperatures to maximize the steam-cycle efficiency. Where copious supplies of water exist the condenser is cooled by 'once-through' circulation of sea- or river-water. Where water is more restricted in availability, e.g. away from the coasts, the condensate is circulated in cooling towers in which it is sprayed in nozzles into a rising volume of air. Some of the water is evaporated, providing cooling. The latent heat of water is $2 \times 10^6$ J/kg compared with a sensible heat of 4200 J/kg per degree C in 'once-through' cooling.

A disadvantage of such towers is the increase in humidity produced in the local atmosphere.

Dry cooling towers in which the water flows through enclosed channels (similar to a car radiator), past which air is blown, avoid local humidity problems, but at a much higher cost than 'wet towers'. Cooling towers emit evaporated water to the atmosphere in the order of 75 000 litres/min for a 1000 MW(e) plant. A crucial aspect of once-through cooling in which the water flows directly into the sea or river is the increased temperature of the latter due to the large volume per minute (typically 360 m$^3$/s for a coolant rise of 2.4°C for a 2.4 GW nuclear station) of heated coolant. Because of their lower thermal efficiency, nuclear power stations require more cooling water than fossil-fuelled plants.

The chemical reaction rate doubles for each 10°C rise in temperature, causing an increased demand for oxygen, but the ability of the water to dissolve oxygen is less at the higher temperatures. Hence extreme care must be taken to safeguard marine life, although the higher temperatures can be used effectively for marine farming if conditions can be controlled.

### 1.6.3 Electromagnetic radiation from overhead lines, cables, and equipment

The biological effects of electromagnetic radiation have produced considerable concern among the general public as to the possible hazards in the home and the workplace. Proximity of dwellings to overhead lines and even buried cables has also led to concerns of possible cancer-inducing effects, with the consequence that research effort has been needed to allay such fears to show that they are unfounded. Of course, it must be remembered that radiologists have employed ionizing and very-high-frequency radiation for medical treatment over many years, but the power frequencies used (50 or 60 Hz) are not considered harmful. The electric field and magnetic field strengths below typical H.V. transmission lines are given in Table 1.2.

Considerable international research and cooperative investigation has now been proceeding for over 20 years into low-frequency electric and magnetic field exposure produced by household appliances, video display terminals, and local power lines. To date there is no convincing evidence that they pose any demonstrable health hazards. Epidemiologic findings of an association between electric and magnetic fields and childhood leukaemia or other childhood or adult cancers are inconsistent and inconclusive; the same is likely to be true of birth defects or other reproductive problems.

It is important for the electrical and electronic engineer to put the human effects of radiation into perspective and to remember that many other environmental agents can produce far greater damage to health than does electromagnetic radiation from power equipment.

**Table 1.2**  Likely maximum electric and magnetic field strengths directly under overhead lines. Note that magnetic field is dependent upon current carried

| Line voltage (V) | Electric field strength (V/m) | Magnetic flux density ($\mu$T) |
|---|---|---|
| 400 000 | 11 000 | 40 |
| 275 000 | 6000 | 40 |
| 132 000 | 2000 | 11 |
| 33 000 | 350 | 7 |
| 11 000 | 120 | 7 |
| 415 | < 1 | 1 |
| National Radiation Protection Board Guidelines for Safety | 10 000–15 000 | 1600 |
| Earth's magnetic field | — | 40–50 |

(*Source: Electricity Association Services Ltd, London 1996*)

Further aspects of fields produced by power lines are discussed in Chapter 3.

### 1.6.4 Visual and audible noise impacts

The presence of overhead lines constitutes an environmental problem (perhaps the most obvious one) on several counts.

1. Space is used which could be used for other purposes. The land allocated for the line is known as the right of way (or wayleave in Britain). The area used for this purpose is already very appreciable.

2. Lines are considered by many to mar the landscape. This is, of course, a subjective matter, but it cannot be denied that several lines converging on a substation or plant, especially from different directions, is offensive to the eye.

3. Radio interference (RI), audible noise (AI), and safety considerations must also be considered.

Although most of the above objections could be overcome by the use of underground cables, these are not free of drawbacks. The limitation to cable transmitting current because of temperature-rise considerations coupled with high manufacture and installation costs results in the ratio of the cost of transmitting energy underground to that for overhead transmission being between 10 and 20, which increases with operating voltage. With novel cables, such as superconducting, still under development it is hoped to reduce this disadvantage. Estimates made of the increased costs resulting from the under-grounding of whole or large parts of a power system indicate that an intolerable financial burden would result on the utilities and hence the consumer. However, it may be expected that in future a larger proportion of circuits will be carried underground, especially in suburban areas. In large urban areas circuits are invariably underground, thereby posing increasing problems as load intensities increase.

Reduction of RI can be achieved in the same way that electromagnetic effects are reduced, but audible noise is a function of the line design. Careful attention to the tightness of joints, the avoidance of sharp or rough edges, and the use of earth screen shielding can reduce AI to acceptable levels at a distance dependent upon voltage.

Safety clearances dependent upon International Standards are, of course, extremely important and must be maintained in adverse weather conditions. The most important of these is the increased sag due to ice and snow build up in winter.

## 1.7 Transmission and Distribution Systems

### 1.7.1 Representation

Modern electricity supply systems are invariably three-phase. The design of distribution networks is such that normal operation is reasonably close to balanced three-phase working, and often a study of the electrical conditions in one phase is sufficient to give a complete analysis. Equal loading on all three phases of a network is ensured by allotting, as far as possible, equal domestic loads to each phase of the low-voltage distribution feeders; industrial loads usually take three-phase supplies.

A very useful and simple way of graphically representing a network is the schematic or line diagram in which three-phase circuits are represented by single lines. Certain conventions for representing items of plant are used and these are shown in Figure 1.26. A typical line or schematic diagram of a part of a power system is shown in Figure 1.27. In this, the generator is star connected, with the star point connected to earth through a resistance. The nature of the connection of the star point of rotating machines and transformers to earth is of vital importance when considering faults which produce electrical imbalance in the three phases. The generator feeds two three-phase lines (overhead or underground). The line voltage is increased from that at the generator term-inals by transformers connected as shown. At the end of the lines the voltage is reduced for the secondary distribution of power. Two lines are provided to improve the *security* of the supply, i.e. if one line develops a fault and has to be switched out the remaining one still delivers power to the receiving end. It is

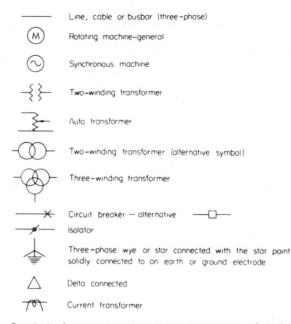

| | |
|---|---|
| —— | Line, cable or busbar (three-phase) |
| (M) | Rotating machine-general |
| (∼) | Synchronous machine |
| —⊰⊱— | Two-winding transformer |
| | Auto transformer |
| —(○○)— | Two-winding transformer (alternative symbol) |
| —(⊕)— | Three-winding transformer |
| ——✕ | Circuit breaker — alternative    —▭— |
| —⧵— | Isolator |
| | Three-phase wye or star connected with the star point solidly connected to an earth or ground electrode |
| △ | Delta connected |
| ⁓⊙⁓ | Current transformer |

**Figure 1.26** Symbols for representing the components of a three-phase power system

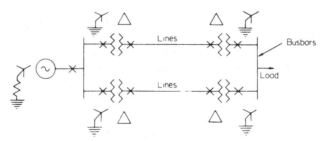
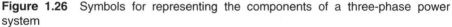

**Figure 1.27** Line diagram of a simple system

not necessary in straightforward current and voltage calculations to indicate the presence of switches, etc., on the diagrams, but in some cases, such as stability calculations, the location of switches, current transformers, and protection is very useful.

Although the use of jargon is avoided as much as possible, certain contantly recurring terms may be rather vague to the newcomer to the subject. A short list of such terms follows, with explanations.

*Systems*    This is used to describe the complete electrical network: generators, loads, and prime movers.

| | |
|---|---|
| *Load* | This may be used in a number of ways: to indicate a device or collection of devices which consume electricity; to indicate the power required from a given supply circuit; to indicate the power or current being passed through a line or machine. |
| *Busbar* | This is an electrical connection of zero impedance joining several items, such as lines, loads, etc. Often this takes the form of actual busbars of copper or aluminium. |
| *Earthing* (*Grounding*) | This is connection of a conductor or frame of a device to the main body of the earth. This must be done in such a manner that the resistance between the item and the earth is below prescribed limits. This often entails the burying of the large assemblies of conducting rods in the earth and the use of connectors of large cross-sectional area. |
| *Fault* | This is a malfunctioning of the network, usually due to the short-circuiting of two conductors or live conductors connecting to earth. |
| *Outage* | Removal of a circuit either deliberately or inadvertently. |
| *Security of supply* | Provision must be made to ensure continuity of supply to consumers, even with certain items of plant out of action. Usually, two circuits in parallel are used and a system is said to be secure when continuity is assured. This is obviously the item of first priority in design and operation. |

## 1.7.2 Transmission

Transmission normally implies the bulk transfer of power by high-voltage links between main load centres. Distribution, on the other hand, is mainly concerned with the conveyance of this power to consumers by means of lower voltage networks.

Generators usually produce voltages in the range 11–25 kV, which is increased by transformers to the main transmission voltage. At substations the connections between the various components of the system, such as lines and transformers, are made and the switching of these components is carried out. Large amounts of power are transmitted from the generating stations to the load-centre substations at 400 kV and 275 kV in Britain, and at 345 kV, 765 kV, and 500 kV in the U.S.A. The network formed by these very-high-voltage lines is sometimes referred to as the *supergrid*. Most of the large and efficient stations feed through transformers directly into this network. This grid, in turn, feeds a subtransmission network operating at 132 kV in Britain and 115 kV in the U.S.A. Some of the older and sometimes less efficient stations feed into this system which, in turn, supplies networks which are concerned with distribution to consumers in a given area. In Britain these networks operate at 33 kV, 11 kV, or 6.6 kV and supply the final consumer feeders at 400–415 V three-phase, giving 230–240 V per phase. Other voltages exist in isolation in various places, e.g. the 66 and 22 kV London cable system. A typical part of a supply network is shown schematically in Figure 1.28. The

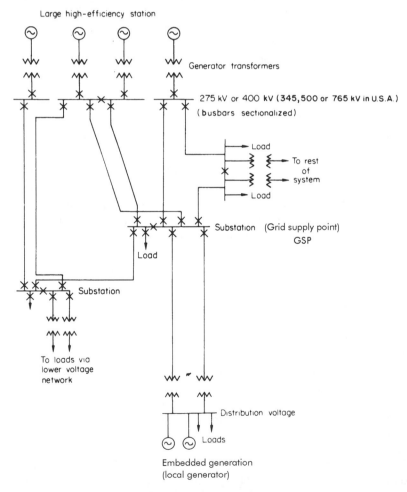

**Figure 1.28** Part of a typical power system

power system is thus made up of networks at various voltages. There exist, in effect, voltage tiers as represented in Figure 1.29.

Summarizing transmission networks deliver to wholesale outlets at 132 kV and above; subtransmission networks deliver to retail outlets at voltages from 115 kV or 132 kV, and distribution networks deliver to final step-down transformers at voltages below 132 kV, usually operated as radial systems.

## Reasons for interconnection

Many generating sets are large and 2000 + MW capacity stations are available to provide base or intermediate load. With CCGT units, their high efficiency and cheap long-term gas contracts, regardless of geographical position, mean that it is more economical to use these efficient stations to full capacity 24 h a day and transmit energy considerable distances than to use less efficient more local stations. The main base load therefore is met by these high-efficiency stations which must be interconnected so that they feed into the general system and not into a particular load.

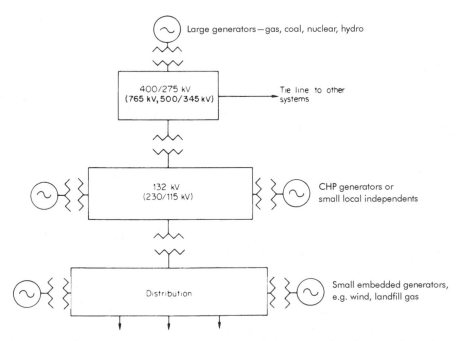

**Figure 1.29** Schematic diagram of the constituent networks of a supply system. U.S.A. voltages in parentheses

To meet sudden increases in load a certain amount of generating capacity, known as the 'spinning reserve', is required. This consists of generators running at normal speed and ready to supply power instantaneously. If the machines are stationary a reasonable time is required (especially for steam turboalternators) to run up to speed; this can approach 1 h although gas turbines can be started and loaded in 3 min or less. Hydro generators can be even quicker. It is more economical to have certain stations serving only this function than to have each station carrying its own spinning reserve.

The electricity supplies over the entire country are synchronized and a common frequency exists: 50 Hz in Europe, 60 Hz in N. America.

Interconnection also allows for alternative paths to exist between generators and bulk supply points supplying the distribution systems. This provides security of supply should any one path fail.

### 1.7.3 Distribution systems

Although in this book there is a bias towards transmission networks and the associated problems, the general area of distribution should not be considered to be of secondary importance. Distribution networks differ from transmission networks in several ways, quite apart from voltage magnitude. The number of branches and sources is much higher in distribution networks and the general structure or topology is different. A typical system consists of a step-down (e.g. 132/11 kV) on-load tap-changing transformer at a bulk supply point feeding a number of lines which can vary in length from a few hundred metres to several kilometres. A series of step-down three-phase transformers, e.g. 11 kV/415 V in Britain or 4.16 kV/220 V in the U.S.A., are spaced along the route and from these are supplied the consumer three-phase, four-wire networks which give 240 V, or, in the U.S.A., 110 V, single-phase supplies to houses and similar loads.

### Rural systems

In rural systems, loads are relatively small and widely dispersed (5–50 kVA per consumer group is usual). It is a predominantly overhead line system at 11 kV, three-phase, with no neutral or single phase for spurs from the main system. Pole-mounted transformers (5–200 kVA) are installed, protected by fuses which require manual replacement after operation; hence rapid access is desirable by being situated as close to roads as possible. Essentially, a radial system is supplied from one step-down point; distances up to 10–15 miles (16–24 km) are feasible with total loads of 500 kVA or so (see Figure 1.30), although in sparsely populated areas, distances of 50 miles may be fed by 11 kV. Single-

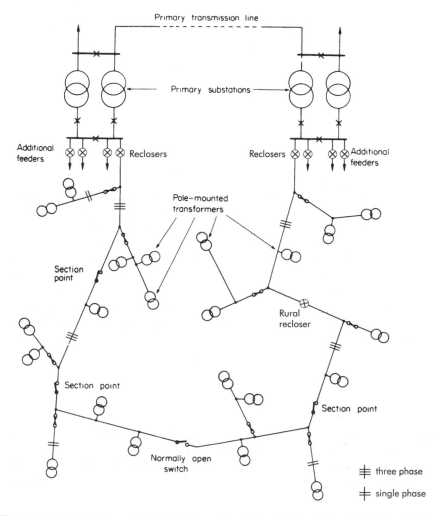

**Figure 1.30** Typical rural distribution system (*Permission of the Electricity Council*)

phase earth-return systems operating at 20 kV are now being used increasingly in developing countries.

Over 80 per cent of faults are transitory due to flashover following some natural or man-made cause. This produces unnecessary fuse-blowing unless autoreclosers are employed on the main supply; these have been used with great success in either single- or three-phase form. The principle, as shown in Figure 1.31, is to open on fault before the fuse has time to operate and to reclose after 1–2 s. If the fault still persists, a second attempt is made to clear, followed by another reclose. Should the fault still not be cleared, the recloser

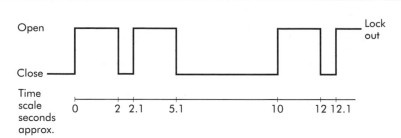

**Figure 1.31**   Sequence of operations for a recloser

remains closed for a longer period to blow the appropriate protective fuse (e.g. on a spur line or a transformer). If the fault is still not cleared, then the recloser opens and locks out to await manual isolation of the faulty section. This process requires the careful coordination of recloser operation and fuse-blowing characteristics where time grading is important. Section switches operated by radio or tele-command enable quick resupply routes to be established following a faulty section isolation.

Good earthing at transformer star points is required to prevent overvoltages at consumers' premises. Surge protection using diverters or arcing horns is essential in lightning-prone areas.

### Surburban systems

These are a development of the rural system into ring mains, with much of it underground for amenity reasons. The rings are sectionalized so that simple protection can be provided. Loads range between 2 and $10\,\text{MW/mile}^2$ (0.8– $4.0\,\text{MW/km}^2$) but diversity factor is wide (2–4) (see Section 1.8.1 for definition of diversity factor).

In high-density housing areas, the practice is to run the L.V. mains on either side of each road, interconnected at junctions by links for sectionalizing (see Figure 1.32). At appropriate points this network is fed by a step-down transformer of 200–500 kVA rating connected to an H.V. cable or overhead line network. Reinforcement is provided by installing further step-down transformers tapped from the H.V. network. It is very rare to up-rate L.V. cables as the load grows.

Short-circuit levels are fairly low due to long feeders from the bulk supply points. With the increasing cost of cables and undergrounding works, but with improved transformer efficiency and lower costs due to rationalization and standardization, an economic case can be made for reducing the L.V. network and extending the H.V. network so that fewer consumers are supplied from each transformer.

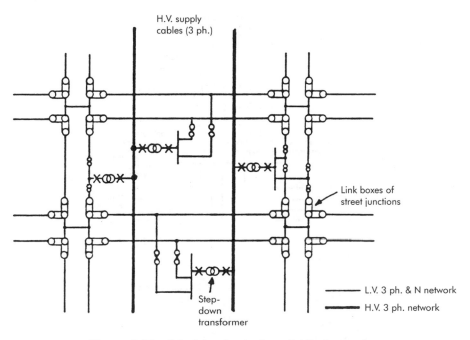

H.V. supply
cables (3 ph.)

Link boxes of
street junctions

Step-
down
transformer

———— L.V. 3 ph. & N network
━━━━ H.V. 3 ph. network

**Figure 1.32** Principle of suburban distribution system

## Urban (town or city) systems

Very heavy loadings (up to $100\,\text{MW/mile}^2$ or $40\,\text{MW/km}^2$) are usual, especially where high-rise buildings predominate. Extensive heating and air-conditioning loads as well as many small motors predominate. Fluorescent lighting reduces the power factor and leads to some waveform distortion but rectifier-fed loads and power electronic motor drives now cause considerable harmonics on all types of network.

Again, a basic L.V. grid, reinforced by extensions to the H.V. network, as required, produces minimum costs overall. The H.V. network is usually in the form of a ring main fed from two separate sections of a double busbar sub-station where 10–60 MVA transformers provide main supply from the trans-mission system (see Figure 1.33). The H.V. network is sectionalized to contain short-circuit levels and to ease protection problems. A high security of supply is possible by overlapping H.V. rings so that the same L.V. grid is fed from several transformers supplied over different routes. Failure of one portion of the H.V. system does not affect consumers, who are then supported by the L.V. network from adjacent H.V. supplies.

In the U.K. transformers of 500 or 750 kVA rating are now standard, with one H.V. circuit breaker or high rupturing capacity (HRC) fuse and two isolators either side to enable the associated H.V. cable to be isolated

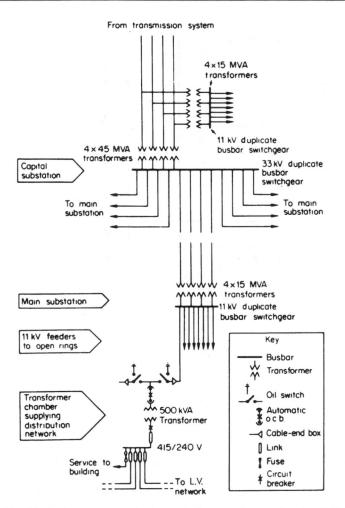

**Figure 1.33** Typical arrangement of supply to an urban network—British practice (Reproduced by permission of the *Institution of Electrical Engineers*)

manually in the event of failure. Average H.V. feeder length is less than 1 mile and restoration of H.V. supplies is usually obtainable in under 1 h. Problems may arise due to back-feeding of faults on the H.V. system by the L.V. system, and in some instances reverse power relay protection is necessary.

In new developments it is essential to acquire space for transformer chambers and cable access before plans are finalized. High-rise buildings may require substations situated on convenient floors as well as in the basement.

## Industrial systems

Apart from the supply of new industrial and housing estates or the electrification of towns and villages, a large part of the work of a planning engineer is involved with the up-rating of existing supplies. This requires good load forecasting over a period of 2–3 years to enable equipment to be ordered and access to sites to be established. One method of forecasting is to survey the demand on transformers by means of a maximum-demand indicator and to treat any transformer which has an average load factor of 70 per cent or above as requiring up-rating over the next planning period. Another method is to analyse consumers' bills, sectionalized into areas, and distributors and, by surveys and computer analysis, to relate energy consumed to maximum demand. This method could be particularly useful and quite economic where computerized billing is used.

In practice, good planning requires sufficient data on load demands, energy growth, equipment characteristics, and protection settings. All this information can be stored and updated from computer files at periodic intervals and provides the basis for the installation of adequate equipment to meet credible future demands without unnecessary load shedding or dangerous overloading.

## 1.7.4 Typical power systems

Throughout the world the general form of power systems follows the same pattern. Magnitudes of voltage vary from country to country, the differences originating mainly from geographical and historical reasons.

Several frequencies have existed, although now only two values—50 Hz and 60 Hz—remain: 60 Hz is used on the American continent, whilst most of the rest of the world uses 50 Hz, although Japan still has 50 Hz for the main island and 60 Hz for the northern islands. The value of frequency is a compromise between higher generator speeds (and hence higher output per unit machine volume) and the disadvantage of high system reactance at higher frequencies. Historically, the lower limit was set by the need to avoid visual discomfort from electric lamps.

The distance that a.c. trnasmission lines can transfer power is limited by the maximum permissible peak voltage between conductor and ground. As voltages increase, more and more clearance must be allowed in air to prevent the possibility of flashover or danger to people or animals on the ground. Unfortunately, the critical flashover voltage increases non-linearly with clearance, such that, with long clearances, proportionally less peak voltages can be safely used. Figure 1.34 shows this effect where critical flashover voltage is 4–6 times the rated voltage of the system.

An additional limitation is the power loading that can be applied to a transmission line dependent upon its surge impedance loading (SIL). The

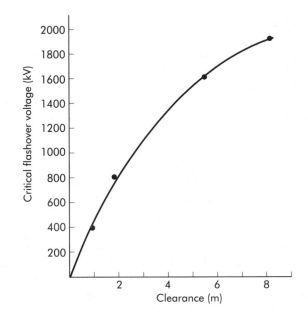

**Figure 1.34**   Critical flashover voltage for V-string insulators in a window tower

surge impedance is given by $Z_0 = \sqrt{L/C}$, where $L$ is the series inductance and $C$ is the shunt capacitance per unit length of the line (see Chapter 5). The curve and data given in Figure 1.35 show the capability for various transmission line-to-line voltages and conductor configurations.

As a consequence of these limitations, the highest a.c. voltage used for transmission is 750 kV, although experimental lines above 1000 kV have been built.

The topology of the power system and the voltage magnitudes used are greatly influenced by geography. Very long lines are to be found in North and South American nations and Russia. This has resulted in higher transmission voltages, e.g. 765 kV, with the possibility of voltages in the range 1000–1500 kV. In some South American countries, e.g. Brazil, vast amounts of hydroelectric resources are in the process of being developed, resulting in very long transmission links. In highly developed countries the available hydro resources have been utilized and a considerable proportion of new generation is nuclear. In geographically smaller countries, as exist in Europe, the degree of interconnection is much tighter, with shorter transmission distances, the upper voltage being about 420 kV.

Systems are universally a.c. with the use of high-voltage d.c. links used for specialist purposes, e.g. very long circuits, submarine cable connections, and back-to-back converters for variable-angle operation. The use of d.c. has been limited by the high cost of the conversion equipment. This requires overhead

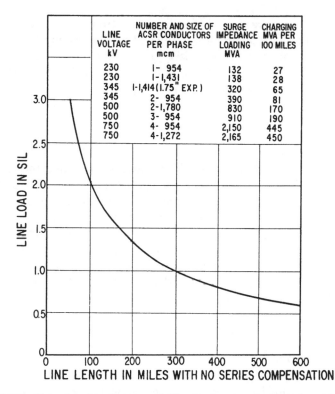

| LINE VOLTAGE kV | NUMBER AND SIZE OF ACSR CONDUCTORS PER PHASE mcm | SURGE IMPEDANCE LOADING MVA | CHARGING MVA PER 100 MILES |
|---|---|---|---|
| 230 | 1- 954 | 132 | 27 |
| 230 | 1-1,431 | 138 | 28 |
| 345 | 1-1,414 (1.75" EXP.) | 320 | 65 |
| 345 | 2- 954 | 390 | 81 |
| 500 | 2-1,780 | 830 | 170 |
| 500 | 3- 954 | 910 | 190 |
| 750 | 4- 954 | 2,150 | 445 |
| 750 | 4-1,272 | 2,165 | 450 |

**Figure 1.35** Transmission-line capability in terms of surge-impedance loading (*Permission of Edison Electric Institute, U.S.A.*)

line lengths of a few hundred kilometres or cables of 30–100 km to enable the reduced circuit costs to offset the conversion costs. Below are presented summaries of the systems of some countries.

## U.S.A.

The systems in the U.S.A. are based on a comparatively few investor-owned generation/transmission utilities responsible for bulk transmission as well as operating and constructing new generation facilities when needed. Delivery is to smaller municipal or cooperative rural distribution companies or to the distribution arm of the transmission utility. The industry is heavily regulated by Federal (state) Agencies and the investor profits are carefully controlled.

The loads differ seasonally from one part of the country to another and load diversity occurs because of the time zones. Generally, the summer load is the highest due to extensive use of air-conditioning. Consumers are supplied at either 110 V, 60 Hz, single phase for lighting and small-consumption appliances but at 220 V for loads above about 3 kW (usually cookers, water-heaters and air-conditioners). This type of supply normally uses a 220 V centre-tapped transformer secondary to provide the 110 V supply.

## U.K.

Here, the frequency (as in the whole of Europe) is 50 Hz and the residential supply is 230–240 V single phase. The European standard is now 230 V, but with ±10 per cent tolerance this voltage can vary between a maximum of 253 V and a minimum of 207 V; equipment and appliance designers must take this allowed variation into account or beware! Most commercial and industrial loads are supplied at 400–415 V or higher voltages, three phase. The winter load produces the highest peak because of the preponderance of heating appliances, although it is noted that the summer proportion is growing due to increasing use of air-conditioning in commercial and industrial premises.

## Continental Europe

Most continental countries still mainly have combined generator/transmission utilities, many of which cover the whole country and which are overseen by government control. Increasingly these systems are being 'unbundled', that is they are being separated into different functions (generation, transmission, distribution), each individually accountable such that private investors can enter into the electricity market (see Chapter 12). Interconnection across national boundaries enables electrical energy to be traded under agreed tariffs, and limited system support is available under disturbed or stressed operating conditions.

The daily load variation tends to be much flatter than in the U.K. because of the dominance of industrial loads with the ability to vary demand and because there is less reliance on electricity for heating in private households. Many German and Scandinavian cities have CHP plants with hot water distribution mains for heating purposes. Transmission voltages are 380–400 kV, 220 kV, and 110 kV with household supplies at 220–230 V, single phase.

## Japan and the Pacific Rim

The fastest-growing systems are generally found in the Far East, particularly China, Indonesia, the Philippines and Malaysia. Voltages up to 500 kV are used for transmission and 220–240 V are employed for households, etc. Hydroelectric potential is still quite large (particularly in China) but with gas and oil still being discovered and exploited, CCGT plant developments are underway. Increasing interconnection, perhaps by direct current, is a possibility.

# 1.8  Utilization

## 1.8.1  Loads

The major consumption groups are industrial, residential (domestic), and commercial. Industrial consumption accounts for up to 40 per cent of the total in many industrialized countries and a significant item is the induction motor. The percentage of electricity in the total industrial use of energy is expected to continue to increase due to greater mechanization and the growth of energy-intensive industries, such as chemicals and aluminium. In the U.S.A. the following six industries account for over 70 per cent of the industrial electricity consumption: metals (25 per cent), chemicals (20 per cent), paper and products (10 per cent), foods (6 per cent), petroleum products (5 per cent), and transportation equipment (5 per cent). Over the past 25 years the amount of electricity per unit of industrial output has increased annually by 1.5 per cent, but this is dependent on the economic cycle. Consumption of electricity in industrialized countries since about 1980 has been no more than 2 per cent per year due largely to the contraction of energy-intensive industries (e.g. steel manufacturing) combined with efforts to load manage and to make better use of electricity (see Chapter 12). Consequently, a tailing off of demand growth is seen, as shown in Figure 1.1. Residential loads are largely made up of refrigerators, fridge-freezers, freezers, cookers (including microwave ovens), space heating, water heating, lighting, and (increasingly in Europe) air-conditioning. Together these loads in the U.K. amount to around 40 per cent of total sales.

The commercial sector comprises offices, shops, schools, etc. The consumption here is related to personal consumption for services, traditionally a relatively high-growth quantity. In this area, however, conservation of energy measures are particularly effective and so modify the growth rate.

Quantities used in measurement of loads are defined as follows:

*Maximum load*   The average load over the half hour of maximum output.

*Load factor*   The units of electricity exported by the generators in a given period divided by the product of the maximum load in this period and the length of the period in hours. The load factor should be high; if it is unity, all the plant is being used over all of the period. It varies with the type of load, being poor for lighting (about 12 per cent) and high for industrial loads (e.g. 100 per cent for pumping stations). Figure 1.36 illustrates the calculation of load factor.

***Diversity factor***    This is defined as the sum of individual maximum demands of the consumers, divided by the maximum load on the system. This factor measures the diversification of the load and is concerned with the installation of sufficient generating and transmission plant. If all the demands occurred simultaneously, i.e. unity diversity factor, many more generators would have to be installed. Fortunately, the factor is much higher than unity, especially for domestic loads.

A high diversity factor could be obtained with four consumers by compelling them to take load as shown in Figure 1.36. Although compulsion obviously cannot be used, encouragement can be provided in the form of tariffs. An example is the two-part tariff in which the consumer has to pay an amount dependent on the maximum demand required, plus a charge for each unit of energy consumed. Sometimes the charge is based on kilovolt-amperes instead of power to penalize loads of low power factor.

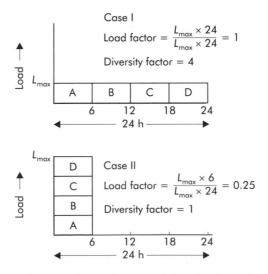

**Figure 1.36**  Two extremes of load factor and diversity factor in a system with four consumers

## Load loss factor

The average losses in the system, e.g. transmission line losses, depend upon the load curves for any given period. The *load loss factor* is defined as the ratio of the actual losses over a period to the losses obtaining if the maximum load is maintained. This is an increasingly important factor because modern methods of charging for metered supplies depend upon accounting for losses between the generator output and the load input.

Referring to Figure 1.36, let us assume that in Case I, $L_{max}$ is 25 kW and in Case II, $L_{max}$ is 100 kW. Case I load is constant for 24 h but Case II is only 6 h duration and is zero for the rest of the day. Then the load factors are as given in the figure.

Now losses in the system, for most practical purposes, are given by the square of the current. But at nominal voltage and unity power factor the current is directly proportional to power in kilowatts. Hence the load loss factor (LLF) over 24 h of operation would be given by the time required for the peak load to produce the same losses as are produced by the actual load over the 24 h period. Thus

$$\text{LLF} = \frac{[\text{average demand (kW)}]^2}{[\text{peak demand (kW)}]^2}$$

In the example cases:

$$\text{Case I:LLF} = \frac{25^2}{25^2} = 1$$

$$\text{Case II:LLF} = \frac{25^2}{100^2} = 0 \cdot 0625$$

because the average demand = 100 ÷ diversity factor = 25 kW and the peak demand = 100 kW.

(Note that average demand can also be calculated as area under the demand–time curve divided by number of hours.)

In practice, the LLF varies with load factor in a quadratic manner dependent upon the daily load curve variation. It can be calculated by carrying out analytical studies on the load pattern over particular networks of known parameters. Typically, in the U.K. the LLF is given approximately by

$$\text{LLF} = 0.2 \text{ (load factor)} + 0.8 \text{ (load factor)}^2$$

## Tariffs

The setting of charges for the supply of electrical energy to consumers is one of the most important aspects of electricity economics. With the advent of competition in both generation and supply markets, the correct and prompt billing of customers is top priority in providing good customer services. Since such billing should now itemize the various charges imposed by the generators and the delivery networks (transmission and distribution systems), to achieve *fair* and *transparent* pricing from source to consumer requires a considerable communications and measurement infrastructure heavily dependent on efficient and compatible computer networking. Here is an excellent example of the application of Information Technology (IT) without which a modern competitive and economically efficient market in electricity supply could not be set up.

The setting of tariffs for generation, supply, and use-of-system (UoS) will be dealt with in more detail in Chapter 12. Briefly, it is worth noting here that two types of charge will be required, namely:

1. A *capacity charge*, to pay for the installed equipment required to produce electrical energy and to deliver it to the customers. This charge is often called the *standing charge* and is dependent, in principle, on the installed capacity assets of the various utilities handling the electrical power from generating sources, through transmission and distribution systems, and supplying the customer's premises. It is designated in monetary units per kilowatt of peak demand (£/kW or $/kW).

2. An *energy charge* to reflect the costs of the primary energy input (coal, oil, gas, hydro, etc.) and the cost of converting it to electrical energy, plus the losses incurred in delivering it to the customer. This charge is dependent upon the energy metered into the consumer's premises and is designated in monetary units per kilowatt-hour (p/kWh or c/kWh).

It should be remarked upon here that whereas the energy consumed is a cumulative quantity measured over a number of hours, days, or years, the peak demand is a 'spot' value which is often recorded only once a year. In practice, and so that the overload capability of equipment is taken into account, the peak kilowatt value is taken as the kilowatt-hour value consumed over, say, 30 min, divided by 0.5 (i.e. half-hour). Naturally, a high load factor (above 0.6) will lead to better utilization of system equipment than a low load factor as this often determines the fuel type which can attract investment.

### Example 1.2

Compare the overall cost per unit generated by coal, oil, gas, and nuclear power stations given the following data:

| Fuel type | Capital cost (£/kW) | Fuel cost (p/kWh) | Efficiency (%) |
|-----------|---------------------|-------------------|----------------|
| Coal | 600 | 1.0 | 38 |
| Oil | 500 | 2.0 | 40 |
| Gas | 300 | 1.3 | 56 |
| Nuclear | 1500 | 0.2 | 30 |

Assume interest on capital and depreciation per annum is 8 per cent and that all have similar life times.

## Solution

It is convenient to work on the basis of 1 kW output and over 1 year. The effect of load factor can be explored by repeating the calculation over a range of load factors. Table 1.3 has been compiled, using coal-fired generation data as an example.

Capital cost on an annual basis for each kW installed

$$= 600 \times \frac{8}{100} = £48$$

Energy generated per kW per annum

$$= LF \times 365 \times 24 = 0.4 \times 8760 = 3504 \text{kWh}$$

Cost of 3504 kWh is given by

$$\frac{\text{Cost per unit}}{\text{Efficiency of conversion}} \times 3504 = \frac{1.0}{0.38} \times \frac{3504}{100} = £92.21$$

Total cost per annum to produce 3504 kWh is capital cost + generating cost

$$= 48 + 92.21 = £140.21$$

Hence, cost per unit generated is

$$\frac{140.21}{3504} = £0.0400 \quad \text{or 4p per unit}$$

**Table 1.3**  Cost in pence per unit at various load factors

| Fuel type | LF | | | | |
|-----------|------|------|------|------|------|
|           | 0.4 | 0.5 | 0.6 | 0.7 | 0.8 |
| Coal | 4.00 | 3.73 | 3.54 | 3.41 | 3.32 |
| Oil | 6.14 | 5.91 | 5.76 | 5.65 | 5.57 |
| Gas | 3.00 | 2.87 | 2.78 | 2.71 | 2.66 |
| Nuclear | 4.09 | 3.41 | 2.95 | 2.62 | 2.38 |

It is worth noting from Table 1.3 how nuclear generation is heavily dependent upon having a high LF to be able to compete with gas (using CCGT) and coal. With the prices quoted, gas is cheaper than coal at any load factor, hence emphasizing the importance of keeping coal costs down through long-term contracts.

## Load management

Attempts to modify the shape of the load curve to produce economy of operation have already been mentioned. These have included tariffs, pumped storage, and the use of seasonal or daily diversity between interconnected systems. A more direct method would be the control of the load either through tariff structure or direct electrical control of appliances, the latter, say, in the form of remote on/off control of electric water-heaters where inconvenience to the consumer is least. For many years this has been achieved with domestic time switches, but some schemes use switches radio-controlled from the utility to give greater flexibility. This permits load reductions almost instantaneously and defers the hot-water and heating of air-conditioning load until after system peaks. This topic is considered further in Chapter 12.

## Load forecasting

It is evident that load forecasting is a crucial activity in electricity supply. Forecasts are based on the previous year's loading for the period in question, updated by factors such as general load increases, major new loads, and weather trends. Both power demand and energy (kWh) forecasts are used, the latter often being the more readily obtained. Demand values may be determined from energy forecasts. Energy trends tend to be less erratic than peak power demands and are considered better growth indicators; however, load factors are also erratic in nature.

As weather has a much greater influence on residential than on industrial demands it may be preferable to assemble the load forecast in constituent parts to obtain the total. In many cases the seasonal variations in peak demand are caused by weather-sensitive domestic appliances, e.g. heaters and air-conditioning. A knowledge of the increasing use of such appliances is therefore essential. Several techniques are available for forecasting. These range from simple curve fitting and extrapolation to stochastic modelling and are given in detail in Sullivan, 1977. The many physical factors affecting loads, e.g. weather, national economic heath, popular TV programmes, public holidays, etc., make forecasting a complex process demanding experience and high analytical ability using probabilistic techniques including neural networks.

# Problems

**1.1** In the U.S.A. in 1971 the total area of right of ways for H.V. overhead lines was $16\,000\,\text{km}^2$. Assuming a growth rate for the supply of electricity of 7 per cent per annum calculate what year the whole of the U.S.A. will be covered with transmission systems (assume area to approximate $4800 \times 1600\,\text{km}$). Justify any assumptions made and discuss critically why the result is meaningless.
(Answer: 91.25 years)

**1.2** The calorific value of natural gas at atmospheric pressure and temperature is $12 \times 10^6\,\text{J/m}^3$. Calculate the energy transfer in a pipe of $1\,\text{m}$ diameter with gas at $60\,\text{atm}$ (gauge) flowing at $5\,\text{m/s}$. If *liquid* hydrogen was transferred at the same velocity calculate the energy transfer if the calorific value of liquid hydrogen is 650 times the value for gas at $1\,\text{atm}$.
(Answer: $2.8\,\text{GW}$, $30.6\,\text{GW}$)

**1.3** (a) An electric car has a steady *output* of $10\,\text{kW}$ over its range of $100\,\text{km}$ when running at a steady $40\,\text{km/h}$. The efficiency of the car (including batteries) is 65 per cent. At the end of the car's range the batteries are recharged over a period of $10\,\text{h}$. Calculate the average charging power if the efficiency of the battery charger is 90 per cent.

(b) The calorific value of gasoline (petrol) is roughly $16\,500\,\text{kJ/gallon}$. By assuming an average filling rate at a pump of 10 gallon/minute, estimate the rate of energy transfer on filling a gasoline-driven car. What range and what cost/km would the same car as (a) above produce if driven by gasoline with a 7 gallon tank? (Assume internal combustion engine efficiency is 60% and gasoline costs £3 per gallon)
(Answer: (a) $4.3\,\text{kW}$; (b) $2.75\,\text{MW}$, $76.8\,\text{km}$, 27p/km)

**1.4** Discuss in a critical manner the possible impact of solar energy on the electrical supply industry ($1\,\text{therm} = 105.5 \times 10^3\,\text{kJ}$)

(a) as a means of central generation;
(b) as a domestic energy source.

A solar panel ($5\,\text{m} \times 2\,\text{m}$) is used for heating domestic water. The absorbing panel has an absorbtivity of 0.93, a top-plate emissivity of 0.9, and a back-plate emissivity of 0.05. The transmittance through the glass top cover is 0.85.

With a peak radiation of $700\,\text{W/m}^2$ the absorber temperature is $100°\text{C}$ and the top cover $60°\text{C}$. If all the absorber heat is transferred to the water which flows through ducts in the absorber with negligible heat-transfer resistance, determine the water mass flow rate (kg/s) necessary to raise the water temperature from $15°\text{C}$ (inlet) to $60°\text{C}$ at peak radiation. State clearly any assumptions made. (Stefan–Boltzmann constant $= 5.7 \times 10^{-8}\,\text{W/m}^2\,\text{K}^4$.)
(Answer: $0.092\,\text{kg/s}$)

**1.5** The variation of load ($P$) with time ($t$) in a power supply system is given by the expression,

$$P\,(\text{kW}) = 4000 + 8t - 0.00091t^2$$

where $t$ is in hours over a total period of 1 year.
This load is supplied by three $10\,\text{MW}$ generators and it is advantageous to fully load a machine before connecting the others. Determine:

(a) the load factor on the system as a whole;
(b) the total magnitude of installed load if the diversity factor is equal to 3;
(c) the minimum number of hours each machine is in operation;
(d) the approximate peak magnitude of installed load capacity to be cut off to enable only two generators to be used.
(Answer: (a) 0.74 (b) 65 MW (c) 8760, 7300, 2500 h (d) 5.4 MW)

**1.6**    (a) Explain why economic storage of electrical energy would be of great benefit to power systems.

(b) List the technologies for the storage of electrical energy which are available now and discuss, briefly, their disadvantages.

(c) Why is hydro power a very useful component in a power system?

(d) Explain the action of pumped storage and describe its limitations.

(e) A pumped storage unit has an efficiency of 78 per cent when pumping and 82 per cent when generating. If pumping can be scheduled using energy costing 2.0 p/kWh, plot the gross loss/profit in p/kWh when it generates into the system with a marginal cost between 2p and 6p/kWh.

(f) Explain why out-of-merit generation is sometimes scheduled.
(Answer: (e) Overall efficiency 64 per cent; Profit max. 2.87 p/kWh; Loss max. 1.13 p/kWh)

*(From Engineering Council (E.C.) Examination, 1996)*

# 2

# *Basic Concepts*

Before the modelling and analysis of power systems is pursued in detail, various basic ideas will be outlined. These include the nature of three-phase systems, the per-unit system, and circuit representation. Hopefully, some of this material will be revision for many readers.

## 2.1 Three-Phase Systems

The rotor flux of an alternating current machine induces sinusoidal e.m.f.s, the conductors forming the stator winding, which in a single-phase machine occupies slots over most of the circumference of the stator core. These e.m.f.s are not in phase and net winding voltage is less than the arithmetic sum of the individual conductor voltages. If this winding is replaced by three separate identical windings, as shown in Figure 2.1(a), each occupying one-third of the available slots, then the effective contribution of all the conductors is greatly increased, yielding a greatly enhanced power capability for a given machine size.

These three windings give voltages displaced in time or phase by 120°, as indicated in Figure 2.1(b). Also, because the voltage in the (a) phase reaches its peak 120° before the (b) phase and 240° before the (c) phase, the order of phase voltages reaching their maxima or *phase sequence* is a–b–c. Most countries use a,b, and c to denote the phases; however, R (Red), Y (Yellow), and B (Blue) is often also used. It is seen that the algebraic sum of the winding or phase voltages (and currents if the winding currents are equal) at every instant in time is zero. Hence, if the 'starts' of each winding are connected, then the electrical situation is unchanged and the three return lines can be dispensed with, yielding a three-phase, three-wire system, as shown in Figure 2.2(a). If the currents from the windings are not equal, then it is usual to connect a fourth wire (neutral) to the common connection or neutral point, as shown in Figure 2.2(b).

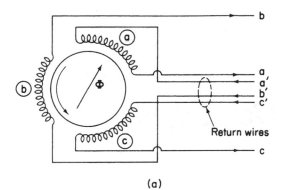

(a)

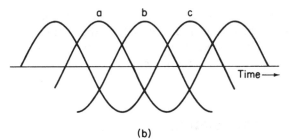

(b)

**Figure 2.1**    (a) Synchronous machine with three separate stator windings a, b, and c displaced physically by 120°. (b) Variation of e.m.f.s developed in the windings with time

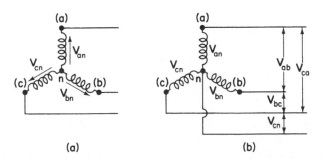

**Figure 2.2**    (a) Wye or star connection of windings. (b) Wye connection with neutral line

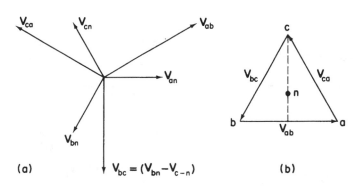

**Figure 2.3** (a) Phasor diagram for wye connection. (b) Alternative arrangement of line-to-line voltages. Neutral voltage is at n, geometric centre of equilateral triangle

This type of winding connection is called 'wye' or 'star' and two sets of voltages exist:

1.  the winding, phase, or line-to-neutral voltage, i.e. $V_{an}$, $V_{bn}$, $V_{cn}$; and

2.  the line-to line voltages, $V_{ab}$, $V_{bc}$, $V_{ca}$. The subscripts here are important, $V_{ab}$, means the voltage of line or terminal (a) with respect to (b) ($V_{ba} = V_{ab}$). The corresponding phasor diagram is shown in Figure 2.3(a) and it can be shown that $V_{ab} = V_{bc} = V_{ca} = \sqrt{3}$ (phase voltage).

The phase rotation of a system is very important. Consider the connection through a switch of two voltage sources of equal magnitude and both of rotation a–b–c. When the switch is closed no current flows. If, however, one source is of reversed rotation (easily obtained by reversing two wires), as shown in Figure 2.4, i.e. a–c–b, a large voltage ($\sqrt{3} \times$ phase voltage) exists across the switch contacts c'b and b'c, resulting in very large currents if the switch is closed. Also, on reversed phase rotation the rotating magnetic field set up by a three-phase winding is reversed in direction and an induction motor will rotate in the opposite direction, often with disastrous results to its mechanical load, e.g. a pump.

A three-phase load is connected in the same way as the machine windings. The load is *balanced* when each phase takes equal currents, i.e. has equal impedance. With the wye system the phase currents are equal to the current in the lines. The *four-wire* system is of particular use for low-voltage distribution networks in which consumers are supplied with a single-phase supply taken between a line and neutral. This supply is often 230 V or 240 V and the line-to-line voltage is 400 V or 415 V. In the U.S.A. the 220 V supply often comes into a house from a centre-tapped transformer, as shown in Figure 2.5, which in effect gives a choice of 220 V (large domestic appliances) or 110 V (lights, etc.).

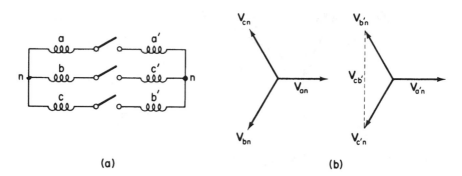

**Figure 2.4**    (a) Two generators connected by switch; phase voltages equal for both sets of windings. (b) Phasor diagrams of voltages. $V_{cb'}$ = voltage across switch; $V_{a'a} = 0$

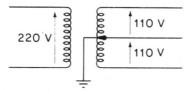

**Figure 2.5**    Tapped single-phase supply to give 220/110 V (centre-tap grounded)

The system planner will endeavour to connect the single-phase loads such as to provide balanced (or equal) currents in the supply lines. At any instant in time it is highly unlikely that consumers will take equal loads, and at this level of supply considerable unbalance occurs, resulting in currents in the neutral line. If the neutral line has zero impedance, this unbalance does not affect the load voltages. Lower currents flow in the neutral and it is usually of smaller cross-sectional area than the main line conductors. The combined or statistical effect of the large number of loads on the low-voltage network is such that when the next higher distribution voltage is considered (say 11 kV (line to line)), which supplies the lower voltage network, the degree of unbalance is small. This and the fact that at this higher voltage, large three-phase, balanced motor loads are supplied, allows the three-wire system to be used. It is used exclusively up to the highest transmission voltages, resulting in much reduced line costs and environmental impact.

In a balanced three-wire system a hypothetical neutral line may be considered and the conditions in only one phase determined. This is illustrated by the phasor diagram of line-to-line voltages shown in Figure 2.3(b). As the system is balanced the magnitudes so derived will apply to the other two phases but the relative phase angles must be adjusted by 120° and 240°. This single-phase approach is very convenient and widely used.

An alternative method of connection is shown in Figure 2.6. The individual phases are connected (taking due cognizance of winding polarity in machines and transformers) to form a closed loop. This is known as the mesh or delta connection. Here the line-to-line voltages are identical to the phase voltages, i.e. $V_{\text{line}} = V_{\text{phase}}$. The line currents are as follows:

$$\mathbf{I_a} = \mathbf{I_{ab}} - \mathbf{I_{ca}} \qquad \mathbf{I_b} = \mathbf{I_{bc}} - \mathbf{I_{ab}} \qquad \mathbf{I_c} = \mathbf{I_{ca}} - \mathbf{I_{bc}}$$

(Note: Bold symbols are complex (phasor) quantities requiring complex addition and subtraction.)

For balanced currents in each phase it is readily shown from a phasor diagram that $I_{\text{line}} = \sqrt{3} \times I_{\text{phase}}$. Obviously a fourth or neutral line is not possible with this connection. It is seldom used for rotating-machine stator windings, but is frequently used for the windings of one side of transformers for reasons to be discussed later. A line-to-line voltage transformation ration of $1 : \sqrt{3}$ is obtained when going from a primary mesh to a secondary wye connection with the same number of turns per phase. Under balanced conditions the idea of the hypothetical neutral and single-phase solution may still be used (the mesh can be converted to a wye using the $\Delta - Y$ transformation).

It should be noted that three-phase systems are usually described by the *line-to-line voltage*.

### 2.1.1  Analysis of simple three-phase circuits

#### Four-wire systems

If the impedance voltage drops in the lines are negligible, then the voltage across each load is the source line or phase voltage.

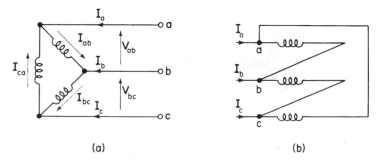

(a)                              (b)

**Figure 2.6**  (a) Mesh or delta-connected load–current relationships. (b) Practical connections

## *Example 2.1*

For the network of Figure 2.7(a), draw the phasor diagram showing voltages and currents and write down expressions for total line currents $\mathbf{I_a}$, $\mathbf{I_b}$, $\mathbf{I_c}$, and the neutral current $\mathbf{I_n}$.

Solution

Note that the power factor of the three-phase load is expressed with respect to the phase current and voltage. The phasor diagram is shown in Figure 2.7(b). The line currents are as follows:

$$\mathbf{I_a} = \frac{V}{\sqrt{3}} \cdot \frac{1}{\mathbf{Z_a}} + I_m(\cos \phi - j \sin \phi)$$

$$\mathbf{I_b} = \frac{V}{\sqrt{3}}(-0.5 - j0.866)\frac{1}{\mathbf{Z_b}} + I_m[-\cos (60 - \phi) - j \sin (60 - \phi)]$$

$$\mathbf{I_c} = \frac{V}{\sqrt{3}}(-0.5 - j0.866)\frac{1}{\mathbf{Z_c}} + I_m[-\cos (60 - \phi) - j \sin (60 - \phi)]$$

The neutral current

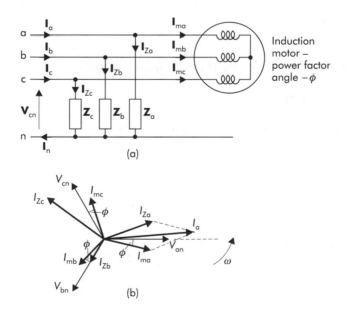

**Figure 2.7** (a) Four-wire system with single-phase unbalanced loads and three-phase balanced motor load. (b) Phasor diagram: only $\mathbf{I_{za}}$ and $\mathbf{I_{ma}}$ have been phasorially added to show $\mathbf{I_a}$. $\mathbf{I_b}$, and $\mathbf{I_c}$ can be found in a similar way. Note: $\mathbf{V_{an}}$ is reference direction. Rotation is anticlockwise

$$I_n = I_a + I_b + I_c = \frac{V}{\sqrt{3}}[Y_a + Y_b\,(0.5 - j0.866) + Y_c(-0.5 + j0.866)]$$

since $I_m$ does not contribute to $I_n$.

## Three-wire balanced

The system may be treated as a single-phase system using phase voltages. It must be remembered, however, that the total three-phase power and reactive power are three times the single-phase values.

## Three-wire unbalanced

In more complex networks the method of symmetrical components is used (see Chapter 7), but in simple situations conventional network theory can be applied. Consider the source load arrangement shown in Figure 2.8, in which $Z_a \neq Z_b \neq Z_c$. From the mesh method of analysis, the following equation is obtained.

$$\begin{bmatrix} Z_a + Z_b & -Z_b \\ -Z_b & Z_b + Z_c \end{bmatrix} \begin{bmatrix} i_1 \\ i_2 \end{bmatrix} = \begin{bmatrix} V \\ V(-0.5 - j0.866) \end{bmatrix}$$

Hence $i_1$ and $i_2$ are determined and from these are found $I_a$, $I_b$, and $I_c$ and the voltage of the neutral point (n) from ground (see Figure 2.8(b)). Depending on the severity of the imbalance, the voltage difference from (n) to ground can attain values exceeding the phase voltage, i.e. (n) can lie outside the triangle of line voltages. Such conditions produce damage to connected equipment and show the importance of earthing (grounding) the neutral.

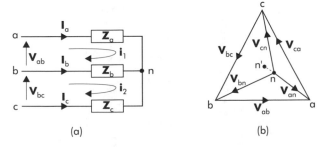

**Figure 2.8** (a) Three-wire system with unbalanced load, $i_1$ and $i_2$ are loop currents. (b) Phasor diagram—voltage difference of neutral connection (n) from ground = n'n; n' = neutral voltage with balance

The power and reactive power consumed by a *balanced* three-phase load for both wye or mesh connections are given by

$$P = \sqrt{3} \cdot V \cdot I \cdot \cos \phi \qquad Q = \sqrt{3} \cdot V \cdot I \cdot \sin \phi$$

where

$V$ = line-to-line voltage;
$I$ = line current;
$\phi$ = angle between load phase current and load *phase voltage*.

Alternatively

$$P = 3V_{\text{ph}} \cdot I_{\text{ph}} \cdot \cos \phi \qquad \text{and} \qquad Q = 3V_{\text{ph}} \cdot I_{\text{ph}} \cdot \sin \phi$$

As the powers in each phase of a balanced load are equal, a single wattmeter may be used to measure total power. For unbalanced loads, two wattmeters are sufficient with three-wire circuits.

## Example 2.2

A three-phase wye-connected load is shown in Figure 2.9. It is supplied from a 200 V, three-phase, four-wire supply of phase sequence a–b–c and the neutral line (n) has a resistance of 5 Ω. Two wattmeters are connected as shown. Calculate the power recorded on each wattmeter.

## Solution

The mesh method will be used to determine the currents in the lines and hence in the wattmeter current coils.

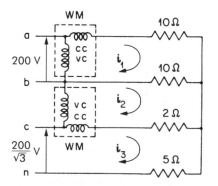

**Figure 2.9** Use of two wattmeters to measure three-phase power. Arrangement for Example 2.2; cc = current coil, vc = voltage coil

The line voltage $V_{ab}$ is used as a reference phasor. Hence,

$$V_{bc} = (-0.5 - j0.866)200$$
$$V_{ca} = (-0.5 + j0.866)200$$

and

$$V_{cn} = +j\frac{200}{\sqrt{3}}$$

From Figure 2.9,

$$\begin{bmatrix} 20 & -10 & 0 \\ -10 & 12 & -2 \\ 0 & -2 & 7 \end{bmatrix}\begin{bmatrix} i_1 \\ i_2 \\ i_3 \end{bmatrix} = \begin{bmatrix} 200 \\ -100 - j173.2 \\ j\dfrac{200}{\sqrt{3}} \end{bmatrix}$$

Eliminate $i_1$,

$$\begin{bmatrix} 7 & -2 \\ -2 & 7 \end{bmatrix}\begin{bmatrix} i_2 \\ i_3 \end{bmatrix} = \begin{bmatrix} j173.2 \\ j\dfrac{200}{3} \end{bmatrix}$$

From which

$$i_3 = j2.66A$$
$$i_2 = -j24.0A$$
$$i_1 = 10 - j12A$$

Power in wattmeter (1) = real part of $V_{ab} \cdot I_a^* = 200(10 + j12)$
$$= 2 \text{ kW}$$
(see Section 2.5 for use of the conjugate $I_a^*$)

Power in wattmeter (2) = real part of $V_{cb} \cdot I_c^* = (-V_{bc})I_c^*$
$$= (100 + j173.2)(i_3 - i_2)^*$$
$$= (100 + j173.2)(-j26.66)$$
$$= 4.612 \text{ kW}$$

Actual power consumed $= I_a^2 \cdot 10 + I_b^2 \cdot 10 + I_c^2 \cdot 2 + I_n^2 \cdot 5$
$$= 6.337 \text{ kW}$$

As the loads are not balanced and there is a return current in the neutral, the measured power is *not* the sum of the two wattmeter readings.

## 2.2   Three-Phase Transformers

The usual form of the three-phase transformer, i.e. the core type, is shown in Figure 2.10. If the magnetic reluctances of the three limbs are equal, then the sum of the fluxes set up by the three-phase magnetizing currents is zero. In fact, the core is the magnetic equivalent of the wye-connected winding. It is apparent

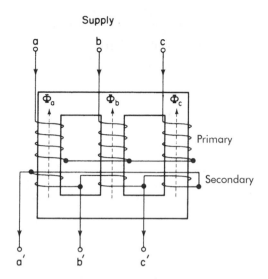

**Figure 2.10**   Three-phase core-type transformer. Primary connected in wye, secondary connected in mesh

from Figure 2.10 that the magnetic reluctances are not exactly equal, but in an introductory treatment may be so assumed. An alternative to the three-limbed core is the use of three separate single-phase transformers. Although more expensive (about 20 per cent extra), this has the advantage of smaller loads for transportation, and this aspect is crucial for large sizes. Also, with the installation of four single-phase units, a spare is available at reasonable cost.

The wound core, as shown in Figure 2.10, is placed in a steel tank filled with insulating oil or synthetic liquid. The oil acts both as electrical insulation and as a cooling agent to remove the losses from the windings and core. The low-voltage winding is situated over the core limb and the high-voltage winding is wound over the low-voltage one. The core comprises steel laminations insulated on one side (to reduce eddy losses) and clamped together.

Note that the voltages across the windings are phase voltages and it is these which are related by the turns ratio, but the secondary voltages are 30° out of phase with the primary voltages.

### 2.2.1  *Autotransformers* (Figure 2.11)

In the autotransformer only one winding is used per phase, the secondary voltage being tapped off the primary winding. There is obviously a saving in size, weight, and cost over a two-windings per phase transformer. It may be shown that the ratio of the weight of conductor on an autotransformer to that on a double-wound one is given by $(1 - N_2/N_1)$. Hence, maximum advantage

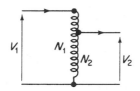

**Figure 2.11** One phase of an autotransformer: $V_2/V_1 = N_2/N_1$

is obtained with a relatively small difference between the voltages on the two sides. There are, however, several requirements in a power system where connections from one voltage level to another do not entail large transformer ratios, e.g. 500 kV/345 kV, 500 kV/725 kV, and then the auto transformer is used. The effective reactance is reduced compared with the equivalent two-winding transformer and this can give rise to high short-circuit currents. The general constructional features of the core and tank are similar to those of double-wound transformers, but note that the primary and secondary voltages are now in-phase.

## 2.3 Harmonics in Three-Phase Systems

Harmonics may exist in three-phase systems and are created by non-linear loads, rectifiers, etc. Consider the instantaneous values of phase voltages in a balanced system containing harmonics up to the third:

$$v_a = V_1 \sin \omega t + V_2 \sin 2\omega t + V_3 \sin 3\omega t$$

$$v_b = V_1 \sin\left(\omega t - \frac{2\pi}{3}\right) + V_2 \sin 2\left(\omega t - \frac{2\pi}{3}\right) + V_3 \sin 3\left(\omega t - \frac{2\pi}{3}\right)$$

$$v_c = V_1 \sin\left(\omega t - \frac{4\pi}{3}\right) + V_2 \sin 2\left(\omega t - \frac{4\pi}{3}\right) + V_3 \sin 3\left(\omega t - \frac{4\pi}{3}\right)$$

Here $V_1$, $V_2$ and $V_3$ are the magnitude of the harmonic voltages. The fundamental terms have the normal phase shifts (as do the fourth, seventh and tenth, etc., harmonics). However, the second harmonic terms possess a reversed phase-sequence (also fifth, eighth, eleventh, etc.), and the third harmonic terms are all in phase (also all multiples of three).

When substantial harmonics are present the $\sqrt{3}$ relation between line and phase quantities no longer holds. Harmonic voltages in successive phases are $2\pi n/3$ out of phase with each other and the resulting line voltage becomes

$$v_{ab} = V_n \sin n\omega t - V_n \sin n\left(\omega t - \frac{2\pi}{3}\right)$$

i.e.

$$\mathbf{v}_{ab} = 2V_n \cos n\left(\omega t - \frac{\pi}{3}\right) \sin\left(\frac{n\pi}{3}\right)$$

When $n = 3$, 6, 9, etc., this becomes zero, i.e. no triple harmonics exist in the line voltages. The mesh connection forms a complete path for the triple harmonic currents which flow in phase around the loop.

When analysing the penetration of harmonics into the power network it is usual to assume that the effective reactance for the $n$th harmonic is $n$ times the fundamental value.

## 2.4  Multiphase Systems

Although the three-phase system is universally used, attention has been given in recent years to the use of more than three phases for transmission purposes. In particular, six- and twelve-phase systems have been studied. Advantages relative to three-phase systems are as follows:

1.  Increased thermal loading capacity of lines.

2.  The higher the number of phases, the smaller the line-to-line voltage becomes relative to the phase voltage, resulting in increased utilization of rights of way because of less phase-to-phase insulation requirement.

3.  For a given conductor size and tower configuration the stress on the conductor surface decreases with the number of phases, leading to reduced corona effects.

4.  The transmission efficiency is higher. Existing double-circuit lines (two three-phase circuits on each tower) could be converted to single-circuit six-phase lines. It is advantageous to describe multiphase systems in terms of the phase voltage rather than line-to-line, as is the case for three-phase systems.

A six-phase supply is obtainable by suitable arrangement of the secondary windings of a three-phase transformer. Consider Figure 2.12. The windings on the three limbs of the transformer are centretapped with the taps mutually connected. This gives the phasor diagram shown in Figure 2.12(b). The line-to-line voltages are equal to the phase voltage in magnitude, i.e. $\mathbf{V}_{af} = \mathbf{V}_{an} - \mathbf{V}_{fn} = \mathbf{V}_{ph}$.

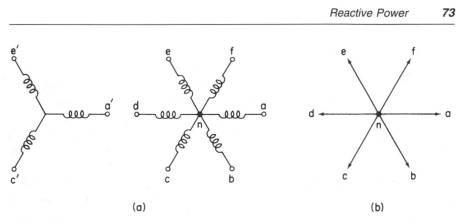

**Figure 2.12**   Six-phase system. (a) Transformer connections. (b) Phasor diagrams

## 2.5   Reactive Power

In the circuit shown in Figure 2.13, let the instantaneous values of voltage and current be

$$e = \sqrt{2}E \sin(\omega t + \phi)$$

and

$$i = \sqrt{2}I \sin(\omega t)$$

The instantaneous power

$$p = ei = EI \cos \phi - EI \cos(2\omega t + \phi)$$

also

$$-EI \cos(2\omega t + \phi) = -EI(\cos 2\omega t \cos \phi - \sin 2\omega t \cdot \sin \phi)$$

$$\therefore \; p = ei = (EI \cos \phi - EI \cos 2\omega t \cos \phi) + (EI \sin 2\omega t \sin \phi)$$
$$= (\text{instantaneous real power}) + (\text{instantaneous reactive power})$$

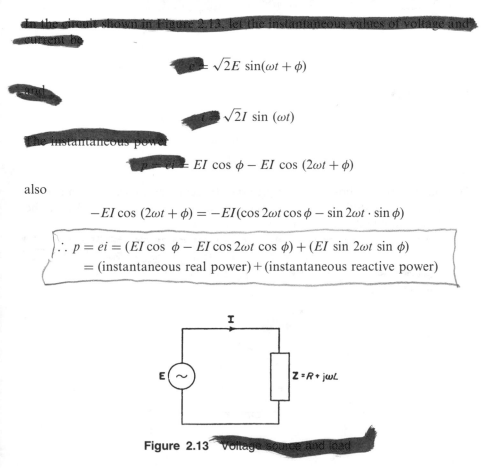

**Figure 2.13**   Voltage source and load

The mean power = $EI \cos \phi$.

The mean value of $EI \sin \phi \sin 2\omega t = 0$, but the maximum value = $EI \sin \phi$.

The voltage source is supplying energy to the load in one direction only. At the same time an interchange of energy is taking place between the source and the load of average value zero, but of peak value $EI \sin \phi$. This latter quantity is known as the *reactive power* (*Q*) and the unit is the VAr (taken from the alternative name, volt-ampere reactive). The interchange of energy between the source and the inductive and capacitive elements (i.e. the magnetic and electric fields) takes place at twice the supply frequency. Therefore, it is possible to think of a power component *P* (watts) of magnitude $EI \cos \phi$ and a reactive power component *Q* (VAr) equal to $EI \sin \phi$, where $\phi$ is the power factor angle, i.e. the angle between **E** and **I**. It should be stressed, however, that the two quantities *P* and *Q* are physically quite different.

The quantity **S** (volt-amperes), known as the complex power, may be found by multiplying **E** by the conjugate of **I** or vice versa. Consider the case when **I** lags **E** and assume $\mathbf{S} = \mathbf{E}^*\mathbf{I}$. Referring to Figure 2.14,

*Magnitude of instance Power (p t)*

$$\mathbf{S} = E\, e^{-j\phi_1} \times I\, e^{j\phi_2} = EI\, e^{-(\phi_1-\phi_2)}$$
$$= EI\, e^{-\phi}$$
$$= P - jQ$$

$$\boxed{P(\text{watts}) = EI \cos\phi \\ Q(\text{VAr}) = EI \sin\phi}$$

Next assume

$$\mathbf{S} = EI^*$$
$$= EI\, e^{j(\phi_1-\phi_2)}$$
$$= P + jQ.$$

$\left( I \text{ lags } E \right)$

Obviously both the above methods give the correct magnitudes for *P* and *Q* but the sign of *Q* is different. The method used is arbitrarily decided and the convention to be adopted in this book is as follows:

The volt-amperes reactive absorbed by an inductive load shall be considered positive, and by a capacitive load negative, hence $\mathbf{S} = \mathbf{E}\mathbf{I}^*$. This convention is recommended by the International Electrotechnical Commission.

**Figure 2.14** Phasor diagram $\left( I \text{ lags } E \right)$

In a network the net stored energy is the sum of the various inductive and capacitive stored energies present. The net value of reactive power is the sum of the vars absorbed by the various components present, taking due account of the sign. Vars can be considered as being either produced or absorbed in a circuit: a capacitive load can be thought of as generating vars. Assuming that an inductive load is represented by $R + jX$ and that the $EI^*$ convention is used, then an inductive load *absorbs* positive vars and a capacitive load *produces* vars. (The reactive power flow towards a busbar is positive when the load is lagging; vars exported from a busbar are considered negative for a lagging power factor and positive for leading one. In this book, reference to the var will imply a lagging power factor, and inductive and capacitive vars will be distinguished by the sign.)

The various elements in a network are characterized by their ability to generate or absorb reactive power. Consider a synchronous generator which can be represented by the simple equivalent circuit shown in Figure 2.15. When the generator is overexcited, i.e. its generated e.m.f. is high, it produces vars and a complex power, $P - jQ$. When the machine is underexcited the generated current leads $V$ and the generator produces $P + jQ$. It can also be thought of as absorbing vars. The reactive power characteristics of various power-system components are summarized as follows: reactive power is generated by overexcited synchronous machines, capacitors, cables, and lightly loaded overhead lines, and absorbed by underexcited synchronous machines, induction motors, inductors, transformers, and heavily loaded overhead lines.

*Handwritten annotations:*

Load (load) source

Synch gen

Generator → $P - jQ$
produces negative vars
E large → over excited gen
E large

Source Only → Produces Neg Vars
$I$ lags $V$ (lagging PF)   E   $(Q \to -)$
$I$ lags $V_s = IX_s$
$IX_s$
Load Lagging

nerator → $P + jQ$
sorbs pos vars
– Small → underexcited generator

Source Only → Absorbs Positive Vars $(Q \to +)$
$I$ leads $V$ (leading PF)
$I$ lags $V_s = IX_s$
Load Leading

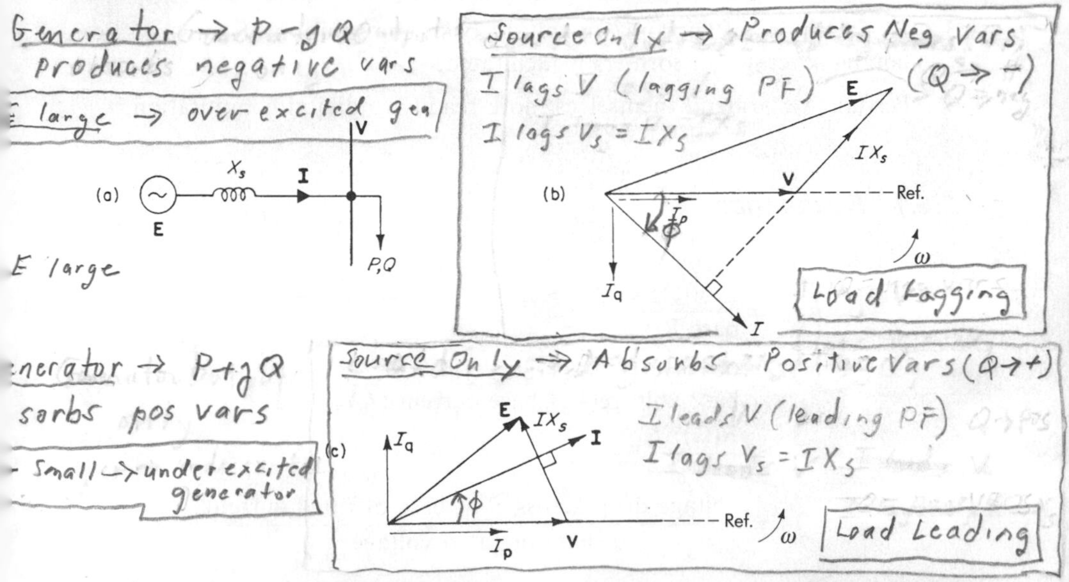

**Figure 2.15** (a) Line diagram of system. (b) Overexcited generator–phasor diagram. (c) Underexcited generator–phasor diagram.

The reactive power absorbed by a reactance of $X_L(\Omega) = I^2 X_L$ where $I$ is the current. Note that at a node, $\Sigma P = 0$ and $\Sigma Q = 0$ in accordance with Kirchhoff's law. A capacitance with a voltage $V$ applied produces $V^2 B$ VAr, where $B = 1/X_c = \omega C$ if $X_c = 1/\omega C$.

## 2.6    The Per-Unit System

In the analysis of power networks, instead of using actual values of quantities it is usual to express them as fractions of reference quantities, such as rated or full-load values. These fractions are called per unit (denoted by p.u.) and the p.u. value of any quantity is defined as

$$\frac{\text{actual value (in any unit)}}{\text{base or reference value in the same unit}}$$

Some authorities express the p.u. value as a percentage. Although the use of p.u. values may at first sight seem a rather indirect method of expression there are, in fact, great advantages; they are as follows:

1.   The apparatus considered may vary widely in size; losses and volt drops will also vary considerably. For apparatus of the same general type the p.u. volt drops and losses are in the same order, regardless of size.

2.   As will be seen later, the use of $\sqrt{3}$'s in three-phase calculations is reduced.

3.   By the choice of appropriate voltage bases the solution of networks containing several transformers is facilitated.

4.   Per-unit values lend themselves more readily to digital computation.

### 2.6.1    Resistance

$$R_{\text{p.u.}} = \frac{R(\Omega)}{\text{base } R(\Omega)}$$

$$= \frac{R(\Omega)}{\text{base voltage}(V_b)/\text{base current } (I_b)}$$

$$= \frac{R(\Omega) \cdot I_b}{V_b}$$

$$= \frac{\text{voltage drop across } R \text{ at base or rated current}}{\text{base or rated voltage}}$$

Strictly, $R_{\text{p.u.}}$ is a non-dimensional ratio, but it is often given a unit (such as p.u. ohm) to provide a check on calculations.

Also

$$R_{\text{p.u.}} = \frac{R(\Omega)I_b^2}{V_b I_b}$$

$$= \frac{\text{power loss at base current}}{\text{base power or volt-amperes}}$$

$\therefore$ the power loss (p.u.) at base or rated current $= R_{\text{p.u.}}$

Power loss (p.u.) at $I_{\text{p.u.}}$ current $= R_{\text{p.u.}} \cdot I_{\text{p.u.}}^2$

Similarly

$$\text{p.u. impedance} = \frac{\text{impedance in ohms}}{\left(\dfrac{\text{base voltage}}{\text{base current}}\right)} = \frac{Z(\Omega).I_b}{V_b}$$

## *Example 2.3*

A d.c. series machine rated at 200 V, 100 A has an armature resistance of 0.1 $\Omega$ and field resistance of 0.15 $\Omega$. The friction and windage loss is 1500 W. Calculate the efficiency when operating as a generator.

## *Solution*

Total series resistance in p.u. is given by

$$R_{\text{p.u.}} = \frac{0.25}{200/100} = 0.125 \text{ p.u.}$$

where

$$V_{\text{base}} = 200V \qquad \text{and} \qquad I_{\text{base}} = 100 \text{ A}$$

Friction and windage loss

$$= \frac{1500}{200 \times 100} = 0.075 \text{ p.u.}$$

At the rated load, the series-resistance loss

$$= 1^2 \times 0.125 \text{ p.u.}$$

and the total loss

$$= 0.125 + 0.075 = 0.2 \text{ p.u.}$$

As the output $=^{\gamma} 1$ p.u., the efficiency

$$= \frac{1}{1 + 0.2} = 0.83 \text{ p.u.}$$

## Three-phase circuits

A p.u. phase voltage has the same numerical value as the corresponding p.u. line voltage. With a line voltage of 100 kV and a rated line voltage of 132 kV, the p.u. value is 0.76. The equivalent phase voltages are $100/\sqrt{3}$ kV and $132/\sqrt{3}$ kV and hence the p.u. value is again 0.76. The actual values of $R$, $X_{\mathrm{L}}$, and $X_{\mathrm{c}}$ for lines, cables, and other apparatus are phase values. When working with ohmic values it is less confusing to use the phase values of all quantities. In the p.u. system, three-phase values of voltage, current, and power can be used without undue anxiety about the result being a factor of $\sqrt{3}$ incorrect.

It is convenient in a.c. circuit calculations to work in terms of base volt-amperes, base $VA$. Thus,

$$\text{base } VA = V_{\mathrm{base}} \times I_{\mathrm{base}} \times \sqrt{3}$$

when $V_{\mathrm{base}}$ is the line voltage and $I_{\mathrm{base}}$ is the line current in a three-phase system.

Hence

$$I_{\mathrm{base}} = \frac{\text{base } VA}{\sqrt{3} \cdot V_{\mathrm{base}}} \tag{2.1}$$

Expression (2.1) shows that if base $VA$ and $V_{\mathrm{base}}$ are specified, then $I_{\mathrm{base}}$ is also specified. Consequently, only *two* base quantities can be chosen from which all other quantities in a three-phase system can be calculated. Thus,

$$Z_{\mathrm{base}} = \frac{V_{\mathrm{base}}/\sqrt{3}}{I_{\mathrm{base}}} = \frac{V_{\mathrm{base}}/\sqrt{3}}{VA_{\mathrm{base}}/(\sqrt{3} \cdot V_{\mathrm{base}})} = \frac{V_{\mathrm{base}}^2}{VA_{\mathrm{base}}} \tag{2.2}$$

Hence

$$Z_{\mathrm{p.u.}} = \frac{Z(\Omega)}{Z_{\mathrm{base}}} = \frac{Z(\Omega) \times VA_{\mathrm{base}}}{V_{\mathrm{base}}^2} \tag{2.3}$$

## Transformers

Consider a single-phase transformer in which the total series impedance of the two windings referred to the primary is $Z_1$ (Figure 2.16).

Then the p.u. impedance $= I_1 Z_1 / V_1$, where $I_1$ and $V_1$ are the rated or base values. The total impedance referred to the secondary

$$= Z_1 N^2$$

and this in p.u. notation is

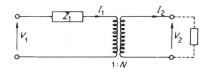

**Figure 2.16** Equivalent circuit of single-phase transformer

$$= Z_1 N^2 \left( \frac{I_2}{V_2} \right)$$

$$= Z_1 N^2 \frac{I_1}{N} \cdot \frac{1}{V_1 N} = \frac{Z_1 I_1}{V_1}$$

Hence the p.u. impedance of a transformer is the same whether considered from the primary or the secondary side. But expression (2.3) shows that $Z_{p.u.}$ is directly proportional to the base $VA$ and inversely proportional to the base voltage squared. Although the VA rating of a transformer is independent of the side to which it refers, the voltage rating is different. Consequently if we wish to vary the voltage base dependent upon which part of the network to which we are referring, so that all voltages are around 1 p.u., we must recalculate the corresponding value of $Z_{p.u.}$, as follows:

$$Z_{p.u.}(\text{new base}) = Z_{p.u.} \text{ (original base)} \times \left( \frac{\text{base V}_{\text{old}}}{\text{base V}_{\text{new}}} \right)^2 \qquad (2.4)$$

Note that the $VA$ base is unchanged. If we wish to calculate $Z_{p.u.}$ to a new $VA$ base, then $Z_{p.u.}$ (new base) $= Z_{p.u.}$ (old base) $\times \left( \dfrac{\text{base }VA_{\text{new}}}{\text{base }VA_{\text{old}}} \right)$ $\qquad (2.5)$

## *Example 2.4*

In the network of Figure 2.17, two single-phase transformers supply a 10 kVA resistance load at 200 V. Show that the p.u. load is the same for each part of the circuit and calculate the voltage at point D.

## *Solution*

The load resistance is $(200^2/10 \times 10^3)$, i.e. 4 $\Omega$.
    In each of the circuits A, B, and C a different voltage exists, so that each circuit will have its own base voltage, i.e. 100 V in A, 400 V in B, and 200 V in C.

    Although it is not essential for rated voltages to be used as bases, it is essential that the *voltage bases used be related by the turns ratios of the transformers*. If this is not so

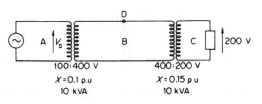

**Figure 2.17**   Network with two transformers—p.u. approach

the whole p.u. framework breaks down. The same volt-ampere base is used for all the circuits as $V_1 I_1 = V_2 I_2$ on each side of a transformer and is taken in this case as 10 kVA. The base impedance in C

$$= \frac{200^2}{10\,000} = 4\ \Omega$$

The load resistance (p.u.) in C

$$= \frac{4}{4} = 1\ \text{p.u.}$$

In B the base impedance

$$= \frac{400^2}{10\,000} = 16\ \Omega$$

and the load resistance referred to B

$$= 4 \times 2^2 = 16\ \Omega$$

Hence the p.u. load referred to B

$$= 1\ \text{p.u.}$$

Similarly, the p.u. load resistance referred to A is also 1 p.u. Hence, if the voltage bases are related by the turns ratios the load p.u. value is the same for all circuits. An equivalent circuit may be used as shown in Figure 2.18. Let the volt-ampere base be 10 kVA; the voltage across the load ($V_R$) is 1 p.u. (as the base voltage in C is 200 V, if the load voltage had been maintained at 100 V, then $V_R$ would be 0.5 p.u.). The base current

$$= \frac{(VA)_b}{V_b} = \frac{10\,000}{200} = 50\ \text{A in C}$$

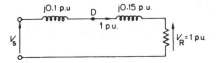

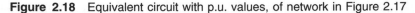

**Figure 2.18**   Equivalent circuit with p.u. values, of network in Figure 2.17

The corresponding currents in the other circuits are 25 A in B, and 100 A in A. The actual load current $= 50/50 = 1$ p.u. (in phase with $V_R$, the reference phasor). Hence the supply voltage $V_s$

$$= 1(j0.1 + j0.15) + 1 \text{ p.u.}$$
$$\therefore V_s = 1.03 \text{ p.u.}$$
$$= 1.03 \times 100 = 103 \text{ V}$$

The voltage at point D in Figure 2.17

$$= 1 + j0.15 \times 1 = 1 + j0.15$$
$$= 1.012 \text{ p.u.modulus}$$
$$= 1.012 \times 400 = 404.8 \text{ V}$$

It is a useful exercise to repeat this example using ohms, volts, and amperes.

## *Example 2.5*

Figure 2.19 shows the schematic diagram of a radial transmission system. The ratings and reactances of the various components are shown, along with the nominal transformer line voltages. A load of 50 MW at 0.8 p.f. lagging is taken from the 33 kV substation which is to be maintained at 30 kV. It is required to calculate the terminal voltage of the synchronous machine. The line and transformers may be represented by series reactances. The system is three-phase.

## *Solution*

It will be noted that the line reactance is given in ohms; this is usual practice. The voltage bases of the various circuits are decided by the nominal transformer voltages, i.e. 11, 132, and 33 kV. A base of 100 MVA will be used for all circuits. The reactances (resistance is neglected) are expressed on the appropriate voltage and MVA bases.

Base impedance for the line

$$= \frac{132^2 \times 10^6}{100 \times 10^6}$$
$$= 174\Omega$$

Hence the p.u. reactance

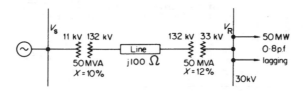

**Figure 2.19**   Line diagram of system for Example 2.5

$$= \frac{j100}{174} = j0.575$$

Per unit reactance of the sending-end transformer

$$= \frac{100}{50} \times j0.1 = j0.2$$

Per unit reactance of the receiving-end transformer

$$= j0.12 \times \frac{100}{50} = j0.24$$

Load current

$$= \frac{50 \times 10^6}{\sqrt{3} \times 30 \times 10^3 \times 0.8} = 1203 \text{ A}$$

Base current for 33 kV, 100 MVA

$$= \frac{100 \times 10^6}{\sqrt{3} \times 33 \times 10^3}$$
$$= 1750 \text{ A}$$

Hence the p.u. load current

$$= \frac{1203}{1750} = 0.687$$

Voltage of load busbar

$$= \frac{30}{33} = 0.91 \text{ p.u.}$$

The equivalent circuit is shown in Figure 2.20.
Also,

$$V_s = 0.687(0.8 - j0.6)(j0.2 + j0.575 + j0.24) + (0.91 + j0)$$
$$= 1.328 + j0.558 \text{ p.u.}$$
$$\therefore V_s = 1.44 \text{ p.u. or } 1.44 \times 11 \text{ kV} = 15.84 \text{ kV}$$

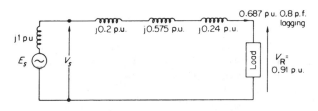

**Figure 2.20**  Equivalent circuit for Example 2.5

## 2.7   Power Transfer and Reactive Power

The circuit shown in Figure 2.21 is very important as it represents the simplest electrical model for a synchronous generator (with an e.m.f. ($E$) feeding into a power system represented by a voltage source ($V$)).

$$\mathbf{S}_A = \mathbf{E}\mathbf{I}^* = \mathbf{E}\left(\frac{\mathbf{E} - \mathbf{V}}{\mathbf{Z}}\right)^*$$

$$= E\, e^{j\delta}\left(\frac{E\, e^{-j\delta} - V}{Z\, e^{-j\theta}}\right)$$

$$S_A = \frac{E^2}{Z}e^{j\theta} - \frac{EV}{Z}e^{(j\theta+\delta)}$$

$$\left.\begin{aligned}P_A &= \frac{E^2}{Z}\cos\theta - \frac{EV}{Z}\cos(\theta+\delta)\\[2mm] Q_A &= \frac{E^2}{Z}\sin\theta - \frac{EV}{Z}\sin(\theta+\delta)\end{aligned}\right\} \tag{2.6}$$

Similarly,

$$\mathbf{S}_B = V\left(\frac{V - Ee^{-j\delta}}{Ze^{-j\theta}}\right)$$

$$\left.\begin{aligned}P_B &= \frac{V^2}{Z}\cos\theta - \frac{EV}{Z}\cos(\theta-\delta)\\[2mm] Q_B &= \frac{V^2}{Z}\sin\theta - \frac{EV}{Z}\sin(\theta-\delta)\end{aligned}\right\} \tag{2.7}$$

The power output to $V$ is a maximum when $\cos(\theta - \delta) = 1$, i.e. $\theta = \delta$.

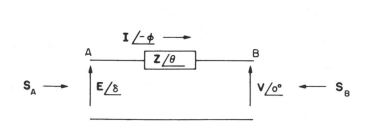

**Figure 2.21**   Power transfer between sources

### 2.7.1 Calculation of sending and received voltages in terms of power and reactive power

The determination of the voltages and currents in a network can obviously be achieved by means of complex notation, but in power systems usually power ($P$) and reactive power ($Q$) are specified and often the resistance of lines is negligible compared with reactance. For example, if $R = 0.1X$, the error in neglecting $R$ is 0.49 per cent, and even if $R = 0.4X$ the error is 7.7 per cent.

A simple transmission link is shown in Figure 2.22(a). It is required to establish the equations for $E$, $V$, and $\delta$. From Figure 2.22(b).

$$E^2 = (V + \Delta V_{\mathrm{p}})^2 + \Delta V_{\mathrm{q}}^2$$
$$= (V + RI\cos\phi + XI\sin\delta)^2 + (XI\cos\phi - RI\sin\phi)^2$$
$$\therefore E^2 = \left(V + \frac{RP}{V} + \frac{XQ}{V}\right)^2 + \left(\frac{XP}{V} - \frac{RQ}{V}\right)^2 \tag{2.8}$$

Hence,

$$\Delta V_{\mathrm{p}} = \frac{RP + XQ}{V} \tag{2.9}$$

and

$$\Delta V_{\mathrm{q}} = \frac{XP - RQ}{V} \tag{2.10}$$

If

$$\Delta V_{\mathrm{q}} \ll V + \Delta V_{\mathrm{p}}$$

then

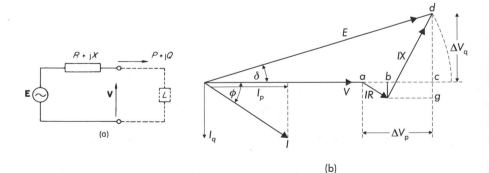

(a)

(b)

**Figure 2.22**    Phasor diagram for transmission of power through a series impedance

$$E^2 = \left(V + \frac{RP + XQ)}{V}\right)^2$$

and

$$E - V = \frac{RP + XQ}{V} = \Delta V_{\mathrm{p}}$$

Hence the arithmetic difference between the voltages is given **approximately** by

$$\frac{RP + XQ}{V}$$

If

$$R = 0$$

then

$$E - V = \frac{XQ}{V} \qquad (2.11)$$

i.e. *voltage depends only on* $Q$, and from equation (2.8) this is valid if $PX \ll V^2 + QX$. The angle of transmission $\delta$ is obtained from $\sin^{-1}$ $(\Delta V_{\mathrm{q}}/E)$, *and depends only on* $P$.

  Equations (2.9) and (2.10) will be used wherever possible because of their great simplicity.

## *Example 2.6*

Consider a 275 kV line of length 160 km ($R$ and $X$ per km $= 0.034$ and $0.32$ $\Omega$, respectively). Obviously, $R \ll X$.

  Compare the sending end voltage $E$ in Figure 2.22(a) when calculated with the accurate and the approximate formulae. Assume a load of 600 MW, 300 MVAr and take a base of 100 MVA. (Note that 600 MVA is nearly the maximum rating of a $2 \times 258$ mm$^2$ line at 275 kV.)

## *Solution*

The base impedance is

$$\frac{275^2 \times 10^6}{100 \times 10^6} = 756 \ \Omega$$

For line of length 160 km,

$$X \ (\Omega) = 160 \times 0.32 = 51.2 \ \Omega$$

$$X = \frac{51.2}{756} = 0.0677 \ \text{p.u.} \ (\Omega)$$

and let the received voltage be

$$V = 275 \text{ kV} = 1 \text{ p.u.}$$

then          $$PX = 6 \times 0.0677 = 0.406 \text{ p.u.}(V)$$
and

$$V^2 + QX = 1 + 3 \times 0.0677 = 1.203 \text{ p.u. } (V)$$

Hence the use of equation (2.9) will involve some inaccuracy
    Using equation (2.8) with $R$ neglected.

$$E^2 = \left(1 + \frac{0.0677 \times 3}{V}\right)^2 + \left(\frac{0.0677 \times 6}{V}\right)^2 = 1.203^2 + 0.406^2$$

$$E = 1.270 \text{ p.u. } (V)$$

An approximate value of $E$ can be found from equation (2.11) as

$$E - V = \Delta V = \frac{0.0677 \times 3}{1} = 0.203$$

$$\therefore E = 1.203 \text{ p.u. } (V)$$

i.e. an error of 5.6 per cent.
    With a 80 km length of this line at the same load (still neglecting $R$) gives

$$PX = 6 \times 0.0338 \quad \text{and} \quad QX = 3 \times 0.0388$$
$$= 0.203 \qquad\qquad\qquad = 0.101$$

The accurate formula gives

$$E = \sqrt{(1 + 0.101)^2 + 0.203^2} = 1.120 \text{ p.u. } (V)$$

and the approximate one gives

$$E = 1 + 0.101 = 1.101 \text{ p.u. } (V)$$

i.e. an error of 1.7 per cent.

    In the solution above, it should be noted that p.u. values for power system calculations are conveniently expressed to three decimal places, implying a measured value to 0.1 per cent error. In practice, most measurements will have an error of at least 0.2 per cent and possibly 0.5 per cent.
    With 132 kV lines on a 100 MVA base, the reactance of a 160 km 258 mm$^2$ line is $160 \times 0.4/174$ or 0.37 p.u. and the rated power is 1.5 p.u. ($Q = 0.75$ p.u. for same power factor). Hence

$$PX = 1.5 \times 0.37 \quad \text{and} \quad V^2 + XQ = 1 + 0.277$$
$$= 0.555 \qquad\qquad\qquad = 1.277$$

Hence, an error in the order of 5 per cent could be expected.
    If $E$ is specified and $V$ is required, the phasor diagram in Figure 2.23 is used. From this,

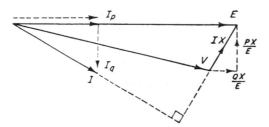

**Figure 2.23**   Phasor diagram when **E** is specified

$$V = \sqrt{\left[\left(E - \frac{QX}{E}\right)^2 + \left(\frac{PX}{E}\right)^2\right]} \qquad (2.12)$$

when $R = 0$.
   If

$$\frac{PX}{E} \ll \frac{E^2 - QX}{E}$$

then

$$V = E - \frac{QX}{E} \qquad (2.13)$$

## 2.8   Useful Network Theory

### 2.8.1   *Four-terminal networks*

A lumped-constant circuit, provided that it is passive, linear, and bilateral, can be represented by the four-terminal network shown in the diagram in Figure 2.24. The complex parameters **A**, **B**, **C**, and **D** describe the network in terms of the sending- and receiving-end voltages and currents as follows:

$$\begin{aligned}
\mathbf{V_S} &= \mathbf{AV_R} + \mathbf{BI_R} \\
\mathbf{I_S} &= \mathbf{CV_R} + \mathbf{DI_R}
\end{aligned} \qquad (2.14)$$

and it can be readily shown that **AD**–**BC** = 1.
   **A**, **B**, **C**, and **D** may be obtained by measurement and certain physical interpretations can be made, as follows:

1.   Receiving-end short-circuited:

$$\mathbf{V_R} = 0, \qquad \mathbf{I_S} = \mathbf{DI_R} \qquad \text{and} \qquad \mathbf{D} = \frac{\mathbf{I_S}}{\mathbf{I_R}}$$

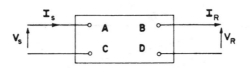

**Figure 2.24**   Representation of a four-terminal (two-port) network

Also,

$$V_S = BI_R \quad \text{and} \quad B = \frac{V_S}{I_R} = \text{short-circuit transfer impedance}$$

2.   Receiving-end open-circuited:

Here

$$I_R = 0, \quad I_S = CV_R \quad \text{and} \quad C = \left(\frac{I_S}{V_R}\right)$$

$$V_S = AV_R, \quad A = \left(\frac{V_S}{V_R}\right)$$

Expressions for the constants can also be found by complex (i.e. magnitude *and* angle) measurements carried out solely at the sending end with the receiving end open and short-circuited.

Often it is useful to have a single four-terminal network for two or more items in series or parallel, e.g. a line and two transformers in series.

It is shown in most texts on circuit theory that the generalized constants for the combined network, $A_0, B_0, C_0$, and $D_0$ for the two networks (1) and (2) in cascade are as follows:

$$A_0 = A_1A_2 + B_1C_2 \qquad B_0 = A_1B_2 + B_1D_2$$
$$C_0 = A_2C_1 + C_2D_1 \qquad D_0 = B_2C_1 + D_1D_2$$

For two four-terminal networks in parallel it can be shown that the parameters of the equivalent single four-terminal network are:

$$A_0 = \frac{A_1B_2 + A_2B_1}{B_1 + B_2} \qquad B_0 = \frac{B_1B_2}{B_1 + B_2} \qquad D_0 = \frac{B_2D_1 + B_1D_2}{B_1 + D_2}$$

$C_0$ can be found from $A_0D_0 - B_0C_0 = 1$.

## Delta–star transformation

The delta network connected between the three terminals A, B, and C of figure 2.25 can be replaced by a star network such that the impedance measured

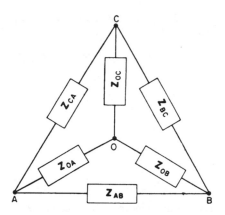

**Figure 2.25** Star–delta, delta–star transformation

between the terminals is unchanged. The equivalent impedances can be found as follows:

Impedance between terminals AB with C open-circuited is given by

$$Z_{OA} + Z_{OB} = Z_{AB}//(Z_{CA} + Z_{BC})$$

Similarly

$$Z_{OB} + Z_{OC} = Z_{BC}//(Z_{AB} + Z_{CA})$$

and

$$Z_{OC} + Z_{OA} = Z_{CA}//(Z_{BC} + Z_{AB})$$

From these three equations $Z_{OA}$, $Z_{OB}$, and $Z_{OC}$, the three unknowns, can be determined as

$$Z_{OA} = \frac{Z_{AB}Z_{CA}}{Z_{AB} + Z_{CA} + Z_{BC}}$$

$$Z_{OB} = \frac{Z_{AB}Z_{BC}}{Z_{AB} + Z_{CA} + Z_{BC}} \qquad (2.15)$$

$$Z_{OC} = \frac{Z_{BC}Z_{CA}}{Z_{AB} + Z_{CA} + Z_{BC}}$$

## Star–delta transformation

A star-connected system can be replaced by an equivalent delta connection if the elements of the new network have the following values (Figure 2.25):

$$Z_{AB} = \frac{Z_{OA}Z_{OB} + Z_{OB}Z_{OC} + Z_{OC}Z_{OA}}{Z_{OC}}$$

$$Z_{BC} = \frac{Z_{OA}Z_{OB} + Z_{OB}Z_{OC} + Z_{OC}Z_{OA}}{Z_{OA}} \qquad (2.16)$$

$$Z_{CA} = \frac{Z_{OA}Z_{OB} + Z_{OB}Z_{OC} + Z_{OC}Z_{OA}}{Z_{OB}}$$

## Problems

**2.1**  The star-connected secondary winding of a three-phase transformer supplies 415 V (line to line) at a load point through a four-conductor cable. The neutral conductor is connected to the winding star point which is earthed. The load consists of the following components:

| | |
|---|---|
| Between a and b conductors | a 1 Ω resistor |
| Between a and neutral conductors | a 1 Ω resistor |
| Between b and neutral conductors | a 2 Ω resistor |
| Between c and neutral conductors | a 2 Ω resistor |

Connected to the a, b, and c conductors is an induction motor taking a balanced current of 100 A at 0.866 power factor (p.f.) lagging. Calculate the currents in the four conductors and the total power supplied.

Take the 'a' to neutral voltage as the reference phasor. The phase sequence is a–b–c. (Answer: 350 kW)

**2.2**  The wye-connected load shown in Figure 2.26 is supplied from a transformer whose secondary-winding star point is solidly earthed. The line voltage supplied to the load is 400 V. Determine (a) the line currents, and (b) the voltage of the load star point with respect to ground. The phase sequence is a–b–c.
(Answer: (a) $I_a(200 - j69)$ A; (b) $V_{n-g} - 47$ V)

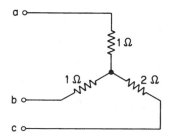

**Figure 2.26**  Circuit for Problem 2.2

**2.3**  Two capacitors, each of 10 $\mu$F, and a resistor R, are connected to a 50 Hz three-phase supply, as shown in Figure 2.27. The power drawn from the supply is the same whether the switch S is open or closed. Find the resistance of R.
(Answer: 143 Ω)

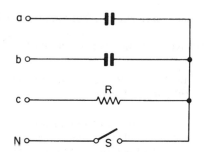

**Figure 2.27**   Circuit for Problem 2.3

**2.4**   The network of Figure 2.28 is connected to a 400 V three-phase supply, with phase sequence a–b–c. Calculate the reading of the wattmeter W.
(Answer: 2.24 kW)

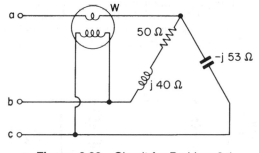

**Figure 2.28**   Circuit for Problem 2.4

**2.5**   A 400 V three-phase supply feeds a delta-connected load with the following branch impedances:

$$\mathbf{Z}_{RY} = 100 \ \Omega \qquad \mathbf{Z}_{YB} = j100 \ \Omega \qquad \mathbf{Z}_{BR} = -j100 \ \Omega$$

Calculate the line currents for phase sequences (a) RYB; (b) RBY.
(Answer: (a) 7.73, 7.37, 4 A: (b) 2.07, 2.07, 4 A)

**2.6**   A synchronous generator, represented by a voltage source in series with an inductive reactance $X_1$, is connected to a load consisting of a fixed inductive reactance $X_2$ and a variable resistance $R$ in parallel. Show that the generator power output is a maximum when

$$1/R = 1/X_1 + 1/X_2$$

**2.7**   A single-phase voltage source of 100 kV supplies a load through an impedance j100 Ω. The load may be represented in either of the following ways as far as voltage changes are concerned:

(a) by a constant impedance representing a consumption of 10 MW, 10 MVAr at 100 kV; or

(b) by a constant current representing a consumption of 10 MW, 10 MVAr at 100 kV.

Calculate the voltage across the load using each of these representations.
(Answer: (a) $(90 - j8.2)$ kV; (b) $(90 - j10)$ kV)

**2.8** Show that the p.u. impedance (obtained from a short-circuit test) of a star–delta three-phase transformer is the same whether computed from the star-side parameters or from the delta side. Assume a rating of $G$ (volt-amperes), a line-to-line input voltage to the star-winding terminals of $V$ volts, a turns ratio of $1:N$ (star to delta), and a short-circuit impedance of $Z$ (ohms) per phase referred to the star side.

**2.9** A 11 kV/132 kV, 50 MVA, three-phase transformer has an inductive reactance of $j0.005$ $\Omega$ referred to the primary (11 kV). Calculate the p.u. value of reactance based on the rating. Neglect resistance.
(Answer: 0.002 p.u.)

**2.10** Express in p.u. all the quantities shown in the line diagram of the three-phase transmission system in Figure 2.29. Construct the single-phase equivalent circuit. Use a base of 100 MVA. The line is 80 km in length with resistance and reactance of 0.1 and 0.5 $\Omega$, respectively, and a capacitive susceptance of 10 $\mu$S per km (split equally between the two ends).

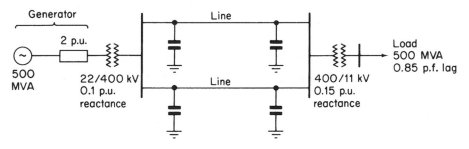

**Figure 2.29** Circuit for Problem 2.10

**2.11** A wye-connected load is supplied from three-phase 220 V mains. Each branch of the load is a resistor of 20 $\Omega$. Using 220 V and 10 kVA bases calculate the p.u. values of the current and power taken by the load.
(Answer: 0.24 p.u.; 0.24 p.u.)

**2.12** A 440 V three-phase supply is connected to three star-connected loads in parallel, through a feeder of impedance $(0.1 + j0.5)$ $\Omega$ per phase. The loads are as follows: 5 kW, 4 kVAr; 3 kW, 0 kVAr; 10 kW, 2 kVAr. Determine:

(a) line current;
(b) power and reactive power losses in the feeder per phase;
(c) power and reactive power from the supply and the supply power factor.
(Answer: (a) 24.9 A; (b) 62 W, 310 VAR; (c) 18.06 kW, 6.31 kVAr, 0.94)

**2.13** Two transmission circuits are defined by the following **ABCD** constants: 1, 50, 0, 1, and $0.9 \angle 2°$, $150 \angle 79°$, $9 \times 10^{-4} \angle 91°$, $0.9 \angle 2°$. Determine the **ABCD** constants of the circuit comprising these two circuits in series.
(Answer: $\mathbf{A} = 45 \angle 89.86$; $\mathbf{B} = 165.7 \angle 63.6$)

**2.14** A 132 kV overhead line has a series resistance and inductive reactance per phase per kilometre of 0.156 and 0.4125 $\Omega$, respectively. Calculate the magnitude of the sending-end voltage when transmitting the full line capability of 125 MVA when the power factor is 0.9 lagging and the received voltage is 132 kV, for 16 km and 80 km lengths of line. Use both accurate and approximate methods.
(Answer: 136.92 kV (accurate), 136.85 kV; 157.95 kV (accurate), 156.27 kV)

**2.15** A synchronous generator may be represented by a voltage source of magnitude 1.7 p.u. in series with an impedance of 2 p.u. It is connected to a zero-impedance voltage source of 1 p.u. The ratio of $X/R$ of the impedance is 10. Calculate the power generated and the power delivered to the voltage source if the angle between the voltage sources is 30°.
(Answer: 0.49 p.u.; 0.44 p.u.)

**2.16** A three-phase star-connected 50 Hz generator generates 240 V per phase and supplies three delta-connected load coils each having a resistance of 10 $\Omega$ and an inductance of 47.75 mH.

Determine:

    (a) the line voltage and current;
    (b) the load current;
    (c) the total real power and reactive power dissipated by the load.

Determine also the values of the three capacitors required to correct the overall power factor to unity when the capacitors are (i) star- and (ii) delta-connected across the load.
    State an advantage and a disadvantage of using the star connection for power factor correction.
(Answer: (a) 416 V, 39.96 A; (b) 23.07 A; (c) 15.77 kW, 24.09 kVAr; (i) 443.8 $\mu$F; (ii) 147.8 $\mu$F

(*From E.C. Examination, 1996*)

# 3

# Components of a Power System

## 3.1  Introduction

In this chapter the essential characteristics of the components of a power system will be discussed. It is essential that these be fully understood before the study of large systems of interconnection of these components is undertaken. In all cases the simplest representation employing an equivalent circuit will be used not only to make the principles clearer but also because these simple models are used in practice. For more sophisticated treatments, especially of synchronous machines, the reader is referred to the more advanced texts given in the reference list at the end of the book.

It is assumed that the reader has completed introductory courses in circuit theory and machines (or electromechanical energy conversion). A knowledge of basic control theory will be needed for the section on the dynamic response of machines, although this may be omitted without essential loss of continuity in the text.

Loads are considered as components even though their exact composition and characteristics are not known with complete certainty. When designing a supply system or extending an existing one the prediction of the loads to be expected is required, statistical methods being used. This is the aspect of power supply known with least precision.

It is stressed that most of the equivalent circuits used are single phase and employ phase-to-neutral values. This assumes that the loads are balanced three-phase which is reasonable for normal steady-state operation. When unbalance exists between the phases, full treatment of all phases is required, and special techniques for dealing with this are given in Chapter 7.

# Synchronous Machines

## 3.2 Characteristics

In this chapter those characteristics of the synchronous machine pertinent to power systems will be discussed. It is assumed that the reader has a basic knowledge of synchronous-machine theory. The two forms of rotor construction produce characteristics that influence the operation of the system to varying extents. In the round or cylindrical rotor the field winding is placed in slots cut axially along the rotor length as illustrated in Figure 3.1(a). The diameter is relatively small (1–1.5 m) and the machine is suitable for operation at high speeds (3000 or 3600 r.p.m.) driven by a steam or gas turbine. Hence it is known as a turbogenerator. In a salient pole rotor the poles project as shown in Figure 3.1(b) and low-speed operation driven by hydraulic turbines, wind, or diesel sets is usual. The frequency of the generated e.m.f. and speed are related by

$$f = \frac{np}{60}$$

where $n$ is speed in r.p.m. and $p$ is the number of *pairs* of poles. A hydraulic turbine rotating at 150–300 r.p.m., depending on type, thus needs many pole pairs to generate at normal frequencies.

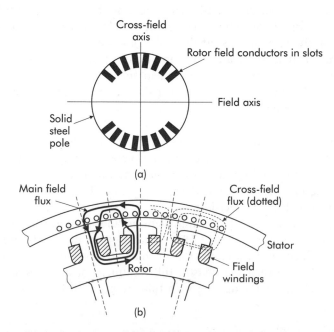

**Figure 3.1**   (a) Cylindrical rotor. (b) Salient pole rotor, and stator with flux paths

The three-phase currents in the stator winding or armature generate a rotating magnetic field which rotates with the synchronously rotating rotor and its field. The effect of the stator winding field on the rotor field is referred to as 'armature reaction'. Figure 3.1(b) indicates the paths of the main field fluxes in a salient pole machine from which it can be seen that these paths are mainly in iron, apart from crossing the air gap between stator and rotor. On the other hand, the cross-field flux has a longer path in air and hence the reluctance of its path is less than for the main field flux. Consequently, any e.m.f.s produced by the main field flux in the stator are larger than those produced by any cross-field fluxes.

Machine designers shape the poles in a salient machine and distribute the windings in a round-rotor machine to obtain maximum fundamental sinusoidal voltage on no-load in the carefully designed stator windings. Also, an acceptably low harmonic content is achieved in contrast to the cross-field flux which often has considerable harmonic content. However, the stator winding is designed and distributed such as to minimize the harmonic voltages induced in it and only the fundamental e.m.f. is significant for power system considerations.

Whether a synchronous machine has a cylindrical or salient-pole rotor, its action is the same when generating power and can be best understood in terms of the 'primitive' representation shown in Figure 3.2(a). The rotor current $I_f$

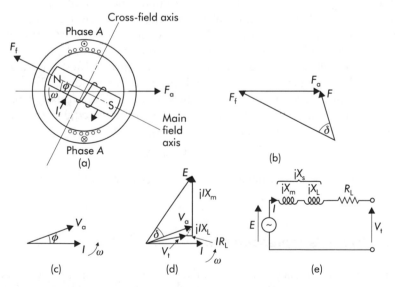

**Figure 3.2** (a) Primitive machine at the instant of maximum current in phase A conductors and field m.m.f. $F_f$. (b) Corresponding space diagram of vector m.m.f.s with resultant $F$. (c) Terminal phasor diagram. (d) Phasor diagram with voltage representation of magnetic conditions. (e) Equivalent circuit for round-rotor machine

produces a magnetomotive force (m.m.f.) $F_f$ ampere turns (AT) rotating with the rotor and producing sinusoidal-induced voltages in the stator windings on no-load. When the stator windings are closed through an external circuit, currents flow, thereby delivering power to the external load. These stator currents produce an m.m.f., shown in Figure 3.2(a) for phase A as $F_a$, which interacts with the rotor m.m.f. $F_f$, thereby producing a retarding force on the rotor that requires torque (and hence power) to be provided from the prime mover. The vector m.m.f.s in Figure 3.2(b) are drawn for a representative instant at which the current $I$ in phase A is a maximum and the voltage is at an angle $\phi$, the power factor angle, past its maximum as shown in Figure 3.2(c). Currents in phases B and C of the stator (equal and opposite at this instant) also contribute along the axis of $F_a$. The resultant air-gap m.m.f. is shown as $F = F_a + F_f$, using vectorial addition as in Figure 3.2(b), and this gives rise to a displaced air-gap flux such that the phase of the stator-induced e.m.f. $E$ is delayed from that produced on no-load for the same rotor position by the torque angle shown as $\delta$ in Figure 3.2(d).

Detailed consideration of generator action taking full account of magnetic path saturation requires a simulation of the fluxes, including all winding currents and eddy currents. However, for power system calculations the m.m.f. summation of Figure 3.2(b) can be transferred to the voltage–current phasor diagram of Figure 3.2(d) by rotation through 90° clockwise and scaling $F$ to $V_a$. Then, $F_f$ becomes $E$ and $F_a$ corresponds to $X_m$ where $X_m$ is an equivalent reactance representing the effect of the magnetic conditions within the machine. Effects due to saturation can be allowed for by changing $X_m$ and relating $E$ to the field current $I_f$ by the open circuit (no-load) saturation curve (see later). The machine terminal voltage $V_t$ is obtained from $V_a$ by recognizing that the stator phase windings have a small resistance $R_L$ and a leakage reactance $X_L$ (about 10 per cent of $X_m$) resulting from flux produced by the stator but not crossing the air gap. The reactances $X_L$ and $X_m$ are usually considered together as the synchronous reactance $X_s$, shown in the equivalent circuit of Figure 3.2(e). The machine phasor equation is:

$$\mathbf{E} = \mathbf{V}_t + \mathbf{I}(R_L + jX_s)$$

In most power system calculations $R_L$ is neglected and a simple phasor diagram results.

### 3.2.1  Two-axis representation

If a machine has saliency, i.e. the rotor has salient poles or the rotor has non-uniform slotting in a solid rotor, then the main field axis and the cross-field axis have different reluctances. The armature reaction m.m.f., $F_a$ in Figure 3.2(b), can be resolved into two components $F_d$ and $F_q$, as shown in Figure 3.3(b) for the primitive machine shown in Figure 3.3(a). The m.m.f./flux/voltage trans-

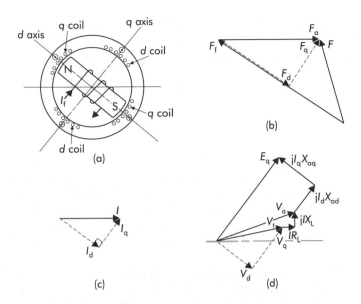

**Figure 3.3** (a) Primitive machine- $d$ and $q$ axes. (b) The m.m.f. vectors resolved in $d$ and $q$ axes. $F_d$ and $F_q$ are components of $F_a$. (c) Current phasors resolved into $I_d$ and $I_q$ components. (d) Phasor diagram of voltages, $E$ of Figure 3.2(d) now becomes $E_q$

formation will now be different on these two component axes (known as the direct and quadrature axes) as less flux and induced e.m.f. will be produced on the $q$ axis. This difference is reflected in the voltage phasor diagram of Figure 3.3(d) as two different reactances $X_{ad}$ and $X_{aq}$, the component of stator current acting on the $d$ axis $I_d$ being associated with $X_{ad}$, and $I_q$ with $X_{aq}$.

The resolution of the stator m.m.f. into two components is represented in Figure 3.3(a) by two coils identified by the axes on which they produce m.m.f. Repeating the previous transformation from m.m.f. $F$ to voltages, Figure 3.3(b) transforms to Figure 3.3(c) by considering the two-axis components of current and then results in Figure 3.3(d). Saturation of the two axes is represented by modifying $X_{ad}$ and $X_{aq}$. Because $I$ has been resolved into two components it is no longer possible for a salient machine to be represented by an equivalent circuit. However, equations can be obtained for the $V_d$ and $V_q$ components of the terminal voltage $V_t$ as:

$$V_q = j\,I_d(X_{ad} + X_L) + I_q R_L - E_q$$
$$V_d = -j\,I_q(X_{aq} + X_L) - I_d R_L$$

It will be seen that the resistive terms (if included) contain the currents from the other axes. The term $E_q$ is the e.m.f. representing the action of the field winding and in many power system studies can be taken as constant if no automatic voltage regulator (AVR) action is assumed.

A single equation can be written in complex form as

$$\mathbf{E}_q = \mathbf{V} + \mathbf{I}R_L + j\mathbf{I}X_L + j(\mathbf{I}_d X_{ad} + \mathbf{I}_q X_{aq})$$
$$= \mathbf{V} + \mathbf{I}R_L + j(\mathbf{I}_d X_d + \mathbf{I}_q X_q)$$

where $\quad X_d = X_{ad} + X_L$

and $\quad X_q = X_{aq} + X_L$

In a salient machine, $X_{ad} \neq X_{aq}$ torque and power can be developed with no field current, the stator m.m.f. 'locking' in with the direct axis as in a reluctance machine. The effect depends upon the saliency but is generally small compared with the maximum power rating with field current. Consequently, for multi-machine power system calculations the equivalent circuit of Figure 3.2(e) can be used.

The transient behaviour of synchronous machines is often expressed in two-axis terms, the basic concept being similar to that shown in Figure 3.3. However, a variety of conventions are employed and care must be taken to establish which direction of current is being taken as positive. In many texts, if a synchronous machine is running as a motor, the current entering the machine is taken as positive. Here we have assumed that current leaving a generator is positive.

To produce accurate voltage and current waveforms under transient conditions the two-axis method of representation with individually coupled coils on the $d$ and $q$ axes is necessary. Since these axes are at right-angles (orthogonal), then fluxes linking coils on the same axis do not influence the fluxes linking coils on the other axis, i.e. no cross-coupling is assumed. Differential equations can be set up to account for flux changes on both axes separately, from which corresponding voltages and currents can be derived through step-by-step integration routines. This form of calculation is left for advanced study. The use of finite elements can extend the accuracy of calculation still further.

### 3.2.2 Effect of saturation on $X_s$—the short-circuit ratio

The open-circuit characteristic is the graph of generated voltage against field current with the machine on open circuit and running at synchronous speed. The short-circuit chracteristic is the graph of stator current against field current with the terminals short-circuited; curves for a modern machine are shown in Figure 3.4. Here, $X_s$ is equal to the open-circuit voltage produced by the same field current that produces rated current on short circuit, divided by this rated stator current. This value of $X_s$ is constant only over the linear part of the open-circuit characteristic (the air-gap line) and ignores saturation. The actual

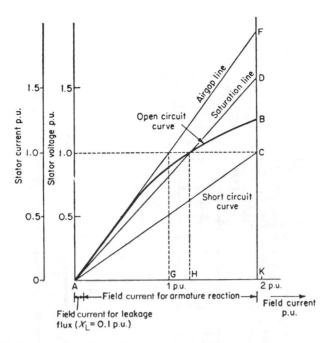

**Figure 3.4** Open- and short-circuit characteristics of a synchronous machine. Unsaturated value of $X_s = FK/CK$. With operation near nominal voltage, the saturation line is used to give a linear characteristic with saturation

value of $X_s$ at full-load current will obviously be less than this value and several methods exist to allow for the effects of saturation.

The short-circuit ratio (SCR) of a generator is defined as the ratio between the field current required to give nominal open-circuit voltage and that required to circulate full-load current in the armature when short-circuited. In Figure 3.4 the short-circuit ratio is AH/AK, i.e. 0.63. The SCR is also commonly calculated in terms of the air-gap line and the short-circuit curve, this giving an unsaturated value. To allow for saturation it is common practice to assume that the synchronous reactance is 1/SCR, which for this machine is 1.58 p.u. Economy demands the design of machines of low SCR and a value of 0.55 is common for modern machines. Unfortunately, transient stability margins are reduced as synchronous reactance increases (i.e. SCR reduces) so a conflict for design between economy and stability arises.

## 3.2.3 Turbogenerators

During the 1970s, the ratings of steam-driven turbine generators were reaching 1000 MW or over to obtain improvements in efficiency made possible by large

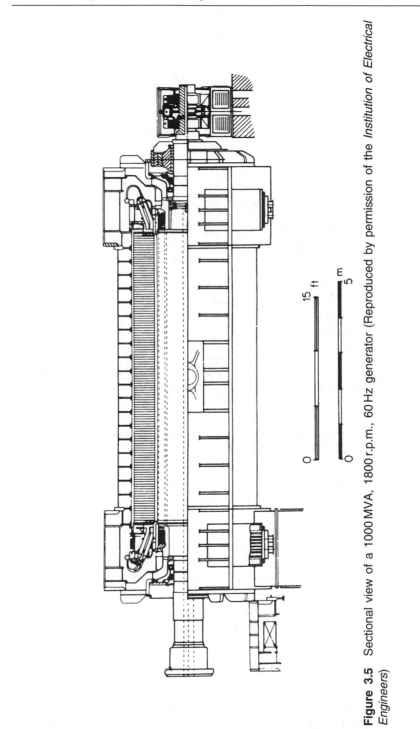

**Figure 3.5**  Sectional view of a 1000 MVA, 1800 r.p.m., 60 Hz generator (Reproduced by permission of the *Institution of Electrical Engineers*)

turbine plant and their lower capital costs per megawatt. Since the increasing availability of improved gas turbines with efficiencies of 55 per cent, even in 100–200 MW sizes, and with the use of gas for firing, air- or hydrogen-cooled generators with five or six separate machines in each power station are becoming common for new installations, particularly by independent plant producers (IPPs).

Large generators are cooled by hydrogen and water pumped through hollow stator conductors, a typical statistic being:

500 MW, 588 MVA    :   winding cooling—rotor, hydrogen; stator, water;

　　　　　　　　　　 :   rotor diameter 1.12 m;

　　　　　　　　　　 :   rotor length 6.2 m;

　　　　　　　　　　 :   total weight 0.63 kg/kVA;

　　　　　　　　　　 :   rotor weight 0.1045 kg/kVA.

A cross-sectional view of a 1000 MVA steam turbogenerator is shown in Figure 3.5.

The most important problems encountered in the development and use of large generators arise from (1) mechanical difficulties due to the rotation of large masses, especially stresses in shafts and rotors, critical speeds, and torsional oscillations; (2) the large acceleration forces produced on stator bars which must be withstood by their insulation (as much as 27 g in some designs); (3) the need for more effective cooling. Mica paper and glass bonded with epoxy resin are at present being used as stator insulation, the maximum permissible temperature being 135°C.

Semiconductor rectifiers are employed for excitation. Traditionally, sets had d.c. exciters mounted on the main shaft. This type was followed by an a.c. exciter from which the current is rectified so that a.c. and d.c. slip-rings are required. More recently, a.c. exciters with integral fused-diode or thyristor rectifiers are employed, thus avoiding any brush gear and consequent maintenance problems, the semiconductors rotating on the main shaft.

Superconducting field windings without ferromagnetic rotors would give magnetic flux densities of about three times the present 2–2.5 T and offer the possibility of reduced machine size and weight and an increase in operating efficiency of about 0.4 per cent; the machine reactance would be smaller (the synchronous reactance may be reduced by as much as four-fifths), but the short-circuit forces would be larger. Superconducting stator windings are impracticable because of the prohibitively large a.c. losses they would produce. Even if such machines are shown to be economic for large sizes their reliability must present major problems. Reliability is all-important, and the loss resulting from an outage of a large conventional generator for 1 day is roughly equivalent to 1 per cent of the initial cost of the machine.

## 3.3 Equivalent Circuit Under Balanced Short-Circuit Conditions

A typical set of oscillograms of the currents in three stator phases when a synchronous generator is suddenly short-circuited is shown in Figure 3.6(a). In all three traces a direct-current component is evident and this is to be expected from a knowledge of transients in *R–L* circuits. The magnitude of direct current in any phase present depends upon the instant at which the short circuit is applied and on the power factor of the circuit. As there are three voltages mutually at 120° it is possible for only one to have the maximum direct-current component. Often, to clarify the physical conditions, the direct component is ignored and a trace of short-circuit current, as shown in Figure 3.6(b), is considered. Immediately after the application of the short circuit the armature current endeavours to create an armature reaction m.m.f., as already mentioned, but the main flux cannot change to a new value immediately as it is linked with low-resistance circuits consisting of: (1) the rotor winding which is effectively a closed circuit; (2) the damper bars, i.e. a winding which consists of short-circuited turns of copper strip set in the poles to dampen oscillatory tendencies, and (3) the rotor body, often of forged steel. As the flux remains unchanged initially, the stator currents are large and can flow only because of the creation of opposing currents in the rotor and damper windings by what is essentially transformer action. Owing to the higher resistance, the current induced in the damper winding decays rapidly and the armature current commences to fall. After this, the currents in the rotor winding and body decay, the armature reaction m.m.f. is gradually established, and the generated e.m.f. and stator current fall until the steady-state condition on short circuit is reached. Here, the full armature reaction effect is operational and the machine is represented by the synchronous reactance $X_s$. These effects are shown in Figure 3.6, with the high initial current due to the damper winding and then the gradual reduction until the full armature reaction is established.

To represent the initial short-circuit conditions two new models must be introduced. If, in Figure 3.6(b), the envelopes of the 50 Hz waves are traced, a discontinuity appears. Whereas the natural envelope continues to point 'a' on the stator current axis, the actual trace finishes at point 'b'; the reasons for this have been mentioned above. To account for both of these conditions, two new reactances are needed to represent the machine, the very initial conditions requiring what is called the *subtransient reactance* $(X'')$ and the subsequent period the *transient reactance* $(X')$. In the following definitions it is assumed that the generator is on no-load prior to the application of the short circuit and is of the round-rotor type. Let the no-load phase voltage of the generator be $E$ volt (r.m.s.)

Then, from Figure 3.6(b), the subtransient reactance

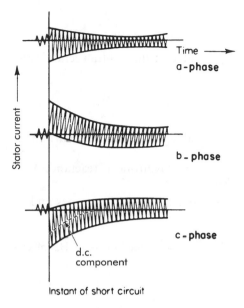

(a)

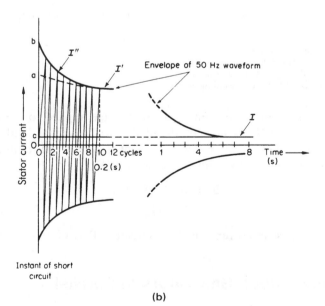

(b)

**Figure 3.6** (a) Oscillograms of the currents in the three phases of a generator when a sudden short circuit is applied. (b) Trace of a short-circuit current when direct-current component is removed

$$(X'') = \frac{E}{0b/\sqrt{2}}$$

where $0b/\sqrt{2}$ is the r.m.s. value of the subtransient current ($I''$); the transient reactance

$$(X') = \frac{E}{0a/\sqrt{2}}$$

where $0a/\sqrt{2}$ is the r.m.s. value of the transient current ($I'$), and finally

$$\frac{E}{0c/\sqrt{2}} = \text{synchronous reactance } X_s$$

Typical values of $X''$, $X'$, and $X_s$ for various types and sizes of machines are given in Table 3.1.

**Table 3.1**  Constants of synchronous machines—60 Hz (all values expressed as per unit on rating)

| Type of machine | $X_s$ (or $X_d$) | $X_q$ | $X'$ | $X''$ | $X_2$ | $X_0$ | $R_L$ |
|---|---|---|---|---|---|---|---|
| Turbo-alternator | 1.2–2.0 | 1–1.5 | 0.2–0.35 | 0.17–0.25 | 0.17–0.25 | 0.04–0.14 | 0.003–0.008 |
| Salient pole (hydroelectric) | 0.16–1.45 | 0.4–1.0 | 0.2–0.5 | 0.13–0.35 | 0.13–0.35 | 0.02–0.2 | 0.003–0.0015 |
| Synchronous compenstor | 1.5–2.2 | 0.95–1.4 | 0.3–0.6 | 0.18–0.38 | 0.17–0.37 | 0.03–0.15 | 0.004–0.01 |

$X_2$ = negative sequence reactance.
$X_0$ = zero sequence reactance.
$X'$ = and $X''$ are the direct axis quantities.
$R_L$ = a.c. resistance of the stator winding per phase.

If the machine is previously on load, the voltage applied to the equivalent reactance, previously $E$, is now modified due to the initial load volt-drop. Consider Figure 3.7. Initially, the load current is $I_L$ and the terminal voltage is $V$. The voltage behind the transient reactance $X'$ is

$$\mathbf{E}' = \mathbf{I_L}(\mathbf{Z_L} + \mathrm{j}\,X')$$
$$= \mathbf{V} + \mathrm{j}\,\mathbf{I_L}\,X'$$

and hence the transient current on short circuit = $\mathbf{E}'/\mathrm{j}\,X'$.

## 3.4  Synchronous Generators in Parallel

Consider two machines A and B (as shown in Figure 3.8(a)), the voltages of which have been adjusted to equal values by means of the field regulators, and the speeds of which are slightly different. In Figure 3.8(b) the phase voltages

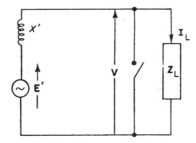

**Figure 3.7** Modification of equivalent circuit to allow for initial load current

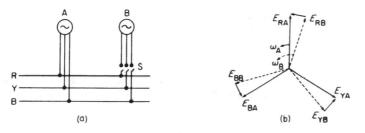

**Figure 3.8** (a) Generators in parallel. (b) Corresponding phasor diagrams

are $E_{RA}$, etc., and the speed of machine A is $\omega_A$ radians per second and of B, $\omega_B$ radians per second.

If voltage phasors of A are considered stationary, those of B rotate at a relative velocity $(\omega_B - \omega_A)$ and hence there are resultant voltages across the switch S of $(\mathbf{E_{RA}} - \mathbf{E_{RB}})$, etc., which reduce to zero during each relative revolution. If the switch is closed at an instant of zero voltage, the machines are connected (synchronized) without the flow of large currents due to the resultant voltages across the armatures. When the two machines are in synchronism they have a common terminal-voltage, speed, and frequency. Any tendency for one machine to accelerate relative to the other immediately results in a retarding or synchronizing torque being set up due to the current circulated.

Two machines operating in parallel are represented by the equivalent circuits shown in Figure 3.9(a) with $\mathbf{E_A} = \mathbf{E_B}$ and on no external load. If A tries to gain speed the phasor diagram in Figure 3.9(b) is obtained and $I = E_R/(\mathbf{Z_A} + \mathbf{Z_B})$. The circulating current $\mathbf{I}$ lags $\mathbf{E_R}$ by an angle $\tan^{-1}(X/R)$ and, as in most machines $X \gg R$, this angle approaches $90°$. This current is a generating current for A and a motoring current for B; hence A is generating power and tending to slow down and B is receiving power from A and speeding up. Therefore, A and B remain at the same speed, 'in step', or in synchronism.

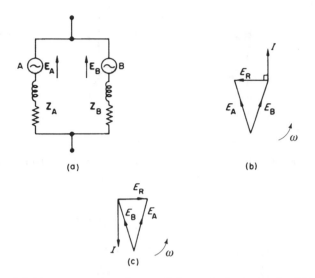

**Figure 3.9**    (a) Two generators in parallel—equivalent circuit. (b) Machine A in phase advance of machine B. (c) Machine B in phase advance of machine A

Figure 3.9(c) shows the state of affairs when B tries to gain speed on A. The quality of a machine to return to its original operating state after a momentary disturbance is measured by the *synchronizing power* and *torque*. It is interesting to note that as the impedance of the machines is largely inductive, the restoring powers and torques are large; if the system were largely resistive it would be difficult for synchronism to be maintained.

Normally, more than two generators operate in parallel and the operation of one machine connected in parallel with many others is of great interest. If the remaining machines in parallel are of such capacity that the presence of the generator under study causes no difference to the voltage and frequency of the other, they are said to comprise an *infinite busbar system*, i.e. an infinite system of generation. In practice, a perfect infinite busbar is never fully realized but if, for example, a 600 MW generator is removed from a 30 000 MW system, the difference in voltage and frequency caused will be very slight.

## 3.5  The Operation of a Generator on Infinite Busbars

In this section, in order to simplify the ideas as much as possible, the resistance of the generator will be neglected, in practice, this assumption is usually reasonable. Figure 3.10(a) shows the schematic diagram of a machine connected to an infinite busbar along with the corresponding phasor diagram. If losses are neglected the power output from the turbine is equal to the power output from

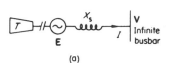
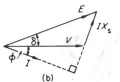

(a)                                          (b)

**Figure 3.10** (a) Synchronous machine connected to an infinite busbar. (b) Corresponding phasor diagram

the generator. T~~he angle δ between the E and V phasors is known as the load angle and is dependent on the power input from the turbine shaft.~~ With an isolated machine supplying its own load the latter dictates the power required and hence the load angle; when connected to an infinite-busbar system, however, the load delivered by the machine is no longer directly dependent on the connected load. B~~y changing the turbine output, and hence δ, the generator can be made to take on any load that the operator desires subject to economic and technical limits.~~

From the phasor diagram in Figure 3.10(b), th~~e power delivered to the infinite busbar~~ $= VI \cos\phi$ ~~per phase,~~ but

$$\frac{E}{\sin(90+\phi)} = \frac{IX_s}{\sin\delta} \qquad \leftarrow \left(\begin{array}{l} Law \ o.f \ sines \\ 90°+\phi \ across \ from \ E \end{array}\right)$$

hence
$$\sin(90+\phi) = \cos\phi$$

$$I\cos\phi = \frac{E}{X_s}\sin\delta$$

$$\therefore \text{ power delivered } = \frac{VE}{X_s}\sin\delta \quad \text{(W)} \qquad (3.1)$$

T~~his expression is of extreme importance as it governs, to a large extent, the operation of a power system. It could have been obtained directly from equation (2.8), putting α = 0.~~

~~Equation (3.1) is shown plotted in Figure 3.11. The maximum power is obtained at δ = 90°. If δ becomes larger than 90° due to an attempt to obtain more than $P_{max}$, increase in δ results in less power output and the machine becomes unstable and loses synchronism. Loss of synchronism results in the interchange of current surges between the generator and network as the poles of the machine pull into synchronism and then out again, i.e. the generator pole slips.~~

~~If the power output of the generator is increased by small increments with the no-load voltage kept constant, the limit of stability occurs at δ = 90° and is known as the steady-state stability limit. There is another limit of stability due to a sudden large change in conditions, such as caused by a fault, and this known as the transient stability limit; it is possible for the rotor to oscillate beyond 90° a number of times. If these oscillations diminish, the machine is~~

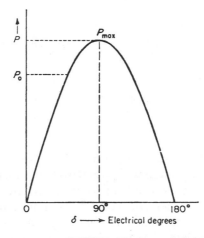

**Figure 3.11**   Power-angle curve of a synchronous machine. Resistance and saliency are neglected

stable. The load angle $\delta$ has a physical significance, it is the angle between like radial marks on the end of the rotor shaft of the machine and on an imaginary rotor representing the system. The marks are in identical physical positions when the machine is on no-load. The synchronizing power coefficient

$$= \mathrm{d}P/\mathrm{d}\delta \qquad \text{watts per radian}$$

and the synchronizing torque coefficient

$$= (1/\omega_s)/(\mathrm{d}P/\mathrm{d}\delta)$$

In Figure 3.12(a) the phasor diagram for the limiting steady-state condition is shown. It should be noted that in this condition, current is always leading. Figures 3.12(b), (c), and (d), show the phasor diagrams for various operational conditions. Another interesting operating condition is variable power and constant excitation. This is shown in Figure 3.13. In this case, as $V$ and $E$ are constant, when the power from the turbine is increased $\delta$ must increase and the power factor changes.

It is convenient to summarize the above types of operation in a single diagram or chart which will enable an operator to see immediately whether the machine is operating within the limits of stability and rating.

### 3.5.1   The performance chart of a synchronous generator

Consider Figure 3.14(a), the phasor diagram for a round-rotor machine ignoring resistance. The locus of constant $IX_s$, $I$, and hence MVA, is a circle, and the locus of constant $E$ is a circle. Hence, provided $V$ is constant, then

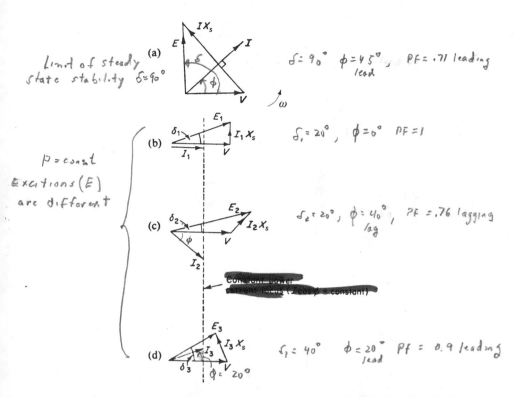

*Limit of steady state stability* $\delta = 90°$

(a)   $\delta = 90°$   $\phi = 45°$,   $PF = .71 \text{ leading}$
lead

$P = const$
*Excitions* $(E)$
*are different*

(b)   $\delta_1 = 20°$,   $\phi = 0°$   $PF = 1$

(c)   $\delta_2 = 20°$,   $\phi = 40°$,   $PF = .76 \text{ lagging}$
lag

Constant power
current leads? ($I \cos \phi$ = constant)

(d)   $\delta_3 = 40°$   $\phi = 20°$   $PF = 0.9 \text{ leading}$
lead

**Figure 3.12** (a) Phasor diagram for generator at limit of steady-state stability. (b) to (d) Phasor diagrams for generator delivering constant power to the infinite busbar system but with different excitations. As $V$ is constant, the in-phase component of $I$ must be constant. As $(EV/X)\sin\delta$ is constant as $E$ changes, $\delta$ must change and      ← *Power = const*

$$\delta_3 > \delta_1 > \delta_2$$

*thus*   $E_3 < E_1 < E_2$   *for Power = const*

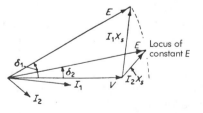

*Locus of constant E*

**Figure 3.13** Operation at variable power and constant excitation

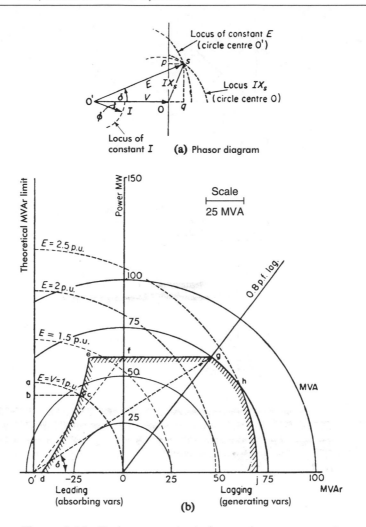

**Figure 3.14** Performance chart of a synchronous generator

$0s$ is proportional to $VI$ or MVA

$ps$ is proportional to $VI \sin \phi$ or MVAr

$sq$ is proportional to $VI \cos \phi$ or MW

To obtain the scaling factor for MVA, MVAr, and MW, the fact that at zero excitation, $E = 0$ and $IX_s = V$, is used, from which $I$ is $V/X_s$ at 90° leading to $00'$, corresponding to VAr/phase.

The construction of a chart for a 60 MW machine follows (Figure 3.14(b)).

## Machine data

60 MW, 0.8 p.f., 75 MVA

11.8 kV, SCR 0.63, 3000 r.p.m.

Maximum exciter current 500 A

$$X_s = \frac{1}{0.63} \text{ p.u.} = 2.94\,\Omega/\text{phase}$$

The chart will refer to complete three-phase values of MW and MVAr. When the excitation and hence $E$ are reduced to zero, the current leads $V$ by $90°$ and is equal to $(V/X_s)$, i.e. $11\,800/\sqrt{3} \times 2.94$. The leading vars correspond to this, given by $3V^2/X = 47.4$ MVAr.

With centre 0 a number of semicircles are drawn of radii equal to various MVA loadings, the most important being the 75 MVA circle. Arcs with $0'$ as centre are drawn with various multiples of $00'$ (or $V$) as radii to give the loci for constant excitations. Lines may also be drawn from 0 corresponding to various power factors, but for clarity only 0.8 p.f. lagging is shown, i.e. the machine is generating vars. The operational limits are fixed as follows. The rated turbine output gives a 60 MW limit which is drawn as shown, i.e. line efg, which meets the 75 MVA line at g. The MVA arc governs the thermal loading of the machine, i.e. the stator temperature rise, so that over portion gh the output is decided by the MVA rating. At point h the rotor heating becomes more decisive and the arc hj is decided by the maximum excitation current allowable, in this case assumed to be 2.5 p.u. The remaining limit is that governed by loss of synchronism at leading power factors. The theoretical limit is the line perpendicular to $00'$ at $0'$ (i.e. $\delta = 90°$), but, in practice, a safety margin is introduced to allow a further increase in load of either 10 or 20 per cent before instability. In Figure 3.14 a 10 per cent margin is used and is represented by ecd: it is constructed in the following manner. Considering point 'a' on the theoretical limit on the $E = 1$ p.u. arc, the power $0'a$ is reduced by 10 per cent of the rated power (i.e. by 6 MW) to $0'b$; the operating point must, however, still be on the same $E$ arc and b is projected to c, which is a point on the new limiting curve. This is repeated for several excitations, finally giving the curve ecd.

The complete operating limit is shown shaded and the operator should normally work within the area bounded by this line, provided the generator is running at rated voltage. If the voltage is different from rated, e.g. at −5 per cent of rated, the whole operating chart will shrink by a pro-rata amount, except for the excitation and turbine maximum power line. Note, therefore, that for the generator to produce rated power at reduced voltage, the stator requires to operate on overcurrent. This may be possible if an overcurrent rating is given.

## Example 3.1

Use the chart of Figure 3.14 to determine the excitation and the load angle at full load (60 MW, 0.8 p.f. lag) and at unity power factor, rated output. Check by calculation.

## Solution

From the diagram, the following values are obtained:

|  | Excitation $E$ | Angle $\delta$ |
|---|---|---|
| 0.8 p.f. lag | 2.3 p.u. | 33° |
| u.p.f. | 1.7 p.u. | 50° |

## Check

$$\text{Power} = \frac{VE}{X_s}\sin\delta$$

$$\therefore 60 \times 10^6 = \frac{11\,800 \times (11\,800 \times 2.3)}{2.94}\sin\delta$$

$$\delta = 33.4° \qquad \text{at 0.8 p.f. lag}$$

$$\delta = 48.2° \qquad \text{at u.p.f.}$$

# 3.6   Salient-Pole Generators

The assumption that a generator has a single impedance $X$, as in Figure 3.2, is rarely true and the two-axis approach of Figure 3.3 with $X_d \neq X_q$ represents generator behaviour better, particularly as it gives a more accurate value for torque angle $\delta$ of a given load. Power as a function of load angle is derived below.

## 3.6.1   Output power

The output power is given by $VI\cos\phi$ where $V$ is shown as $V_t$ in Figure 3.3(d). If we redraw this figure neglecting $R_L$ and splitting $IX_L$ into two components $I_qX_L$ and $I_dX_L$ we obtain Figure 3.15.

From Figure 3.15(b) we see that

$$\mathbf{E} = \mathbf{V} + j\mathbf{I}_d X_d + j\mathbf{I}_q X_q$$
$$= \mathbf{V} + j\mathbf{I}X_q + j\mathbf{I}_d(X_d - X_q) \tag{3.2}$$

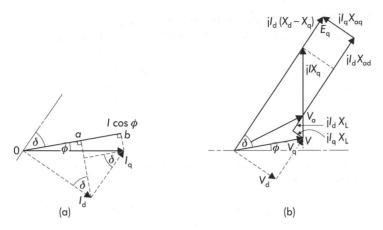

**Figure 3.15** Current and voltage phasor diagrams for salient-pole generator. (a) Current phasors and components. (b) Voltage phasors and components

As $j\mathbf{I}_d(X_d - X_q)$ is in phase with $E$, and we know $V$, $I$, and $\phi$ from the loading condition, $IX_q$ can be drawn at right angles to the current $I$ so that the direction of $E$ can be determined. From this, the $d$ and $q$ axis directions can be drawn on both the current and voltage phasor diagrams and hence the remaining phasors can be established.

From Figure 3.15(a) we have

$$I \cos \phi = \text{oa} + \text{ab} = I_d \sin \delta + I_q \cos \delta$$

But

$$V \sin \delta = I_q X_q \quad \text{and} \quad V \cos \delta = E - I_d X_d$$

Hence

$$\text{Output power} P = V[I \cos \phi]$$

$$= V \left[ \frac{V \sin \delta}{X_q} \cos \delta + \frac{E - V \cos \delta}{X_d} \sin \delta \right]$$

$$= \frac{VE}{X_d} \sin \delta + \frac{V^2(X_d - X_q)}{2X_d X_q} \sin 2\delta \tag{3.3}$$

and

$$\frac{\mathrm{d}P}{\mathrm{d}\delta} = \frac{VE}{X_d} \cos \delta + \frac{V^2(X_d - X_q)}{X_d X_q} \cos 2\delta \tag{3.4}$$

$$= \text{synchronizing power coefficient}$$

The plot of $P$ against $\delta$ is shown in Figure 3.16, from which it is evident that the steady-state stability limit occurs at a $\delta$ of less than 90° when $X_d$ and $X_q$ are very different.

In steady-state calculations, as previously discussed, the effects of saliency may often be neglected, especially compared with saturation effects.

## Example 3.2

Compare the effects of saliency for a generator with the following parameters when producing maximum power output.

$$X_d = 1.2 \, \text{pu}; \qquad X_q = 0.9 \, \text{p.u.}; \qquad \text{excitation} E = 1.5 \, \text{p.u.}$$

Generator is connected to an infinite bus with voltage $V = 1 \, \text{p.u.}$

## Calculation

$$P_{max} \text{ (saliency neglected)} = \frac{EV}{X_d} \sin \delta = \frac{1.5 \times 1.0}{1.2}$$
$$= 1.25 \, \text{p.u.} \qquad (\text{W})$$

$$P_{max} \text{ (with saliency)} = \frac{EV}{X_d} \sin \delta + V^2 \frac{(X_d - X_q)}{2X_d X_q} \sin 2\delta$$
$$= 1.25 \sin \delta + 0.14 \sin 2\delta$$

The quickest way to solve this equation for $P_{max}$ is to try several values of $\delta$, and produce a table as follows:

| $\delta$ (degrees) | $P_{max}$ |
|---|---|
| 90 | 1.25 (already calculated) |
| 80 | 1.279 |
| 70 | 1.26 (less than before) |
| 75 | 1.277 |
| 78 | 1.280 (Max value at 78°) |

$$\text{Error is} \quad \frac{1.28 - 1.25}{1.28} = 2.3 \text{ per cent}$$

Generally, provided that $E > V$, saliency can be neglected for most studies. The reader should try other values of $E$ to check this out, but note that the $P_{max}$ with saliency occurs at angles 10° or more less than without saliency.

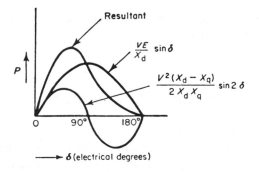

**Figure 3.16** Power-angle curve—salient-pole machine

## 3.7 Automatic Voltage Regulators (AVRs)

Excitation of a synchronous generator is derived from a d.c. supply with variable voltage and requires considerable power to produce the required operating flux. Hence an exciter may require 1 MW or more to excite a large generator's field winding on the rotor. Nowadays, such a power amplifier is best arranged through an a.c. generator driven or overhung on the rotor shaft feeding the rotor through diodes, also mounted on the shaft and rotating with it, or by means of thyristors whose grids can be controlled by a feedback signal (see Figure 3.17).

Originally, the main exciter consisted of a d.c. shunt-controlled generator supplying the generator rotor winding through slip rings on the rotor shaft. The main exciter was itself excited from a pilot exciter, self-excited for reliability. The most important aspect of the excitation system is its speed of response between a change in the reference signal $V_{ref}$ and the change in the excitation current $I_f$. Control system theory using Bode diagrams and phase

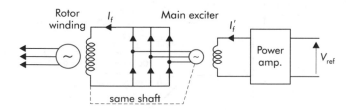

**Figure 3.17** Excitation arrangements for a synchronous generator using diodes. Diodes can be replaced by thyristor controllers and the power amplifier can be eliminated

margins is required to design appropriate responses and stability boundaries of a generator excitation system.

### 3.7.1 Types of automatic voltage regulator

The detailed study of AVRs is a specialist field, and it is not intended here to discuss types in any detail but rather to indicate their general effect. There are two broad divisions of automatic regulators, both of which set out to control the output voltage of the synchronous generator by controlling the exciter. In general the deviation of the terminal voltage from a prescribed value is passed to control circuits and thus the field current is varied, and it is in the manner and speed in achieving this that the division occurs.

The first and older type can be broadly classed as *electromechanical*. A well-known variety of this is the carbon-pile regulator. In this, a voltage proportional to the deviation voltage operates a solenoid assembly to vary the pressure exerted on a carbon-pile resistor in the exciter field, thus varying its resistance. Another type depends upon the conversion of the deviation voltage into a torque by means of a 'torque motor'; according to the angular position of the shaft of the motor, certain resistors in a resistor chain are cut out of the circuit and hence the exciter-field current changes. There are various other types, including the popular vibrating-reed regulator. All these types suffer from the disadvantages of being relatively slow acting and possessing dead bands, i.e. a certain deviation must occur before the mechanism operates; this is illustrated in Figure 3.18.

The other main group of regulators is known as 'continuously acting' as in Figure 3.17 and these are faster than the above types and have no dead bands. A general block diagram of a typical control system is given in Figure 3.19.

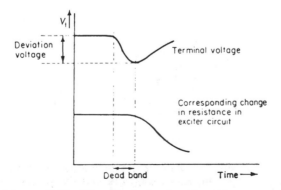

**Figure 3.18**    The effect of a dead band in a regulator

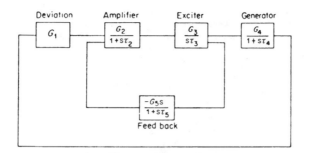

**Figure 3.19** Block diagram of a continuously acting closed-loop automatic voltage regulator

### 3.7.2 Automatic voltage regulators and generator characteristics

The equivalent circuit used to represent the synchronous generator can be modified to account for the action of a regulator. Basically there are three conditions to consider:

1. Operation with fixed excitation and constant no-load voltage ($E$), i.e. no regulator action. This requires the usual equivalent circuit of $E$ in series with $Z_s$.

2. Operation with a regulator which is not continuously acting, i.e. the terminal voltage varies with load changes. This can be simulated by $E'$ and a reactance smaller than the synchronous value. It has been suggested by experience, in practice, that a reasonable value would be the transient reactance, although some authorities suggest taking half of the synchronous reactance. This mode will apply to most modern regulators.

3. Terminal voltage constant. This requires a very-fast-acting regulator and the nearest approach to it exists in the forced-excitation regulators (i.e. AVRs capable of reversing their driving voltage to suppress $I_f$ quickly) used on generators supplying very long lines.

   Each of the above representations will give significantly different values of maximum power output. The degree to which this happens depends on the speed of the AVR, and the effect on the operation chart of the synchronous generator is shown in Figure 3.20, which indicates clearly the increase in operating range obtainable. It should be noted, however, that operation in these improved leading power-factor regions may be limited by the heating of the stator winding. The actual power-angle curve may be obtained by a step-by-step process by using the gradually increasing values of $E$ in $EV/X \sin \delta$.

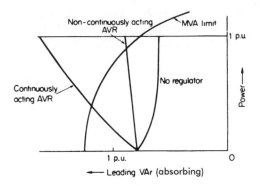

**Figure 3.20** Performance chart as modified by the use of automatic voltage regulators

When a generator has passed through the steady-state limiting angle of $\delta = 90°$ with a fast-acting AVR, it is possible for synchronism to be retained. The AVR, in forcing up the voltage, increases the power output of the machine so that instead of the power falling after $\delta = 90°$, it is maintained and $dP/d\delta$ is still positive. This is illustrated in Figure 3.21(b), where the P–δ relation for the system in Figure 3.21(a) is shown. Without the AVR the terminal voltage of the machine $(V_t)$ will fall with increased δ, the generated voltage $E$ being constant and the power reaching a maximum at 90°. With a perfect AVR, $V_t$ is maintained constant, $E$ being increased with increase in δ. As $P = (EV_s/X)\sin\delta$, it is evident that power will increase beyond 90° until the excitation limit is reached, as shown in Figure 3.21(b).

## Example 3.3

Determine the limiting powers for the system shown in Figure 3.22 for the three types of voltage regulation. All values are on 100 MVA, 132 kV bases. It may be assumed that the lines and transformers are each represented by a single-series reactance.

## Solution

(a) No control, constant excitation voltage.

$$X = 1.5 + 0.15 + 0.1 + \frac{0.4}{2} = 1.95\,\text{p.u.}$$

From equation (2.8)

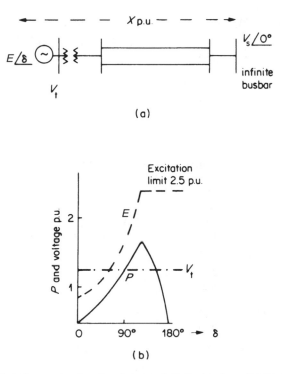

**Figure 3.21** (a) Generator feeding into an infinite busbar. (b) Variations of output power $P$, generated voltage $E$, and terminal voltage $V_t$ with load (transmission) angle $\delta$. Perfect AVR

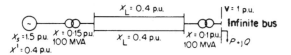

**Figure 3.22** Line diagram of system for Example 3.3. Normal operating load $P = 0.8\,\text{p.u.}$, $Q = 0.5\,\text{p.u.}$

$$E = \sqrt{\left[\left(1 + \frac{0.5 \times 1.95}{1}\right)^2 + \left(\frac{0.8 \times 1.95}{1}\right)^2\right]}$$

$$= 2.52\,\text{p.u.}$$

$$\therefore P_{max} = \frac{EV}{X} = \frac{2.52 \times 1}{1.95} = 1.29\,\text{p.u.}$$

(b) Non-continuously acting AVR.

$$X = 0.4 + 0.15 + 0.1 + 0.2 = 0.85 \, \text{p.u.}$$

$$E' = \sqrt{\left[ \left(1 + \frac{0.5 \times 0.85}{1}\right)^2 + \left(\frac{0.8 \times 0.85}{1}\right)^2 \right]}$$

$$= 1.59 \, \text{p.u.}$$

$$\therefore \ P_{max} = \frac{1.59 \times 1}{0.85} = 1.87 \, \text{p.u.}$$

(c)

$$X = 0.15 + 0.1 + 0.2 = 0.45 \, \text{p.u.}$$

Terminal voltage of generator $V_t$ constant,

$$V_t = \sqrt{\left[ \left(1 + \frac{0.5 \times 0.45}{1}\right)^2 + \left(\frac{0.8 \times 0.45}{1}\right)^2 \right]}$$

$$= 1.28 \, \text{p.u.}$$

$$P_{max} = \frac{1.28 \times 1}{0.45} = 2.85 \, \text{p.u.}$$

It is interesting to compare the power limit when one of the two lines is disconnected; the transient-reactance representation will be used. Now

$$X = 0.4 + 0.15 + 0.1 + 0.4 = 1.05 \, \text{p.u.}$$

$$E' = \sqrt{\left(1 + \frac{0.5 \times 1.05}{1}\right)^2 + \left(\frac{0.8 \times 1.05}{1}\right)^2}$$

$$= 1.74 \, \text{p.u.}$$

$$P_{max} = \frac{1.74 \times 1}{1.05} = 1.65 \, \text{p.u.}$$

# Lines, Cables, and Transformers

## 3.8   Overhead Lines—Types and Parameters

Overhead lines are suspended from insulators which are themselves supported by towers or poles. The span between two towers is dependent upon the allowable sag in the line, and for steel towers with very-high-voltage lines the span is normally 370–460 m (1200–1500 ft). Typical supporting structures are shown in Figures 3.23 and 3.24. There are two main types of tower:

1. Those for straight runs in which the stress due to the weight of the line alone has to be withstood;

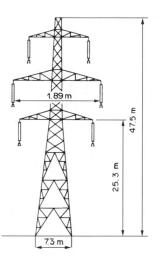

**Figure 3.23** A 400 kV double-circuit overhead-line tower. Two conductors per phase (bundle conductors) (Reproduced by permission of *Institution of Electrical Engineers*)

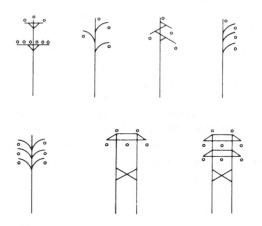

**Figure 3.24** Typical pole-type structures

2. Those for changes in route, called deviation towers; these withstand the resultant forces set up when the line changes direction.

When specifying towers and lines, ice and wind loadings are taken into account, as well as extra forces due to a break in the lines on one side of a tower. For lower voltages and distribution circuits, wooden, reinforced concrete poles or, nowadays, glass fibre tubes are used with conductors supported in horizontal formations.

Conductor tied down
with soft copper wire

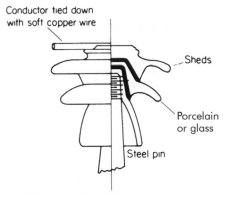

Sheds

Porcelain
or glass

Steel pin

**Figure 3.25**   Pin-type insulators

The live conductors are insulated from the towers by insulators which take
two basic forms: the *pin type* and the *suspension type*. The pin-type insulator is
shown in Figure 3.25 and it is seen that the conductor is supported on the top
of the insulator. This type is used for lines up to 33 kV. The two or three
porcelain 'sheds' or 'petticoats' provide an adequate leakage path from the
conductor to earth and are shaped to follow the equipotentials of the electric
field set up by the conduction–tower system. Suspension insulators (Figure
3.26) consist of a string of interlinking separate discs made of glass or porcelain. A string may consist of many discs depending upon the line voltage; for
400 kV lines, 19 discs of overall length 3.84 m (12 ft 7 in) are used. The conductor is held at the bottom of the string which is suspended from the tower.
Owing to the capacitances existing between the discs, conductor, and tower, the
distribution of voltage along the insulator string is not uniform, the discs
nearer the conductor being the more highly stressed. Methods of calculating
this voltage distribution are available, but are of dubious value owing to the
shunting effect of the leakage resistance (see Figure 3.27). This resistance
depends on the presence of pollution on the insulator surfaces and is considerably modified by rain and fog.

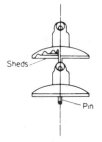

Sheds

Pin

**Figure 3.26**   Suspension-type insulators

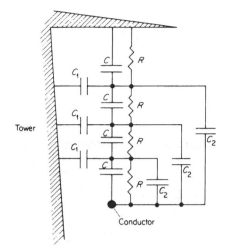

**Figure 3.27** Equivalent circuit of a string of four suspension insulators. $C$ = self-capacitance of disc; $C_1$ = capacitance disc to earth; $C_2$ = capacitance disc to line; $R$ = leakage resistance

### 3.8.1 Parameters

The parameters of interest for circuit analysis are inductance, capacitance, resistance, and leakage resistance. The derivation of formulae for the calculation of these quantities is given in the *E.H.V. Transmission Line Reference Book*. It is intended here merely to quote these formulae and discuss their application.

The inductance of a single-phase two-wire line

$$= \frac{\mu_0}{4\pi}\left[1 + 4\ln\left(\frac{d-r}{r}\right)\right]\text{H/m} \tag{3.5}$$

where $d$ is the distance between the centres, and $r$ is the radius, of the conductors. When performing load flow and balanced-fault analysis on three-phase systems it is usual to consider one phase only, with the appropriate angular adjustments made for the other two phases. Therefore, phase voltages are used and the inductances and capacitances are the equivalent phase or line-to-neutral values. For a three-phase line with equilateral spacing (Figure 3.28) the inductance and capacitance with respect to the hypothetical neutral conductor are used, and this inductance can be shown to be half the loop inductance of the single-phase line, i.e. the inductance of one conductor.

The line–neutral inductance for equilateral spacing

$$= \frac{\mu_0}{8\pi}\left[1 + 4\ln\left(\frac{d-r}{r}\right)\right]\text{H/m} \tag{3.6}$$

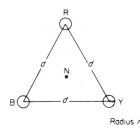

Radius $r$

**Figure 3.28**   Overhead line with equiliateral spacing of the conductors

The capacitance of a single-phase line

$$C = \frac{\pi \varepsilon_0}{\ln[(d-r)/r]} \, \text{F/m} \tag{3.7}$$

With three-phase conductors spaced equiliaterally, the capacitance of each line to be hypothetical neutral is double that for the two-wire circuit, i.e.

$$\frac{2\pi \varepsilon_0}{\ln[(d-r)/r]} \, \text{F/m} \tag{3.8}$$

In practice, the conductors are rarely spaced in the equiliateral formation, and it can be shown that the average value of inductance or capacitance for any formation of conductors can be obtained by the representation of the system by one of equivalent equilateral spacing. The equivalent spacing $d_{\text{eq}}$ between conductors is given by

$$d_{\text{eq}} = \sqrt[3]{(d_{12} \cdot d_{23} \cdot d_{31})}$$

Often, two three-phase circuits are electrically in parallel; if physically remote from each other, the reactances of the lines are identical. When the two circuits are situated on the same towers, however, the magnetic interaction between them should be taken into account. The use of bundle conductors, i.e. more than one conductor per insulator, reduces the reactance; it also reduces conductor–surface voltage gradients and hence corona loss and radio interference. Unsymmetrical conductor spacing results in different inductances for each phase which causes an unbalanced voltage drop, even when the load currents are balanced. The residual or resultant voltage or current induces unwanted voltages into neighbouring communication lines. This can be overcome by the interchange of conductor positions at regular intervals along the route, a practice known as *transposition* (see Figure 3.29). In practice, lines are rarely transposed at regular intervals and transposition is carried out where physically convenient, e.g. at substations. In short lines the degree of unbalance existing without transposition is small and may be neglected in calculations.

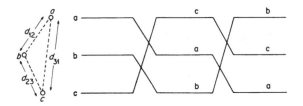

**Figure 3.29** Transposition of conductors

## Resistance

Overhead-line conductors usually comprise a stranded steel core (for mechanical strength) surrounded by aluminium wires which form the conductor. It should be noted that aluminium and ACSR (aluminium conductor steel reinforced) conductors are sometimes described by area of a copper conductor having the same d.c. resistance, i.e. their *copper equivalent*. In Table 3.2(a), $258 \, \text{mm}^2$ ($0.4 \, \text{in}^2$) ACSR conductor implies the copper equivalent. The actual area of aluminium is approximately $430 \, \text{mm}^2$, and including the steel core the overall cross-section of the line is about $620 \, \text{mm}^2$, i.e. a diameter of 28 mm. Electromagnetic losses in the steel increase the effective a.c. resistance, which increases with current magnitude, giving an increase of up to about 10 per cent. With direct current the steel carries 2–3 per cent of the total current. The electrical resistivities of some conductor materials are as follows in ohm-metres at 20°C: copper $1.72 \times 10^{-8}$; aluminium $2.83 \times 10^{-8}$ ($3.52 \times 10^{-8}$ at 80°C); aluminium alloy $3.22 \times 10^{-8}$.

**Table 3.2(a)** Overhead-line constants at 50 Hz (British) (per phase, per km)

| Voltage: | | 132 kV | | 275 kV | | 400 kV | |
|---|---|---|---|---|---|---|---|
| No. and area of conductors (mm)²: | | 1× 113 | 1× 258 | 2× 113 | 2× 258 | 2× 258 | 4× 258 |
| Resistance ($R$) | Ω | 0.155 | 0.068 | 0.022 | 0.034 | 0.034 | 0.017 |
| Reactance ($X_L$) | Ω | 0.41 | 0.404 | 0.335 | 0.323 | 0.323 | 0.27 |
| Susceptance ($1/X_c$) | S × 10⁻⁶ | 7.59 | 7.59 | 9.52 | 9.52 | 9.52 | 10.58 |
| Charging current ($I_c$) | A | 0.22 | 0.22 | 0.58 | 0.58 | 0.845 | 0.945 |
| Surge impedance | Ω | 373 | 371 | 302 | 296 | 296 | 258 |
| Natural load | MW | 47 | 47 | 250 | 255 | 540 | 620 |
| $X_L/R$ ratio | — | 2.6 | 5.9 | 4.3 | 9.5 | 9.5 | 15.8 |
| Thermal rating: | | | | | | | |
|   Cold weather (below 5°C) | MVA | 125 | 180 | 525 | 760 | 1100 | 2200 |
|   Normal (5–18°C) | MVA | 100 | 150 | 430 | 620 | 900 | 1800 |
|   Hot weather (above 18°C) | MVA | 80 | 115 | 330 | 480 | 790 | 1580 |

**Table 3.2(b)** Typical characteristics of bundled-conductor E.H.V. lines

| Number of subconductors in bundle | Country | Line voltage and number of circuits (in parentheses) | Diameter of subconductors | Radius of circle on which subconductors are arranged | Resistance of bundle | Inductive reactance at 50 Hz | Susceptance at 50 Hz |
|---|---|---|---|---|---|---|---|
| | | kV | mm | mm | $\Omega$/km | $\Omega$/km | $\mu$S/km |
| 1 | Japan | 275 (2) | 27.9 | — | 0.0744 | 0.511 | 3.01 |
| | Canada | 300 (2) | 35.0 | — | 0.0451 | 0.492 | 2.33 |
| | Australia | 330 (1) | 45.0 | — | 0.0367 | 0.422 | 2.78 |
| | Russia | 330 (1) | 30.2 | — | 0.065 | 0.404 | 2.82 |
| | Italy | 380 (1) | 50.0 | — | 0.0294 | 0.398 | 2.84 |
| Average (one conductor): | | | | | 0.048 | 0.442 | 2.75 |
| 2 | Japan | 275 (2) | 25.3 | 200 | 0.0444 | 0.374 | 4.35 |
| | Canada | 360 (1) | 28.1 | 299 | 0.04 | 0.314 | 3.57 |
| | Australia | 330 (1) | 31.8 | 190.5 | 0.0451 | 0.341 | 3.34 |
| | U.S.A. | 345 (1) | 30.4 | 228.6 | 0.0315 | 0.325 | 3.55 |
| | Italy | 380 (1) | 31.5 | 200 | 0.0285 | 0.315 | 3.57 |
| | Russia | 330 (1) | 28.0 | 200 | 0.04 | 0.321 | 3.43 |
| Average (two conductors): | | | | | 0.0040 | 0.323 | 3.58 |
| 3 | Sweden | 380 (1) | 31.68 | 260 | 0.018 | 0.29 | 3.85 |
| | Russia | 525 (1) | 30.2 | 230 | 0.0212 | 0.294 | 3.88 |
| 4 | Germany | 380 (2) | 21.7 | 202 | 0.0316 | 0.260 | 4.3 |
| | Germany | 500 (1) | 22.86 | 323 | 0.0285 | 0.272 | 4.0 |
| | Canada | 735 (1) | 35.04 | 323 | 0.0121 | 0.279 | 3.9 |

**Table 3.2(c)**   Overhead-line parameters—U.S.A.

| Line voltage (kV) | 345 | 345 | 500 | 500 | 735 | 735 |
|---|---|---|---|---|---|---|
| Conductors per phase at 18 in spacing | 1 | 2 | 2 | 4 | 3 | 4 |
| Conductor code | Expanded | Curlew | Chuker | Parakeet | Expanded | Pheasant |
| Conductor diameter (inches) | 1.750 | 1.246 | 1.602 | 0.914 | 1.750 | 1.382 |
| Phase spacing (ft) | 28 | 28 | 38 | 38 | 56 | 56 |
| GMD (ft) | 35.3 | 35.3 | 47.9 | 47.9 | 70.6 | 70.6 |
| 60 Hz inductive reactance $\Omega$/mile $\quad X_A$ | 0.3336 | 0.1677 | 0.1529 | 0.0584 | 0.0784 | 0.0456 |
| $X_D$ | 0.4325 | 0.4325 | 0.4694 | 0.4694 | 0.5166 | 0.5166 |
| $X_A + X_D$ | 0.7661 | 0.6002 | 0.6223 | 0.5278 | 0.5950 | 0.5622 |
| 60 Hz capacitive reactance M$\Omega$/miles $\quad X'$ | 0.0777 | 0.0379 | 0.0341 | 0.0126 | 0.0179 | 0.0096 |
| $X'_D$ | 0.1057 | 0.1057 | 0.1147 | 0.1147 | 0.1263 | 0.1263 |
| $X'_A + X'_D$ | 0.1834 | 0.1436 | 0.1488 | 0.1273 | 0.1442 | 0.1359 |
| $Z_0(\Omega)$ | 374.8 | 293.6 | 304.3 | 259.2 | 276.4 | |
| Natural loading MVA | 318 | 405 | 822 | 965 | 1844 | 1955 |
| Conductor d.c. resistance at 25°C ($\Omega$/mile) | 0.0644 | 0.0871 | 0.0510 | 0.162 | 0.0644 | 0.0709 |
| Conductor a.c. resistance (60 Hz) at 50°C ($\Omega$/mile) | 0.0728 | 0.0979 | 0.0599 | 0.179 | 0.0728 | 0.0805 |

$X_A$ = component of inductive reactance due to flux within a 1 ft radius.
$X_D$ = component due to other phases.
Total reactance per phase = $X_A + X_D$.

$$X_A = 0.2794 \log_{10} \left( \frac{1}{[N(GMR)(A)^{N-1}]^{1/N}} \right) \Omega/\text{mile}.$$

$X_D = 0.2794 \log_{10} (GMD) \, \Omega/\text{mile}.$
$X'_A$ and $X'_D$ are similarly defined for capacitive reactance and

$$X'_A = 0.0683 \log_{10} \left( \frac{1}{[Nr(A)^{N-1}]^{1/N}} \right) \text{M}\Omega\text{-mile}; \quad X'_D = 0,0683 \log_{10} (GMD) \, \text{M}\Omega\text{-miles}.$$

$GMR$ = geometric mean radius in feet;
$GMD$ = geometric mean diameter in feet;
$N$ = number of conductors per phase;
$A$ = $S/2 \sin (\pi/N); N > 1$;
$A$ = 0; $N = 1$;
$S$ = bundle spacing in feet;
$r$ = conductor radius in feet.
(*Permission of Edison Electric Institute*)

Leakage resistance is usually negligible for most calculation purposes and is very difficult to assess because of its dependence on the weather. This resistance represents the combined effect of all the various paths to earth from the line. The main path is that presented by the surfaces of the line insulators, the resistance of which depends on the condition of these surfaces. This varies considerably according to location; in industrial areas there will be layers of dirt and soot whilst in coastal districts deposits of salt occur. The bushings of circuit breakers and transformers form other leakage paths. The leakage losses on 132 kV lines vary between 0.3 and 1 kW per mile.

Table 3.2(a) gives the parameters for various overhead-line circuits for the line voltages operative in Britain, Table 3.2(b) gives values for international lines, and Table 3.2(c) relates to the U.S.A.

# 3.9    Representation of Lines

The manner in which lines and cables are represented depends very much on their length and the accuracy required. There are three broad classifications of length: short, medium, and long. The actual line or cable is a distributed-constant circuit, i.e. it has resistance, inductance, capacitance, and leakage resistance distributed evenly along its length, as shown in Figure 3.30. Except for long lines the total resistance, inductance, capacitance, and leakage resistance of the line are concentrated to give a lumped-constant circuit. The distances quoted are only a rough guide.

## 3.9.1    The short line (up to 80 km, 50 miles)

The equivalent circuit is shown in Figure 3.31 and it will be noticed that both shunt capacitance and leakage resistance have been neglected. The four-terminal network constants are (see Section 2.8):

$$A = 1 \qquad B = Z \qquad C = 0 \qquad D = 1$$

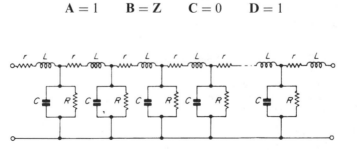

**Figure 3.30**  Distributed constant representation of a line: $L$ = inductance line to neutral per unit length; $r$ = a.c. resistance per unit length; $C$ = capacitance line to neutral per unit length; $R$ = leakage resistance per unit length

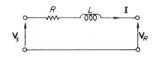

**Figure 3.31** Equivalent circuit of a short line—representation under balanced three-phase conditions

The drop in voltage along a line is important and the *regulation* is defined as

$$\frac{\text{received voltage on no load}(V_0) - \text{received voltage on load } (V_R)}{\text{received voltage on load } (V_R)}$$

It should be noted that if $\mathbf{I}$ is leading $\mathbf{V}_R$ in phase, i.e. a capacitance load, then $V_R > V_S$, as shown in Figure 3.32.

### 3.9.2 Medium-length lines (up to 240 km, 150 miles)

Owing to the increased length the shunt capacitance is now included to form either a $\pi$ or a T network. The circuits are shown in Figure 3.33. Of these two versions the $\pi$ representation tends to be in more general use but there is little difference in accuracy between the two. For the $\pi$ network:

$$\mathbf{V}_S = \mathbf{V}_R + \mathbf{IZ} \qquad \mathbf{I} = \mathbf{I}_R + \mathbf{V}_R \frac{\mathbf{Y}}{2} \qquad \mathbf{I}_S = \mathbf{I} + \mathbf{V}_S \frac{\mathbf{Y}}{2}$$

from which $\mathbf{V}_S$ and $\mathbf{I}_S$ are obtained in terms of $\mathbf{V}_R$ and $\mathbf{I}_R$ giving the following constants:

$$\mathbf{A} = \mathbf{D} = 1 + \frac{\mathbf{ZY}}{2} \qquad \mathbf{B} = \mathbf{Z} \qquad \text{and} \qquad \mathbf{C} = \left(1 + \frac{\mathbf{ZY}}{4}\right)\mathbf{Y}$$

Similarly, for the T network:

$$\mathbf{V}_S = \mathbf{V}_C + \frac{\mathbf{ZI}_S}{2} \qquad \mathbf{V}_C = \mathbf{V}_R + \frac{\mathbf{ZI}_R}{2} \qquad \mathbf{I}_S = \mathbf{I}_R + \mathbf{V}_C \mathbf{Y}$$

giving

$$\mathbf{A} = \mathbf{D} = 1 + \frac{\mathbf{ZY}}{2} \qquad \mathbf{B} = \left(1 + \frac{\mathbf{ZY}}{4}\right)\mathbf{Z} \qquad \text{and} \qquad \mathbf{C} = \mathbf{Y}$$

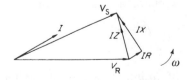

**Figure 3.32** Phasor diagram for short line on leading load

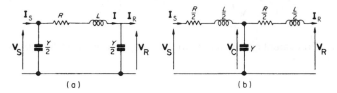

**Figure 3.33**    (a) Medium-length line—$\pi$ equivalent circuit. (b) Medium-length line—T equivalent circuit

### 3.9.3   The long line (above 240 km, 150 miles)

Here the treatment assumes distributed parameters. The changes in voltage and current over an elemental length $\Delta x$ of the line, $x$ metres from the sending end, are determined, and conditions for the whole line are obtained by integration.
Let

| | | | |
|---|---|---|---|
| $R =$ | resistance/unit length | $\mathbf{z} =$ | impedance/unit length |
| $L =$ | inductance/unit length | $\mathbf{y} =$ | shunt admittance/unit length |
| $G =$ | leakage/unit length | $\mathbf{Z} =$ | total series impedance of line |
| $C =$ | capacitance/unit length | $\mathbf{Y} =$ | total shunt admittance of line |

The voltage and current $x$ metres from the sending end are given by

$$\left.\begin{aligned} \mathbf{V}_x &= \mathbf{V}_S \cosh \mathbf{P}x - \mathbf{I}_S \mathbf{Z}_0 \sinh \mathbf{P}x \\ \mathbf{I}_x &= \mathbf{I}_S \cosh \mathbf{P}x - \frac{\mathbf{V}_S}{\mathbf{Z}_0} \sinh \mathbf{P}x \end{aligned}\right\} \tag{3.9}$$

where

$$\mathbf{P} = \text{propagation constant } = (\alpha + j\beta)$$
$$= \sqrt{(R + j\omega L)(G + j\omega C)} = \sqrt{\mathbf{z} \cdot \mathbf{y}}$$

and

$$\mathbf{Z}_0 = \text{characteristic impedance}$$
$$= \sqrt{\frac{R + j\omega L}{G + j\omega C}} \tag{3.10}$$

$\mathbf{Z}_0$ is the input impedance of an infinite length of the line; hence if any line is terminated in $\mathbf{Z}_0$ its input impedance is also $\mathbf{Z}_0$.

The propagation constant **P** represents the changes occurring in the transmitted wave as it progresses along the line; $\alpha$ measures the attentuation, and $\beta$ the angular phase-shift. With a loss-free line, where $R = G = 0$, $\mathbf{P} = j\omega\sqrt{LC}$ and $\beta = \sqrt{(LC)}$. With a velocity of propagation of $3 \times 10^5$ km/s the wavelength of the transmitted voltage and current at 50 Hz is 6000 km. Thus, lines are much shorter than the wavelength of the transmitted energy.

Usually conditions at the load are required when $x = l$ in equations (3.9).

$$\therefore \; \left. \begin{aligned} \mathbf{V_R} &= \mathbf{V_S} \cosh \mathbf{P}l - \mathbf{I_S Z_0} \sinh \mathbf{P}l \\ \mathbf{I_R} &= \mathbf{I_S} \cosh \mathbf{P}l - \frac{\mathbf{V_S}}{\mathbf{Z_0}} \sinh \mathbf{P}l \end{aligned} \right\} \tag{3.11}$$

Alternatively,

$$\left. \begin{aligned} \mathbf{V_S} &= \mathbf{V_R} \cosh \mathbf{P}l + \mathbf{I_R Z_0} \sinh \mathbf{P}l \\ \mathbf{I_S} &= \frac{\mathbf{V_R}}{\mathbf{Z_0}} \sinh \mathbf{P}l + \mathbf{I_R} \cosh \mathbf{P}l \end{aligned} \right\} \tag{3.12}$$

The parameters of the equivalent four-terminal network are thus,

$$\mathbf{A} = \mathbf{D} = \cosh \sqrt{\mathbf{ZY}}$$

$$\mathbf{B} = \sqrt{\frac{\mathbf{Z}}{\mathbf{Y}}} \sinh \sqrt{\mathbf{ZY}}$$

and

$$\mathbf{C} = \sqrt{\frac{\mathbf{Y}}{\mathbf{Z}}} \sinh \sqrt{\mathbf{ZY}}$$

The easiest way to handle the hyperbolic functions is to use the appropriate series.

$$\mathbf{A} = \cosh \sqrt{\mathbf{ZY}} = 1 + \frac{\mathbf{YZ}}{2} + \frac{\mathbf{Y^2Z^2}}{24} + \frac{\mathbf{Y^3Z^3}}{720} + \cdots$$

$$\mathbf{B} = \mathbf{Z}\left(1 + \frac{\mathbf{YZ}}{6} + \frac{\mathbf{Y^2Z^2}}{120} + \frac{\mathbf{Y^3Z^3}}{5040}\right)$$

$$\mathbf{C} = \mathbf{Y}\left(1 + \frac{\mathbf{YZ}}{6} + \frac{\mathbf{Y^2Z^2}}{120} + \frac{\mathbf{Y^3Z^3}}{5040}\right)$$

Usually not more than three terms are required, and for (overhead) lines less than 500 km (312 miles) in length the following expressions for the constants hold approximately:

$$\mathbf{A} = \mathbf{D} = 1 + \frac{\mathbf{ZY}}{2} \qquad \mathbf{B} = \mathbf{Z}\left(1 + \frac{\mathbf{ZY}}{6}\right) \qquad \mathbf{C} = \mathbf{Y}\left(1 + \frac{\mathbf{ZY}}{6}\right)$$

An exact equivalent circuit for the long line can be expressed in the form of the $\pi$ section shown in Figure 3.34. The application of simple circuit laws will show that this circuit yields the correct four-terminal network equations. If only the first term of the expansions is used, then

$$\mathbf{B} = \mathbf{Z} \quad \text{and} \quad \frac{\mathbf{A} - 1}{\mathbf{B}} = \frac{\mathbf{Y}}{2}$$

i.e. the medium-length $\pi$ representation. Figure 3.34 is only for conditions at the ends of the line; if intermediate points are to be investigated, then the full equations must be used.

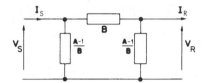

**Figure 3.34** Equivalent circuit to accurately represent the terminal conditions of a long line

## Example 3.4

The conductors of a 1.6 km (1 mile) long, 3.3 kV, overhead line are in horizontal formation with 762 mm (30 in) between centres. The effective diameter of the conductors is 3.5 mm. The resistance per kilometre of the conductors is 0.41 Ω. Calculate the line-to-neutral inductance of the line. If the sending-end voltage is 3.3 kV and the load is 1 MW at 0.8 p.f. lagging, calculate the voltage at the load busbar, the power loss in the line, and the efficiency of transmission.

## Solution

The equivalent equilateral spacing is given by

$$d_e = \sqrt[3]{(d_{12}d_{23}d_{31})}$$

In this case

$$d_e = \sqrt[3]{(762 \times 762 \times 1524)} = 762\sqrt[3]{2} = 960 \, \text{mm}$$

The inductance (line to neutral)

$$= \frac{\mu_0}{2 \times 4\pi} \left[ 1 + 4\ln\left(\frac{d-r}{r}\right) \right] \text{H/m}$$

$$= \frac{4\pi}{10^7 \times 2 \times 4\pi} \left[ 1 + 4\ln\left(\frac{960 - 1.75}{1.75}\right) \right] \text{H/m}$$

$$= \frac{1}{2 \times 10^7} \left(\frac{26.21}{1}\right) \text{H/m} = \frac{13.105}{10^7} \text{H/m}$$

Total inductance

$$= 1.0 \times 5280 \times 0.3048 \times \frac{13.105}{10^7} = 2.11 \times 10^{-3} \text{H}$$

Inductive reactance

$$= 2\pi f L = 2\pi \times 50 \times 2.11 \times 10^{-3} = 0.06 \,\Omega$$

The line is obviously in the category of short and will be treated accordingly.
The load can be expressed as

$$P = 1\,\text{MW} \qquad Q = 0.75\,\text{MVAr}$$

The load can be represented by an equivalent shunt impedance, in which case $P$ and $Q \propto V_R^2$. If $P$ and $Q$ are constant regardless of $V_R$, then an iterative procedure must be used. The latter will be used here. As the load voltage is unknown, to obtain the current the nominal voltage of 3.3 kV will be assumed. The current is then

$$\frac{10^6}{\sqrt{(3)} \times 3300 \times 0.8} = 218\angle - \cos^{-1} 0.8 \,\text{A}$$

$$\mathbf{V_S} = \mathbf{V_R} + \mathbf{IZ} \qquad \text{and} \qquad \mathbf{Z} = (0.41 \times 1.6 + \text{j}0.66)$$

$$\frac{3300}{\sqrt{3}} = \mathbf{V_R} + (218 \times 0.8 - \text{j}218 \times 0.6)(0.66 + \text{j}0.66)$$

$$= \mathbf{V_R} + 115.0 + 86.7 + \text{j}(115.0) - 86.7)$$

$$1900 = \mathbf{V_R} + 201.7 + \text{j}28.3$$

$$\mathbf{V_R} = 1698.3 - \text{j}28.3$$

$$V_R = 1698 \text{ V with negligible phase shift}$$

A quicker method would be to use

$$\Delta V = \frac{RP + XQ}{V_R}$$

$$= \frac{\frac{1}{3} \times 0.66 \times 10^6 + \frac{1}{3} \times 0.66 \times 0.75 \times 10^6}{1900} = \frac{385 \times 10^3}{1.9 \times 10^3} = 202 \text{ V}$$

$$\therefore \ V_R = 1900 - 202 = 1698 \text{ V}$$

However, as the load is 1 MW at the receiving end, $I$ should be recalculated and the above process repeated. As the approximate method is more convenient, it will be used.

$$\Delta V' = \frac{385}{1.698} = 227 \text{ V}$$

$$V_R' = 1900 - 227 = 1673 \text{ V}$$

Iterating again,

$$\Delta V'' = \frac{385}{1.673} = 230 \text{ V}$$
$$V_R'' = 1670 \text{ V}$$

Again,

$$\Delta V''' = \frac{385}{1670} = 230$$

The final value of $V_R$ is therefore 1670 V.

Here, $V_R$ can be obtained analytically without recourse to iterative procedures, but this is more lengthy in calculation. The type of load in this question in which $P$ and $Q$ remain constant regardless of the voltage is called *stiff*.

$$\text{Line loss} = 218^2 \times 0.66 \text{ W per phase}$$
$$= 94 \text{ kW for three phases}$$

$$\text{Efficiency of transmission} = \frac{\text{output}}{\text{output} + \text{losses}} = \frac{1000}{1000 + 94}$$
$$= 91.4 \text{ per cent}$$

This efficiency is typical of a rural distribution line or cable.

## *Example 3.5*

A 150 km long overhead line with the parameters given in Table 3.2(a) for 400 kV, quad conductors is to be used to transmit 1800 MW (normal weather loading) at 0.9 p.f. lag. Calculate the required sending end voltage using three-line representation methods and compare the results.

## *Solution*

From Table 3.2(a)

$$R = 0.017 \, \Omega/\text{km}, \qquad X_L = 0.270 \, \Omega/\text{km}, \qquad 1/X_c = 10.580 \times 10^{-6} \text{S/km}$$

1. Short-line representation:
$$R_{\text{total}} = 0.017 \times 150 = 2.550 \, \Omega$$
$$X_{L\,\text{total}} = 0.270 \times 150 = 40.500 \, \Omega$$
$$\mathbf{Z} = 2.550 + j\,40.500 = 40.580 \,\underline{/86.4°}\, \Omega$$
Receiving end power at nominal voltage

$$= 1800 \times 10^6 \text{ W}$$
$$= \sqrt{3} \times 400 \times 10^6 \times I \times \cos\theta$$

Hence

$$I = 2887 \text{ A at angle} \cos^{-1} 0.9 \text{ lag to } V_R$$

$$= 2887 \underline{/- 25.842^\circ}$$

$$V_S = V_R + IZ = \frac{400 \times 10^3}{\sqrt{3}} + 2887 \underline{/- 25.842} \times 40.580 \underline{/86.4}$$

$$= 230\,940 + 117\,154 \underline{/60.56^\circ}$$

$$= 306\,031 \underline{/19.47^\circ} \ V \text{(phase)}$$

$$= 530.061 \text{ kV (line)}$$

2. Medium-line representation:

$$A = D = 1 + \frac{ZY}{2}; B = Z; Y_{total} = j\,10.58 \times 10^{-6} \times 150 = j\,1587 \times 10^{-6}$$

$$\frac{ZY}{2} = 40.580 \ \underline{/86.4} \times 1587 \times 10^{-6} \underline{/90} = (-0.032 + j\,0.002)$$

Hence

$$A = 1 - 0.032 + j\,0.002 = 0.968 + j\,0.002 = 0.968 \underline{/0.12^\circ}$$

$$V_S = AV_R + BI_R = 0.968 \ \underline{/0.12} \times \frac{400 \times 10^3}{\sqrt{3}} + 40.58 \underline{/86.4^\circ} \times 2887 \underline{/- 25.84^\circ}$$

$$= 223\,549.550 + j\,460.398 + 117\,154.476 + j\,0.030$$

$$= 340\,704.337 \underline{/0.077} \ V \text{ (phase)}$$

$$= 590.117 \text{ kV (line)}$$

(Note: It is important to work to three decimal places as some numbers can become quite small.)

3. Long-line representation:

$$A = D = \cosh \sqrt{ZY} = 1 + YZ + \frac{Y^2 Z^2}{24} + \cdots$$

$$B = \sqrt{\frac{Z}{Y}} \sinh \sqrt{ZY} = Z\left(1 + \frac{YZ}{6} + \frac{Y^2 Z^2}{120} + \cdots\right)$$

$$C = \sqrt{\frac{Y}{Z}} \sinh \sqrt{ZY} = Y\left(1 + \frac{YZ}{6} + \frac{Y^2 Z^2}{120} + \cdots\right)$$

Using three terms of the appropriate series approximation for cosh and sinh, we get

$$YZ = 64400.460 \times 10^{-6} \underline{/176.4^\circ} = 0.0644 \underline{/176.4^\circ}$$

$$Y^2 Z^2 = -1587^2 \times 10^{-12} \times 40.580^2 \underline{/2 \times 86.4^\circ} = -0.004 \ \underline{/172.8^\circ}$$

Hence

$$A = D = 1 + \frac{0.0644}{2}\underline{/176.4} + \frac{-0.004\underline{/172.8°}}{24} = 0.968 + j\,0.002$$

$$B = 40.58\underline{/86.4°}\left(1 + \frac{0.0644}{6}\underline{/176.4°} + \frac{-0.004\underline{/172.8}}{24}\right)$$

$$= 2.495 + j\,40.067 = 40.145\underline{/86.4°}$$

$$V_S = AV_R + BI_R = (0.968 + j\,0.002)\frac{400 \times 10^3}{\sqrt{3}}$$

$$+ 40.145\underline{/86.4°}(2887\underline{/-25.84°})$$
$$= 223\,550.024 + j\,461.880 + 56903.30 + j\,100966.540$$
$$= 298\,231.1\underline{/19.88°}\ \text{V (phase)}$$
$$= 516.551\ \text{kV (line)}$$

Summarizing these results:

| Representation | $V_R$ (kV line) | Error (%) on long-line representation | Overvoltage (%) |
|---|---|---|---|
| Short line | 530 | 2.7 | 32.5 |
| Medium line | 590 | 14.3 | 47.5 |
| Long line | 516 | — | 29.0 |

Line–line voltage at sending end is

$$\sqrt{3} \times 255 = 442\,\text{kV}$$

(Note: This is an extremely high voltage and could not be tolerated as a 400 kV system would be designed for only about 10 per cent steady-state overvoltage, i.e. $400 + 40 = 440$ kV. In practice, an overhead line over 300 km would be equipped with series capacitors to reduce the total series reactance to a value such that the line voltage drop was about 10 per cent nominal phase voltage. In addition, the power factor at the receiving end would be raised to at least 0.95 lag by the use of shunt capacitors or synchronous compensators.)

### 3.9.4 The natural load

The characteristic impedance $Z_0$ (equation 3.10) is also known as the surge impedance. When a line is terminated in its characteristic impedance the power delivered is known as the *natural load*. For a loss-free line under natural-load conditions the reactive power absorbed by the line is equal to the reactive power generated, i.e.

$$\frac{V^2}{X_c} = I^2 X_L$$

and

$$\frac{V}{I} = Z_0 = \sqrt{X_L X_C} = \sqrt{\frac{L}{C}}$$

At this load, $V$ and $I$ are in phase all along the line and optimum transmission conditions are obtained; in practice, however, the load impedances are seldom in the order of $Z_0$. Values of $Z_0$ for various line voltages are as follows (corresponding natural loads are shown in parentheses): 132 kV, 150 Ω (50 MW); 275 kV, 315 Ω (240 MW); 380 kV, 295 Ω (490 MW). The angle of the impedance varies between 0 and $-15°$. For underground cables $Z_0$ is roughly one-tenth of the overhead-line value. Lines are operated above the natural loading, whereas cables operate below this loading.

## 3.10 Parameters of Underground Cables

In cables with three conductors contained within the lead or aluminium sheath, the electric field set up has components tangential to the layers of impregnated-paper insulation in which direction the dielectric strength is poor. Therefore, at voltages over 11 kV, each conductor is separately screened (Höchstadter type) to ensure only radial stress through the paper. Cables using cross-linked polyethylene (XPLE) are now coming into use. The capacitance of single-conductor and individually screened three-conductor cables is readily calculated. For three-conductor unscreened cables one must resort to empirical design data. A typical high-voltage paper insulated cable is shown in Figure 3.35. The capacitance ($C$) of single-core cables may be calculated from design data by the use of the formula

$$C = \frac{2\pi\varepsilon_0\varepsilon_r}{\ln(R/r)} \tag{3.13}$$

where $r$ and $R$ are the inner and outer radii of the dielectric and $\varepsilon_r$ is the relative permittivity of the dielectric, often in the order of 3.5. This also holds for three-core cables with each conductor separately screened.

The various capacitances present in three-core unscreened cables may be represented as shown in Figure 3.36, in which the conductor-to-conductor capacitance is $C_1$ and the conductor-to-sheath capacitance is $C_0$. The equivalent circuit is shown in Figure 3.36(b) and circuits obtained by star–delta or delta–star transformations are shown in Figure 3.36(c) and (d). The value of

$$C_3 = \frac{C_0}{3}$$

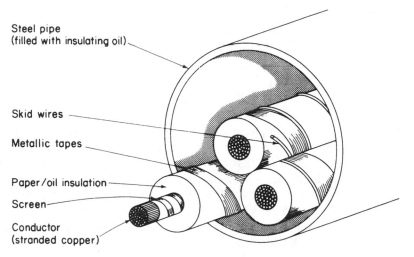

**Figure 3.35** High-pressure, oil-filled, pipe-type high-voltage cable

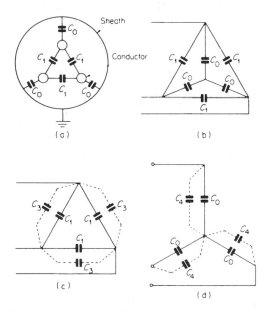

**Figure 3.36** Capacitances in a three-core cable

and, similarly, $C_4 = 3C_1$. The parallel combination of $C_4$ and $C_0$ gives the equivalent line-to-neutral value of the cable capacitance. $C_0$ and $C_1$ may be measured as follows. The three cores are short-circuited and the capacitance between them and the sheath is measured; this measurement gives $3C_0$. Next, the capacitance between two cores is measured, the third being connected to the sheath. It is easily shown that the measured value is now $C_1 + [(C_0 + C_1)/2]/$ hence $C_0$ and $C_1$ are obtained.

Owing to the symmetry of the cable the normal phase-sequence values of $C$ and $L$ are the same as the negative phase-sequence values (i.e. for reversed phase rotation). The series resistance and inductance are complicated by the magnetic interaction between the conductor and sheath.

The effective resistance of the conductor is the direct current resistance modified by the following factors: the skin effect in the conductor; the eddy currents induced by adjacent conductors (the proximity effect); and the equivalent resistance to account for the $I^2 R$ losses in the sheath. The determination of these effects is complicated and is left for advanced study.

The parameters of the cable having been determined, the same equivalent circuits are used as for overhead lines, paying due regard to the selection of the correct model for the appropriate length of cable. Owing to the high capacitance of cables the charging current, especially at high voltages, is an important factor in deciding the permissible length to be used. Table 3.3 gives the charging currents for self-contained low-pressure oil-filled (LPOF) cables.

**Table 3.3** Undergound cable constants at 50 Hz (per km)

|  |  | 132 kV | | 275 kV | | 400 kV |
|---|---|---|---|---|---|---|
| Size | mm$^2$ | 355 | 645 | 970 | 1130 | 1935 |
| Rating (soil $g = 120°C$ cm/W) | A | 550 | 870 | 1100 | 1100 | 1600 |
|  | MVA | 125 | 200 | 525 | 525 | 1100 |
| Resistance ($R$) at 85°C | $\Omega$ | 0.065 | 0.035 | 0.025 | 0.02 | 0.013 |
| Reactance ($X_L$) | $\Omega$ | 0.128 | 0.138 | 0.22 | 0.134 | 0.22 |
| Charging current ($I_c$) | A | 7.90 | 10.69 | 15.70 | 17.77 | 23.86 |
| $X_L/R$ Ratio |  | 2.0 | 4.0 | 8.8 | 6.6 | 16.8 |

## 3.11 Transformers

The equivalent circuit of one phase of a transformer referred to the primary winding is shown in Figure 3.37. The resistances and reactances can be found from the well-known open- and short-circuit tests. In the absence of complete information for each winding, the two arms of the T network can each be assumed to be half the total transformer impedance. Also, little accuracy is lost in transferring the shunt branch to the input terminals, forming a 'cantilever' circuit.

In power transformers the current taken by the shunt branch is usually a very small percentage of the load current and can be neglected.

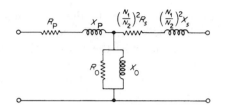

**Figure 3.37**   Equivalent circuit of a two-winding transformer

### 3.11.1   Phase shifts in three-phase transformers

Consider the transformer shown in Figure 3.38(a). The red phases on both sides are taken as reference and the transformation ratio is 1 : $N$. The corresponding phasor diagrams are shown in Figure 3.38(b). Although no neutral point is available in the delta side, the effective voltages from line to earth are denoted by $\mathbf{E}_{R'n}$, $\mathbf{E}_{Y',n}$, and $\mathbf{E}_{B'n}$. Comparing the two phasor diagrams, the following relationships are seen:

$$\mathbf{E}_{R'n} = \text{line–earth voltage on the delta side}$$
$$= N\mathbf{E}_{Rn}\angle 30°,$$

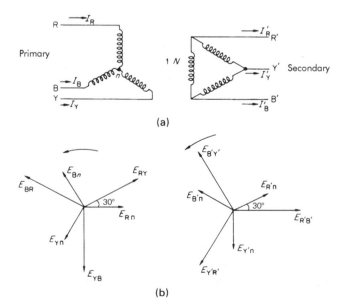

(a)

(b)

**Figure 3.38**   (a) Star–delta transformer with turns ratio 1 : $N$. (b) Corresponding phasor diagrams ($N$ taken as 1 in diagrams)

i.e. the positive sequence or normal balanced voltage of each phase is advanced through 30°. Similarly, it can be shown that the positive sequence currents are advanced through 30°.

By a consideration of the negative phase-sequence phasor diagrams (these are phasors with reversed rotation, i.e. R–B–Y) it will readily be seen that the phase currents and voltages are shifted through −30°. When using the per unit system the transformer ratio does not directly appear in calculations. In star–star and delta–delta connected transformers there are no phase shifts. Hence transformers having these connections and those with star–delta should not be connected in parallel. To do so introduces a resultant voltage acting in the local circuit formed by the usually low transformer impedances. Figure 3.39 shows the general practice on the British network with regard to transformers with phase shifts. It is seen that the reference phasor direction is different at different voltage levels. The larger than 30° phase shifts are obtained by suitable rearrangement of the winding connections.

## Three-winding transformers

Many transformers used in power systems have three windings per phase, the third winding being known as the tertiary. This winding is provided to enable compensation equipment to be connected at an economic voltage (e.g. 13 kV) and to provide a circulating current path for third harmonics so that these currents do not appear outside the transformer.

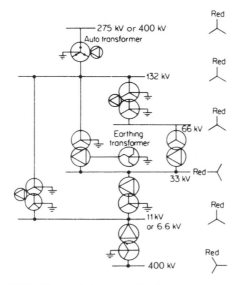

**Figure 3.39** Typical phase shifts in a power system—British

The three-winding transformer can be represented under balanced three-phase conditions by a single-phase equivalent circuit of three impedances star-connected as shown in Figure 3.40. The values of the equivalent impedances $Z_p$, $Z_s$, and $Z_t$ may be obtained by test. It is assumed that the no-load currents are negligible.

Let

$Z_{ps}$ = impedance of the primary when the secondary is short-circuited and the tertiary open

$Z_{pt}$ = impedance of the primary when the tertiary is short-circuited and the secondary open

$Z_{st}$ = impedance of the secondary when the tertiary is short-circuited and the primary open

The above impedances are in ohms referred to the same voltage base. Hence,

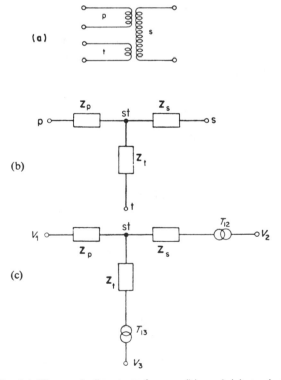

**Figure 3.40**    (a) Three-winding transformer (b) and (c) equivalent circuits

$$Z_{ps} = Z_p + Z_s$$
$$Z_{pt} = Z_p + Z_t$$
$$Z_{st} = Z_s + Z_t$$
$$\left.\begin{array}{l} Z_p = \tfrac{1}{2}(Z_{ps} + Z_{pt} - Z_{st}) \\ Z_s = \tfrac{1}{2}(Z_{ps} + Z_{st} - Z_{pt}) \\ Z_t = \tfrac{1}{2}(Z_{pt} + Z_{st} - Z_{ps}) \end{array}\right\} \qquad (3.14)$$

It should be noted that the star point st in Figure 3.40b is purely fictitious and that the diagram is a single-phase equivalent circuit. In most large transformers the value of $Z_s$ is very small and can be negative. All impedances must be referred to common volt-ampere and voltage bases. The complete equivalent circuit is shown in Figure 3.40(c) where $T_{12}$ and $T_{13}$ provide any off nominal turns ratio.

## Autotransformers

The symmetrical autotransformer may be treated in the same manner as two- and three-winding transformers. This type of transformer shows to best advantage when the transformation ratio is small and it is widely used for the interconnection of the supply networks working at different voltages, e.g 275 kV to 132 kV or 400 kV to 275 kV. The neutral point is solidly grounded, i.e. connected directly to earth without intervening resistance.

## Earthing (grounding) transformers

A means of providing an earthed point or neutral in a supply derived from a delta-connected transformer may be obtained by the use of a zigzag transformer shown schematically in Figure 3.39. By the interconnection of two windings on each limb, a node of zero potential is obtained.

## Harmonics

Due to the non-linearity of the magnetizing characteristic for transformers the current waveform is distorted and hence contains harmonics; these flow through the system impedances and set up harmonic voltages. In transformers with delta-connected windings the third and ninth harmonics circulate round the delta and are less evident in the line currents. An important source of harmonics is a rectifier load; this is becoming more prevalent as electronic devices proliferate.

On occasions, the harmonic content can prove important due mainly to the possibility of resonance occurring in the system, e.g. resonance has occurred with fifth harmonics. Also, the third-harmonic components are in phase in the three conductors of a three-phase line, and if a return path is present these currents add and cause interference in neighbouring communication circuits and increase the neutral return current.

When analysing systems with harmonics it is sufficient to use the normal values for series inductance and shunt capacitance but one must remember to calculate reactance and susceptance at the frequency of the harmonic. The effect on resistance is more difficult to assess: however, it is usually only required to assess the presence of harmonics and the possibility of resonance.

## Tap-changing transformers

A method of controlling the voltages in a network lies in the use of transformers, the turns ratio of which may be changed. In Figure 3.41(a) a schematic diagram of an off-load tap changer is shown; this, however, requires the

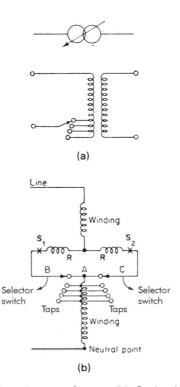

(a)

(b)

**Figure 3.41**   (a) Tap-changing transformer. (b) On-load tap-changing transformer. $S_1$, $S_2$ transfer switches, R centre-tapped reactor

disconnection of the transformer when the tap setting is to be changed. Many transformers now have on-load changers, the basic form of which is shown in Figure 3.41(b). In the position shown, the voltage is a maximum and the current divides equally in the two halves of reactor R, resulting in zero-resultant flux and minimum impedance. To reduce the voltage, $S_1$ opens and the total current passes through the other half of the reactor. Selector switch B then moves to the next contact and $S_1$ closes. A circulating current now flows in R superimposed on the load current. Now $S_2$ opens and C moves to the next tapping; $S_2$ then closes and the operation is complete. Six switch operations are required for one change in tap position. The voltage change between taps is often 1.25 per cent of the nominal voltage. This small change is necessary to avoid large voltage disturbances at consumer busbars. A schematic block diagram of the on-load tap systems is shown in Figure 3.42. The line drop compensator (LDC) is used to allow for the voltage drop along the feeder to the load point, so that the actual load voltage is seen and corrected by the transformer. The total range of tapping varies with the transformer usage, a typical figure for generator transformers is +2 to −16 per cent in 18 steps.

## Typical parameters for transformers

The leakage reactances of two-winding transformers increase slightly with their rating for a given voltage, i.e. from 3.2 per cent at 20 kVA to 4.3 per cent at 500 kVA at 11 kV. For larger sizes, i.e., 20 MVA upwards, 10 per cent is a typical value at all voltages. For autotransformers the impedances are usually less than for double wound. Parameters for large transformers are as follows:

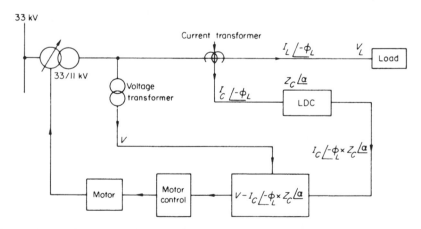

**Figure 3.42** Schematic diagram of a control system for an on-load tap-changing transformer incorporating line drop compensation (LDC)

(1) 400/275 kV autotransformer, 500 MVA, 12 per cent impedance, tap range +10 to −20 per cent; (2) 400/132 kV double wound, 240 MVA, 20 per cent impedance, tap range +5 to −15 per cent. It should be noted that a 10 per cent reactance implies ten times rated current on short circuit. Winding forces depend upon current squared; the transformer must be designed to withstand short circuits.

## Example 3.6

Two three-phase, 275/66/11 kV transformers, are connected in parallel and the windings are employed as follows:

- the 11 kV winding supplies a synchronous compensator (i.e. a synchronous motor running light);
- the 66 kV winding supplies a network with 80 MW of demand;
- the 275 kV winding is connected to the primary transmission system.

The winding data are as shown below. The line diagram is shown in Figure 3.43(a).
    It is required to determine the division of power between the two 275/66 kV windings. Resistance may be neglected.

## Solution

To obtain the equivalent star circuits the reactances will be expressed on common bases of 90 MVA, 275 kV.

| | $X\%$ referred to 90 MVA | | | MVA of windings | | |
|---|---|---|---|---|---|---|
| | 275/ 66 kV | 66/ 11 kV | 11/ 275 kV | 275 kV | 66 kV | 11 kV |
| Transformer (1) | 10.5 | 20 | 8 | 95 | 90 | 60 |
| Transformer (2) | 10.5 | 5 | 15 | 90 | 85 | 45 |

For transformer (1),

$$X_{275} + X_{66} = \frac{10.5}{100} \times \frac{275^2}{90} = 88.4\,\Omega$$

$$X_{66} + X_{11} = \frac{20}{100} \times \frac{275^2}{90} = 168\,\Omega$$

$$X_{11} + X_{275} = \frac{8}{100} \times \frac{275^2}{90} = 67.2\,\Omega$$

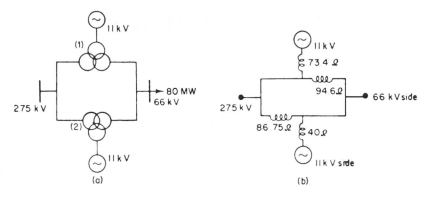

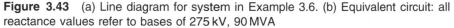

**Figure 3.43** (a) Line diagram for system in Example 3.6. (b) Equivalent circuit: all reactance values refer to bases of 275 kV, 90 MVA

The corresponding quantities for transformer (2) are $88.4\,\Omega$, $42.1\,\Omega$, and $126.3\,\Omega$, respectively.

For transformer (1),

$$
\begin{aligned}
X_{\mathrm{p}}\ (275\ \text{kV side}) &= \tfrac{1}{2}(X_{\mathrm{ps}} + X_{\mathrm{pt}} - X_{\mathrm{st}}) \\
&= \tfrac{1}{2}(88.4 + 67.2 - 168) \\
&= -6.2\,\Omega
\end{aligned}
$$

Similarly,

$$X_{\mathrm{S}}\ (66\ \text{kV side}) = 94.6\,\Omega$$

and

$$X_{\mathrm{T}}\ (11\ \text{kV side}) = 73.4\,\Omega$$

For transformer (2),

$$X_{\mathrm{p}} = 86.75\,\Omega \qquad X_{\mathrm{S}} = 1.65\,\Omega \qquad \text{and} \qquad X_{\mathrm{T}} = 40\,\Omega$$

$X_{\mathrm{p}}$ in transformer (1) and $X_{\mathrm{S}}$ in transformer (2) will be neglected.

The equivalent circuit is shown in Figure 3.43(b) from which it will be seen that two impedances are in parallel between 275 kV and 66 kV sides.

The power will divide according to the reactances of the two transformers. Hence,

$$P_{(1)} = \frac{94.6 \times 80}{94.6 + 86.75} = 41.7\,\text{MW}$$

and

$$P_{(2)} = \frac{86.75 \times 80}{86.75 + 94.6} = 38.3\,\text{MW}$$

## 3.12 Connection of Three-Phase Transformers

In Section 3.11 the phase changes occurring in star–delta transformers are briefly discussed and sufficient treatment is given for the following analysis to be understood. From the practical view, however, the connection of such transformers requires further discussion to be fully appreciated. It is seen that for the connection shown in Figure 3.38(a) the output line voltages on the delta side are 30° in advance of the input line voltages to the star winding for normal a–b–c (R–Y–B) phase notation. Also, the output line voltage is equal to the input line multiplied by $(N/\sqrt{3})$, and impedances transferred from one side to the other are modified by $(N/\sqrt{3})^2$ and vice versa. The true single-phase equivalent circuit is therefore as shown in Figure 3.44.

However, by suitable rearrangement of the winding connections, phase shifts of 30, 90, 150, 210, 270, or 330 electrical degrees may be obtained, and there is a need for standardization of nomenclature and procedure. Should a phase difference exist between the secondary output voltages of transformers connected to a common supply, dangerously large circulating currents will occur and the circuit protection should operate. For uniformity, the transformer windings of Figure 3.38(a) are shown in Figure 3.44 with capitals referring to the H.V. winding. It is seen that with the star–delta a shift of $-30°$ can also be produced by a rearrangement of the terminal connections. British practice* involves the use of a 'vector group reference' number to describe transformer connections. The first number indicates the phase shift, e.g. 1 indicates 0°, $2 = 180°$, $3 = (-30°)$, $4 = (+30°)$; the next number indicates the interphase connection of the secondary, e.g. 1 = star, 2 = delta, 3 = interconnected star. Letters y and D indicate star (y) or delta (D) connection, the H.V. winding coming first. Finally comes the phase displacement angle in the form of a clock-reference, the low-voltage winding-induced voltage being given by the hour hand and the H.V. voltage by the minute hand, e.g. $0° = 12$ o'clock,

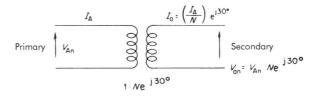

**Figure 3.44** Equivalent single-phase circuit of a three-phase transformer with a phase shift from primary to secondary

*The American Standards Association adopt the standard that the high-voltage reference phase is always 30° ahead of the low-voltage reference phase, regardless of whether the three-phase connections are delta–star or star–delta.

$180° = 6$ o'clock, $+30° = 11$ o'clock and $-30° = 1$ o'clock; $A_2$ in the phasor diagrams is always at 12 o'clock. The application of this principle to the most widely used transformer connections is shown in Figure 3.45.

From an analytical viewpoint it may be desirable to have a phase shift of 90°, thus facilitating the use of operator 'j'. In this type of work the exact arrangement is not important provided the system adopted is used consistently and the winding arrangement is seldom the cause for concern. However, when dealing with the connection of actual equipment in the system, the correct arrangement of the transformer winding is of vital importance.

## 3.13   Voltage Characteristics of Loads

The variation of the power and reactive power taken by a load with various voltages is of importance when considering the manner in which such loads are to be represented in load flow and stability studies. Usually, in such studies, the load on a substation has to be represented and is a composite load consisting of industrial and domestic consumers. A typical composition of a substation load is as follows:

| | | |
|---|---|---|
| Induction motors | 50–70% | (air-conditioners, freezers, washers, fans, pumps, etc.) |
| Lighting and heating | 20–25% | (water heating, resistance heaters, etc.) |
| Synchronous motors | 10% | |
| (Transmission and distribution losses | 10–12%) | |

### 3.13.1   Lighting

This is independent of frequency and consumes no reactive power. The power consumed does not vary as the (voltage)$^2$, but approximately as (voltage)$^{1.6}$. With fluorescent and sodium/mercury lamps, distorted currents are taken.

### 3.13.2   Heating

This maintains constant resistance with voltage change and hence the power varies with (voltage)$^2$.

The above loads may be described as static.

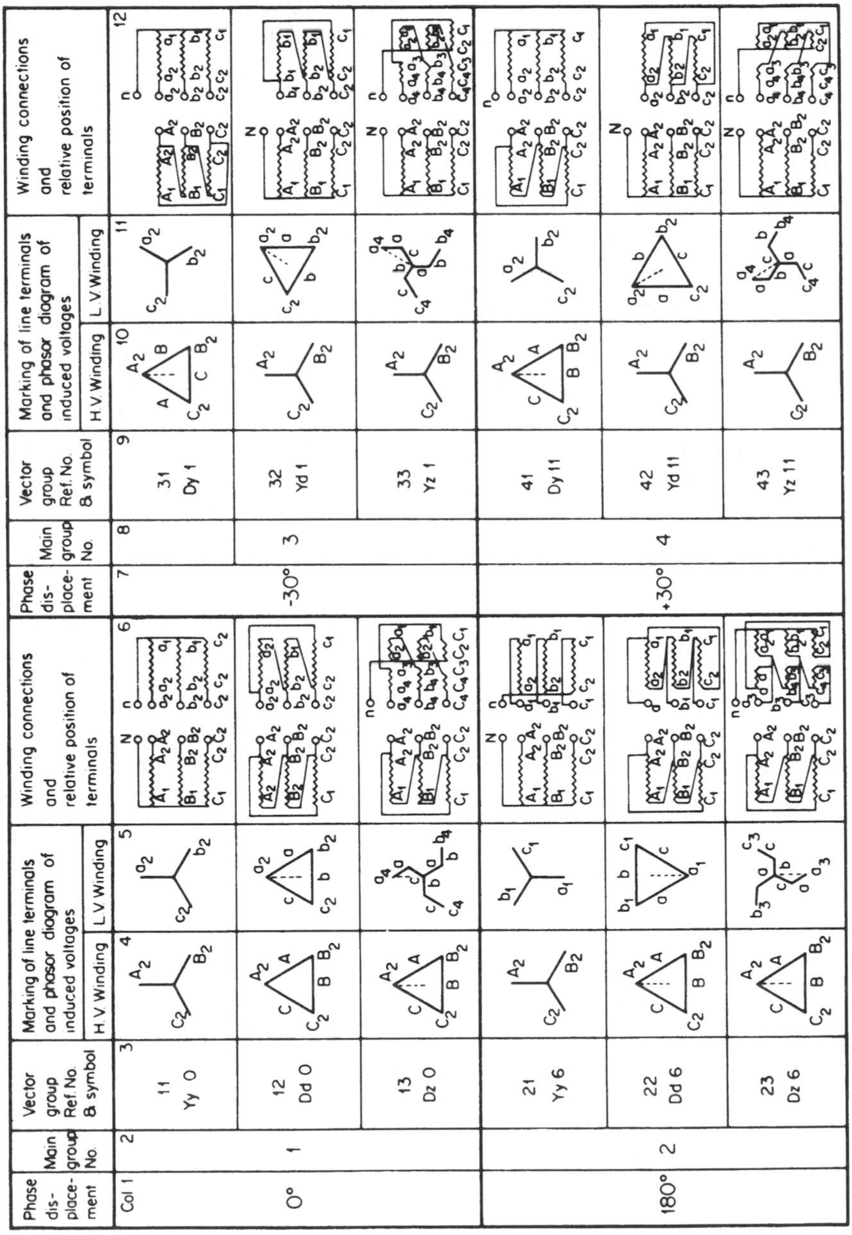

**Figure 3.45**   British Standard for connections of three-phase transformers and resulting phase shifts

### 3.13.3 Synchronous motors

The power consumed is approximately constant. For a given excitation the vars change in a leading direction with voltage reduction. The *P–V*, *Q–V*, generalized characteristics are shown in Figure 3.46.

### 3.13.4 Induction motors

The *P–V*, *Q–V* characteristics may be determined by the use of the simplified circuit shown in Figure 3.47. It is assumed that the mechanical load on the shaft is constant.

The electrical power,

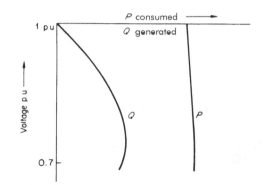

**Figure 3.46**   *P–V* and *Q–V* curves for a synchronous motor

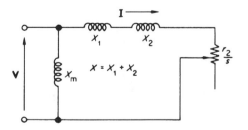

**Figure 3.47**   Equivalent circuit of an induction motor: $X_1$ = stator leakage reactance; $X_2$ = rotor leakage reactance referred to the stator; $X_m$ = magnetizing reactance; $r_2$ = rotor resistance; $s$ = slip p.u. Magnetizing losses have been ignored and the stator losses are lumped in with the line losses

$$P_{\text{electrical}} = P_{\text{mechanical}} = P$$

$$= 3I^2 \frac{r_2}{s} = \text{constant}$$

$$\therefore s = \frac{3I^2 r_2}{P}$$

The reactive power consumed

$$= \frac{3V^2}{X_m} + 3I^2(X_1 + X_2) \tag{3.15}$$

Also, from Figure 3.47

$$P = 3I^2 \frac{r_2}{s} = \frac{3V^2}{(r_2/s)^2 + X^2} \cdot \frac{r_2}{s}$$

$$= \frac{3V^2 r_2 s}{r_2^2 + (sX)^2} \tag{3.16}$$

The graphs of $V$–$P$ and $V$–$Q$ are shown in Figure 3.48. The $P$ and $Q$ scales have been so arranged that $P$ and $Q$ are both equal to 1 p.u. at a voltage of 1 p.u. The effect of shaft torque constant on $P$ is also shown. Similar analysis can be performed for the induction motor with torque $\propto$ (speed)$^2$ and torque $\propto$ speed.

The well-known power-slip curves for an induction motor are shown in Figure 3.49. It is seen that for a given mechanical torque there is a critical voltage and a corresponding critical slip $s_{\text{cr}}$. If the voltage is reduced further the motor becomes unstable and stalls. This critical point occurs when

$$\frac{\mathrm{d}P}{\mathrm{d}s} = 0$$

i.e. when

$$V^2 r_2 \cdot \frac{r_2^2 - (sX)^2}{[r_2^2 + (sX)^2]^2} = 0$$

so that

$$s = \frac{r_2}{X} \quad \text{and} \quad P_{\text{max}} = \frac{V^2}{2X}$$

Alternatively, for a given output power $P$,

$$V_{\text{critical}} = \sqrt{2PX}$$

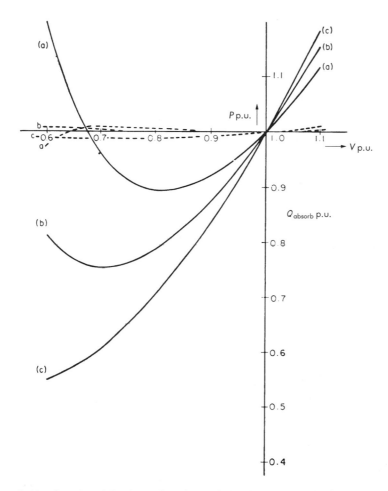

**Figure 3.48** Graphs of *P–V* and *Q–V* for an induction motor—shaft torque constant. Curve (a) full load, (b) 0.75 full load, (c) 0.5 full load; – – – *P*, —— *Q*

### 3.13.5 Composite loads

Except in isolated cases, such as very large induction motor drives in refineries and steel mills, it is the composite load which is of interest. Tests to obtain *P–V* and *Q–V* characteristics of complete substation loads have been few (the consumers object!). The results of one such test on the Polish network are given in Table 3.4. Instead of actual characteristics, values of $\partial P/\partial V$ and $\partial Q/\partial V$ at 1 p.u. voltage are given. As will be seen later, these quantities are of considerable interest. Composite load characteristics can be obtained by summing the characteristics of the constituent loads.

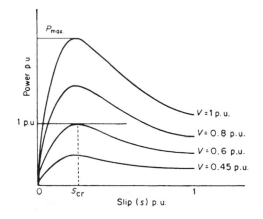

**Figure 3.49**   Power-slip curves for an induction motor. If voltage falls to 0.6 p.u. at full load $P = 1$, the condition is critical (slip $s_{cr}$)

**Table 3.4**   Polish network test results of loads

| $dP/dV$ at nominal voltage | Morning | Evening | Night |
|---|---|---|---|
| Large towns | 0.9–1.2 | 1.5–1.7 | 1.5–1.6 |
| Small towns | 0.6–0.7 | 1.4–1.6 | 1.4–1.6 |
| Villages | 0.5–0.6 | 1.5 | 1.5 |
| Street lighting | — | 1.55 | 1.55 |
| Mines | 0.6–0.76 | 0.75–0.89 | 0.65–0.78 |
| Ironworks | 0.6–0.7 | 0.6–0.75 | 0.6–0.7 |
| Textile industries | 0.5–0.6 | 0.6–0.7 | 0.6–0.65 |
| Large chemical industries | 0.6–0.7 | 0.6–0.75 | 0.6–0.7 |
| Machine industry | 0.5–0.55 | 0.6–0.65 | 0.6–0.65 |
| Junction points in the H.V. network | 0.6–0.7 | 0.8–1.1 | 0.8–1.0 |

| $dQ/dV$ at nominal voltage | Morning | Evening | Night |
|---|---|---|---|
| Large towns | 2.5–3.0 | 2.2–2.5 | 2.4–3.0 |
| Small towns | 2.6–3.5 | 2.0–2.5 | 2.5–3.0 |
| Villages | 2.6–3.5 | 2.0–2.5 | 2.4–3.5 |
| Industry (total) | 1.8–2.7 | 1.8–2.3 | 1.8–2.7 |
| Junction points in 6 kV network | 2.2–2.9 | 1.8–2.5 | 2.0–2.6 |
| Junction points in 110 kV network | 1.8–2.6 | 1.5–2.2 | 1.7–2.3 |

(Source: Boyucki, A., and Wojcik, M., 'Static characteristics of loads', *Energetyka B.,* No. 3 (Polish), 1959.)

In practice, great difficulty is experienced in assessing the degree to which motors are loaded and this materially affects the shape of the characteristics. Loads which require the same power at reduced voltage are called 'stiff'; loads, such as heating, in which the power falls off rapidly at lower voltages are termed 'soft'.

In much analytical work, loads are represented by constant impedances, and hence it is assumed that $P \propto V^2$ and $Q \propto V^2$. If the load is $P + jQ$ p.u. the current at voltage $V$ is

$$\frac{P - jQ}{V} \text{ p.u.} \qquad \text{(remembering } \mathbf{VI^*} = \mathbf{S}\text{)}$$

Hence the impedance to represent this load

$$= \frac{\mathbf{V}}{\mathbf{I}} = \frac{V^2}{P - jQ} \text{ p.u.}$$

The $P$–$V$, $Q$–$V$ characteristics can be used if they are known, but this process is rather tedious. Some authorities suggest that the representation of the load by a constant-current consuming device gives a good approximation to practical conditions. When digital methods are used the correct representation of loads presents no great difficulties. When extensive reduction of a network to a simpler form is necessary, then the constant impedance representation is usually resorted to, although the load this represents seldom occurs in practice. However, it does usually give a worst-case result.

### 3.13.6    Load–frequency characteristics

Often, voltage changes are accompanied by frequency changes. Again, information regarding the characteristics of composite loads with frequency is scarce. With the small frequency changes allowed in practice, the effects on the power and reactive power consumed are usually neglected in calculations.

## Problems

(*Note*: All machines, etc., are three-phase unless stated otherwise.)

**3.1**  When two four pole, 50 Hz synchronous generators are paralleled their phase displacement is 2° mechanical. The synchronous reactance of each machine is 10 Ω/ phase and the common busbar voltage is 6.6 kV. Calculate the synchronizing torque. (Answer: 973.5 Nm)

**3.2**  A synchronous generator has a synchronous impedance of 2 p.u. and a resistance of 0.01 p.u. It is connected to an infinite busbar of voltage 1 p.u. through a transformer of reactance j0.1 p.u. If the generated (no-load) e.m.f. is 1.1 p.u. calculate the current

and power factor for maximum output.
(Answer: 0.708 p.u.; 0.74 leading)

**3.3**   A 6.6 kV synchronous generator has negligible resistance and synchronous reactance of 4 $\Omega$/phase. It is connected to an infinite busbar and gives 2000 A at unity power factor. If the excitation is increased by 25 per cent find the maximum power output and the corresponding power factor. State any assumptions made.
(Answer: 31.6 MW; 0.96 leading)

**3.4**   A synchronous generator whose characteristic curves are given in Figure 3.4 delivers full load at the following power factors: 0.8 lagging, 1.0, and 0.9 leading. Determine the percentage regulation at these loads.
(Answer: 167, 119, 76 per cent)

**3.5**   A salient-pole, 75 MVA, 11 kV synchronous generator is connected to an infinite busbar through a link of reactance 0.3 p.u. and has $X_d = 1.5$ p.u. and $X_q = 1$ p.u., and negligible resistance. Determine the power output for a load angle 30° if the excitation e.m.f. is 1.4 times the rated terminal voltage. Calculate the synchronizing coefficient in this operating condition. All p.u. values are on a 75 MVA base.
(Answer: $P = 0.48$ p.u.; $\partial P/\partial \delta = 0.78$ p.u.)

**3.6**   A synchronous generator of open-circuit terminal voltage 1 p.u. is on no-load and then suddenly short-circuited; the trace of current against time is shown in Figure 3.6(b). In Figure 3.6(b) the current oc $= 1.8$ p.u., oa $= 5.7$ p.u., and ob $= 8$ p.u. Calculate the values of $X_s$, $X'$, and $X''$. Resistance may be neglected. If the machine is delivering 1 p.u. current at 0.8 power factor lagging at the rated terminal voltage before the short circuit occurs, sketch the new envelope of the 50 Hz current waveform.
(Answer: $X_s = 0.8$ p.u.; $X' = 0.25$ p.u.; $X'' = 0.18$ p.u.)

**3.7**   Construct a performance chart for a 22 kV, 500 MVA, 0.9 p.f. generator having a short-circuit ratio of 0.55.

**3.8**   A 275 kV three-phase transmission line of length 96 km is rated at 800 A. The values of resistance, inductance, and capacitance per phase per kilometre are 0.078 $\Omega$, 1.056 mH, and 0.029 $\mu$F, respectively. The receiving-end voltage is 275 kV when full load is transmitted at 0.9 power factor lagging. Calculate the sending-end voltage and current, and the transmission efficiency, and compare with the answer obtained by the short-line representation. Use the nominal $\pi$ and T methods of solution. The frequency is 60 Hz.
(Answer: $V_s = 178$ kV per phase)

**3.9**   A 220 kV, 60 Hz three-phase transmission line is 320 km long and has the following constants per phase per km:

| | |
|---|---|
| Inductance | 1.3 mH |
| Capacitance | 0.22 $\mu$F |
| Resistance | 0.55 $\Omega$ |

Ignore leakage conductance.
If the line delivers a load of 300 A, 0.8 power factor lagging, at a voltage of 220 kV, calculate the sending-end voltage. Determine the $\pi$ circuit which will represent the line.

Calculate the error resulting from the use of the medium-length $\pi$ section.
(Answer: Series $(11 + j\,116)\,\Omega$; shunt $(0.44 + j\,0.72) \times 10^{-3}\,\Omega^{-1}$)

**3.10** In a three-core cable, the capacitance between the three cores short-circuited together and the sheath is $87\,\mu\text{F/km}$, and that between two cores connected together and the third core is $0.84\,\mu\text{F/km}$.

Determine the kVA required to keep 16 km of this cable charged when the supply is 33 kV, three phase, 50 Hz.
(Answer: 270 kVA)

**3.11** Calculate the **A B C D** constants for a 275 kV overhead line of length 83 km. The parameters per kilometre are as follows:

Resistance           $0.078\,\Omega$

Reactance            $0.33\,\Omega$

Admittance (shunt capacitative) $9.53 \times 10^{-6}\,\text{S}$

The shunt conductance is zero.
(Answer: $[A = 0.8615 + j\,0.0317;\ B = 34.0 + j\,153;\ C = (-19.2 + j\,1693 \times 10^{-6})]$

**3.12** A 132 kV, 60 Hz transmission line has the following generalized constants:

$$A = 0.9696\angle 0.49°$$
$$B = 52.88\angle 74.79°\,\Omega$$
$$C = 0.001177\angle 90.15°\,\text{S}$$

If the receiving-end voltage is to be 132 kV when supplying a load of 125 MVA 0.9 p.f. lagging, calculate the sending-end voltage and current.
(Answer: 95.9 kV phase, 166.1 kV line 554 A)

**3.13** Two identical transformers each have a nominal or no-load ratio of 33/11 kV and a leakage reactance of $2\,\Omega$ referred to the 11 kV side; resistance may be neglected. The transformers operate in parallel and supply a load of 9 MVA, 0.8 p.f. lagging. Calculate the current taken by each transformer when they operate five tap steps apart (each step is 1.25 per cent of the nominal voltage). Also calculate the kVAr absorbed by this tap setting.
(Answer: 307 A, 194 A for three-phase transformers, 118 kVAr)

**3.14** An induction motor, the equivalent circuit of which is shown in Figure 3.50 is connected to supply busbars which may be considered as possessing a voltage and frequency which is independent of the load. Determine the reactive power consumed for various busbar voltages and construct the $Q$–$V$ characteristic. Calculate the critical

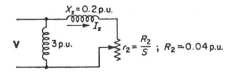

**Figure 3.50** Equivalent circuit of 500 kW, 6.6 kV induction motor in Problem 3.14. All p.u. values refer to rated voltage and power ($P = 1$ p.u. and $V = 1$ p.u.)

voltage at which the motor stalls and the critical slip, assuming that the mechanical load is constant.
(Answer: $s_{cr} = 0.2$, $V_c = 0.63$ p.u.)

**3.15**  A 100 MVA round-rotor generator of synchronous reactance 1.5 p.u. supplies a substation (L) through a step-up transformer of reactance 0.1 p.u., two lines each of reactance 0.3 p.u. in parallel and a step-down transformer of reactance 0.1 p.u. The load taken at L is 100 MW at 0.85 lagging. L is connected to a local generating station which is represented by an equivalent generator of 75 MVA and synchronous reactance of 2 p.u. All reactances are expressed on a base of 100 MVA.

Draw the equivalent single-phase network. If the voltage at L is to be 1 p.u. and the 75 MVA machine is to deliver 50 MW, 20 MVAr, calculate the internal voltages of the machines.
(Answer: $E_1 = 2$ p.u.; $E_2 = 1.72$ p.u.; $\delta_{2L} = 35.45°$)

**3.16**  The following data applies to the power system shown in Figure 3.51.

Generating station A: Four identical turboalternators, each rated at 16 kV, 125 MVA, and of synchronous reactance 1.5 p.u. Each machine supplies a 125 MVA, 0.1 p.u. transformer connected to a busbar sectioned as shown.

Substation B: Two identical 150 MVW, three-winding transformers, each having the following reactances between windings: 132/66 kV windings 10 per cent; 66/11 kV windings 20 per cent; 132/11 kV windings 20 per cent; all on a 150 MVA base.

The secondaries supply a common load of 200 MW at 0.9 p.f. lagging. To each tertiary winding is connected a 30 MVA synchronous compensator of synchronous reactance 1.5 p.u.

Substation C: Two identical 150 MVA transformers, each of 0.15 p.u. reactance, supply a common load of 300 MW at 0.85 p.f. lagging.

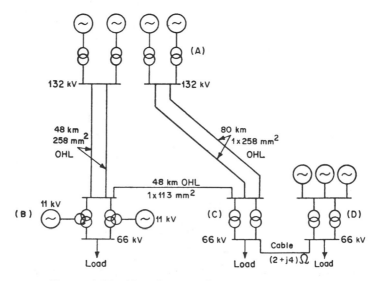

**Figure 3.51**  Line diagram of system in Problem 3.16

Generating station D: Three identical 11 kV, 75 MVA generators, each of 1 p.u. synchronous reactance, supply a common busbar which is connected to an outgoing 66 kV cable through two identical 100 MVA transformers. Load 50 MW, 0.8 p.f. lagging.

Determine the equivalent circuit for balanced operation giving component values on a base of 100 MVA. Treat the loads as impedances.

**3.17** Distinguish between kW, kVA, and kVAr.
Explain why

(a) generators in large power systems usually run overexcited, 'generating' VAr.
(b) remote hydro-generators need an underexcited rating so that they can absorb VAr.
(c) loss of an overexcited generator in a power system will normally cause a drop in voltage at its busbar.

A load of 0.8 p.u. power and 0.4 p.u. VAr lagging is supplied from a busbar of 1.0 p.u. voltage through an inductive line of reactance 0.15 p.u. Determine the load terminal voltage assuming that p.u. current has the same value as p.u. VA.
(Answer: 0.934 p.u.)
(*From E.C. Examination, 1996*)

**3.18** Sketch the performance chart of a synchronous generator indicating the main limits.

Consider a generator with the following nameplate data: 500 MVA, 20 kV, 0.8 p.f. (power factor), $X_s = 1.5$ p.u.

(a) Calculate the internal voltage and power angle of the generator operating at 400 MW with $\cos \phi = 0.8$ (lagging) with a 1 p.u. terminal voltage.
(b) What is the maximum reactive power this generator can absorb from the system?
(c) What is the maximum reactive power this generator can deliver to the system, assuming a maximum internal voltage of 2.25 p.u.
(d) Place the numerical values calculated on the performance chart.

A graphical solution is acceptable.
(Answer: (a) 2.23 p.u., 33.5°; (b) 333 MVAr; (c) 417 MVAr)
(*From E.C. Examination, 1997*)

# 4

# Control of Power and Frequency

## 4.1 Introduction

Although to a certain extent the control of power and frequency is interrelated to the control of reactive power and voltage, it is hoped that by dealing with power and frequency separately from voltage control, a better appreciation of the operation of power systems will be obtained. In a large interconnected system, many generation stations, large and small, are synchronously connected and hence all have the same frequency. The following remarks refer mainly to networks in which the control of power is carried out by the decisions and actions of engineers, as opposed to systems in which the control and allocation of load to machines is effected completely automatically. The latter are sometimes based on a continuous load-flow calculation by analogue or digital computers. The allocation of the required power among the generators has to be decided before the advent of the load, which must therefore be predicted. An analysis is made of the loads experienced over the same period in previous years, account is also taken of the value of the load immediately previous to the period under study and of the weather forecast. The probable load to be expected, having been decided, is allocated to the various turbine-generators.

A daily load cycle is shown in Figure 1.2. It is seen that the rate of rise of the load is very high and varies between 2 MW/min per 1000 MW of peak demand in the early hours and about 8 MW/min per 1000 MW of peak demand between 07.00 and 08.00 hours. Hence the ability of machines to increase their output quickly from zero to full load is important. It is extremely unlikely that the output of the machines at any instant will exactly equal the load on the system. If the output is higher than the demand the machines will tend to increase in speed and the frequency will rise, and vice versa. Hence the frequency is not a constant quantity but continuously varies; these variations are normally small and have no noticeable effect on most consumers. The frequency is continuously monitored against standard time-sources and when long-term tendencies to rise or fall are noticed, the control engineers take appropriate action by regulating the generator outputs.

Should the total generation available be insufficient to meet the demand, the frequency will fall. If the frequency falls by more than 1 Hz the reduced speed of power station pumps, fans, etc., may reduce the station output and a serious situation arises. In this type of situation, although the reduction of frequency will cause a reduction in power demand, voltage must be reduced, and if this is not sufficient then loads will have to be disconnected and continue to be disconnected until the frequency rises to a reasonable level. All utilities have a scheme of planned load shedding based on under-frequency relays set to reduce loads in blocks to prevent complete shut-down in extreme emergencies.

When a permanent increase in load occurs on the system, the speed and frequency of all the interconnected generators fall, since the increased energy requirement is met from the kinetic energy of the machines. This causes an increase in steam or water admitted to the turbines due to the operation of the governors and hence a new load balance is obtained. Initially, the boilers have a thermal reserve by means of which sudden changes can be supplied until the new firing rate has been established. Modern gas turbines have an overload capability for a few minutes which can be usefully exploited in emergency situations.

## 4.2   The Turbine Governor

In the following discussion of governor mechanisms, both steam and water turbines will be considered together and points of difference indicated where required. A simplified schematic diagram of a traditional governor system is shown in Figure 4.1. The sensing device, which is sensitive to change of speed,

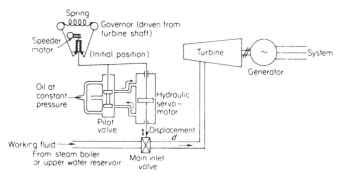

**Figure 4.1**   Governor control system employing the Watt governor as sensing device and a hydraulic servo-system to operate main supply valve. Speeder-motor gear determines the initial setting of the governor position

is the time-honoured Watt centrifugal governor. In this, two weights move radially outwards as their speed of rotation increases and thus move a sleeve on the central spindle. This sleeve movement is transmitted by a lever mechanism to the pilot-valve piston and hence the servo-motor is operated. A dead band is present in this mechanism, i.e. the speed must change by a certain amount before the valve commences to operate, because of friction and mechanical backlash. The time taken for the main steam valve to move due to delays in the hydraulic pilot-valve and servo-motor systems is appreciable, 0.2–0.3 s. The governor characteristic for a large steam turboalternator is shown in Figure 4.2 and it is seen that there is a 4 per cent drop in speed between no load and full load. Because of the requirement for high response speed, low dead band, and accuracy in speed and load control, the mechanical governor has been replaced in large modern turbogenerators by electrohydraulic governing. The method normally used to measure the speed is based on a toothed wheel (on shaft) and magnetic-probe pickup. The use of electronic controls requires an electrohydraulic conversion stage, using conventional servo-valves.

An important feature of the governor system is the mechanism by means of which the governor sleeve and hence the main-valve positions can be changed and adjusted quite apart from when actuated by the speed changes. This is accomplished by the speed changer, or 'speeder motor', as it is sometimes termed. The effect of this adjustment is the production of a family of parallel characteristics, as shown in Figure 4.3. Hence the power output of the generator at a given speed may be adjusted at will and this is of extreme importance when operating at optimum economy.

The torque of the turbine may be considered to be approximately proportional to the displacement $d$ of the main inlet valve. Also, the expression for the change in torque with speed may be expressed approximately by the equation

$$T = T_0(1 - kN) \tag{4.1}$$

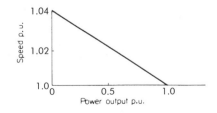

**Figure 4.2** Idealized governor characteristic of a turboalternator with 4 per cent droop from zero to full load

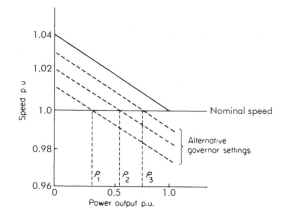

**Figure 4.3** Effect of speeder gear on governor characteristics $P_1, P_2$, and $P_3 =$ outputs at various settings but at same speed

where $T_0$ is the torque at speed $N_0$ and $T$ the torque at speed $N$; $k$ is a constant for the governor system. As the torque depends on both the main-valve position and the speed, $T = f(d, N)$.

There is a time delay between the occurrence of a load change and the new operating conditions. This is due not only to the governor mechanism but also to the fact that the new flow rate of steam or water must accelerate or decelerate the rotor in order to attain the new speed. In Figure 4.4 typical curves are shown for a turbogenerator which has a sudden decrease in the electrical power required, pehaps due to an external power network fault, and hence the retarding torque on the turbine shaft is suddenly much smaller. In the ungoverned

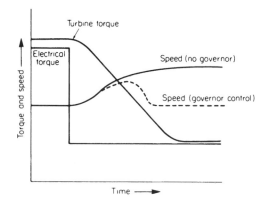

**Figure 4.4** Graphs of turbine torque, electrical torque, and speed against time when the load on a generator suddenly falls

case the considerable time-lag between the load change and the attainment of the new steady speed is obvious. With the regulated or governed machine, due to the dead band in the governor mechanism, the speed–time curve starts to rise, the valve then operates, and the fluid supply is adjusted. It is possible for damped oscillations to be set up after the load change; this will be discussed in Chapter 8.

An important factor regarding turbines is the possibility of overspeeding, when the load on the shaft is lost completely, with possible drastic mechanical breakdown. To avoid this, special valves are incorporated to automatically cut off the energy supply to the turbine. In a turbogenerator normally running at 3000 r.p.m. this overspeed protection operates at about 3300 r.p.m.

### *Example 4.1*

An isolated 75 MVA synchronous generator feeds its own load and operates initially at no-load at 3000 r.p.m., 50 Hz. A 20 MW load is suddenly applied and the steam valves to the turbine commence to open after 0.5 s due to the time-lag in the governor system. Calculate the frequency to which the generated voltage drops before the steam flow meets the new load. The stored energy for the machine is 4 kW-s per kVA of generator capacity.

### *Solution*

For this machine the stored energy at 3000 r.p.m.

$$= 4 \times 75\,000 = 300\,000 \text{ kW–s}$$

Before the steam valves start to open the machine loses $20\,000 \times 0.5 = 10\,000$ kW-s of the stored energy in order to supply the load.

The stored energy $\propto$ (speed)$^2$. Therefore the new frequency

$$= \sqrt{\frac{300\,000 - 10\,000}{300\,000}} \times 50 \text{ Hz}$$

$$= 49.2 \text{ Hz}$$

## 4.3  Control Loops

The machine and its associated governor and voltage-regulator control systems may be represented by the block diagram shown in Figure 4.5. The nature of the voltage control loop has been discussed already in Chapter 3; accurate machine representation is required, involving two-axis expressions, as indicated in Chapter 2.

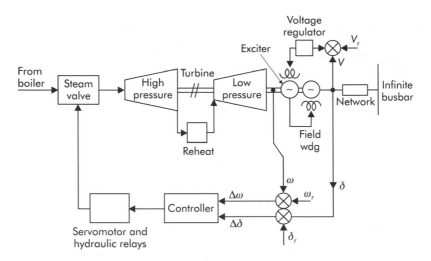

**Figure 4.5** Block diagram of complete turboalternator control systems. The governor system is more complicated that that shown in Figure 4.1 owing to the inclusion of the load-angle $\delta$ in the control loop. Suffixes $r$ refer to reference quantities and $\Delta$ to the error quantities. The controller modifies the error signal from the governor by taking into account the load angle

Two factors have a large influence on the dynamic response of the prime mover: (1) entrained steam between the inlet valves and the first stage of the turbine (in large machines this can be sufficient to cause loss of synchronism after the valves have closed); (2) the storage action in the reheater which causes the output of the low-pressure turbine to lag behind that of the high-pressure side. The transfer function

$$\frac{\text{prime mover torque}}{\text{valve opening}}$$

accounting for both these effects is

$$\frac{G_1 G_2}{(1 + \tau_t s)(1 + \tau_r s)} \tag{4.2}$$

where

$G_1$ = entrained steam constant;
$G_2$ = reheated gain constant;
$\tau_t$ = entrained steam time constant;
$\tau_r$ = reheater time constant.

The transfer function relating steam-valve opening $d$ to changes in speed $\omega$ due to the governor feedback loop is

$$\frac{\Delta d}{\Delta \omega}(s) = \frac{G_3 G_4 G_5}{(1 + \tau_g s)(1 + \tau_1 s)(1 + \tau_2 s)} \qquad (4.3)$$

where

$\tau_g$ = governor-relay time constant

$\tau_1$ = primary-relay time constant;

$\tau_2$ = secondary-valve-relay time constant;

$(G_3 G_4 G_5)$ = constants relating system-valve lift to speed change.

By a consideration of the transfer function of the synchronous generator with the above expressions the dynamic response of the overall system may be obtained.

## 4.4   Division of Load Between Generators

The use of the speed changer enables the steam input and electrical power output at a given frequency to be changed as required. The effect of this on two machines can be seen in Figure 4.6. The output of each machine is not therefore determined by the governor characteristics but can be varied by the operating personnel to meet economical and other considerations. The governor characteristics only completely decide the outputs of the machines when a sudden change in load occurs or when machines are allowed to vary their outputs according to speed within a prescribed range in order to keep the frequency constant. This latter mode of operation is known as *free-governor action*.

It has been shown in Chapter 2 that the voltage difference between the two ends of an interconnector of total impedance $R + jX$ is given by

$$\Delta V = E - V = \frac{RP + XQ}{V}$$

Also the angle between the voltage phasors (i.e. the transmission angle) $\delta$ is given by

$$\sin^{-1} \frac{\Delta V_q}{E}$$

where

$$\Delta V_q = \frac{XP - RQ}{V}$$

When $X \gg R$, i.e. for most transmission networks,

$$\Delta V_q \propto P \qquad \text{and} \qquad \Delta V_p \propto Q$$

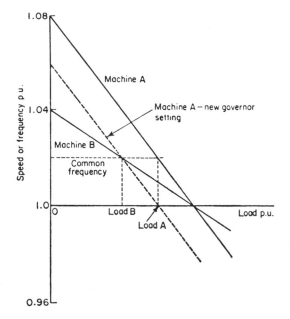

**Figure 4.6** Two machines connected to an infinite busbar. The speeder gear of machine A is adjusted so that the machines load equally

Hence, (1) *the flow of power between two nodes is determined largely by the transmission angle;* (2) *the flow of reactive power is determined by the scalar voltage difference between the two nodes.*

These two facts are of fundamental importance to the understanding of the operation of power systems.

The angular advance of $G_A$ (Figure 4.7) is due to a greater relative energy input to turbine A than to B. The provision of this extra steam (or water) to A is possible because of the action of the speeder gear without which the power outputs of A and B would be determined solely by the nominal governor characteristics. The following simple example illustrates these principles.

## Example 4.2

Two synchronous generators operate in parallel and supply a total load of 200 MW. The capacities of the machines are 100 MW and 200 MW and both have governor droop characteristics of 4 per cent from no load to full load. Calculate the load taken by each machine, assuming free governor action.

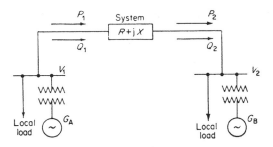

**Figure 4.7** Two generating stations linked by an interconnector of impedance $(R + jX)\,\Omega$. The rotor of A is in phase advance of B and $V_1 > V_2$

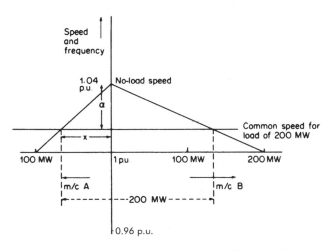

**Figure 4.8**   Speed–load diagram for Example 4.2

## Solution

Let $x$ megawatts be the power supplied from a 100 MW generator. Referring to Figure 4.8,

$$\frac{4}{100} = \frac{\alpha}{x}$$

For the 200 MW machine,

$$\frac{4}{200} = \frac{\alpha}{200 - x}$$

$$\therefore \frac{4x}{100} = \frac{800 - 4x}{200}$$

and $x = 66.6$ MW = load on the 100 MW machine. The load on the 200 MW machine = 133.3 MW.

It will be noticed that when the governor droops are the same the machines share the total load in proportion to their capacities or ratings. Hence it is advantageous for the droops of all turbines to be equal.

## *Example 4.3*

Two units of generation maintain 66 kV and 60 kV (line) at the ends of an interconnector of inductive reactance per phase of 40 $\Omega$ and with negligible resistance and shunt capacitance. A load of 10 MW is to be transferred from the 66 kV unit to the other end. Calculate the necessary conditions between the two ends, including the power factor of the current transmitted.

## *Solution*

$$\Delta V_q = \frac{XP - RQ}{V_R} = \frac{40 \times 3.33 \times 10^6}{60\,000/\sqrt{3}} = 3840\text{V}$$

Also,

$$\frac{\Delta V_q}{66\,000/\sqrt{3}} = \sin\delta = 0.101$$

$$\therefore \delta = 5° \ 44'$$

Hence the 66 kV busbars are $5° \ 44'$ in advance of the 60 kV busbars.

$$\Delta V_p = \frac{66\,000 - 60\,000}{\sqrt{3}} = \frac{RP + XQ}{V_R} = \frac{40Q}{60\,000/\sqrt{3}}$$

Hence

$$Q = 3 \text{ MVAr per phase (9 MVAr total)}$$

The p.f. angle $\phi = \tan^{-1} Q/P = 42°$ and hence the p.f. $= 0.74$.

# 4.5 The Power–Frequency Characteristic of an Interconnected System

The change in power for a given change in the frequency in an interconnected system is known as the *stiffness* of the system. The smaller the change in frequency for a given load change the stiffer the system. The power–frequency characteristic may be approximated by a straight line and $\Delta P/\Delta f = K$, where

$K$ is a constant (MW per Hz) depending on the governor and load characteristics.

Let $\Delta P_G$ be the change in generation with the governors operating 'free acting' resulting from a sudden increase in load $\Delta P_L$. The resultant out-of-balance power in the system

$$\Delta P = \Delta P_L - \Delta P_G \qquad (4.4)$$

and therefore

$$K = \frac{\Delta P_L}{\Delta f} - \frac{\Delta P_G}{\Delta f} \qquad (4.5)$$

$\Delta P_L/\Delta f$ measures the effect of the frequency characteristics of the load and $\Delta P_G \propto (P_T - P_G)$, where $P_T$ is the turbine capacity connected to the network and $P_G$ the output of the associated generators. When steady conditions are again reached, the load $P_L$ is equal to the generated power $P_G$ (neglecting losses): hence, $K = K_1 P_T - K_2 P_L$, where $K_1$ and $K_2$ are the power-frequency coefficients relevant to the turbines and load respectively.

Here, $K$ can be determined experimentally by connecting two large separate systems by a single link, breaking the connection and measuring the frequency change. For the British system, tests show that $K = 0.8 P_T - 0.6 P_L$ and lies between 2000 and 5500 MW per Hz, i.e. a change in frequency of 0.1 Hz requires a change in the range 200–550 MW, depending on the amount of plant connected. In smaller systems the change in frequency for a reasonable load change is relatively large and rapid-response electrical governors have been introduced to improve the power-frequency characteristic.

In 1977, owing to a series of events triggered off by lightning, New York City was cut off from external supplies and the internal generation available was much less than the city load. The resulting fall in frequency with time is shown in Figure 4.9, illustrating the time–frequency characteristics of an isolated power system.

## 4.6 System Connected by Lines of Relatively Small Capacity

Let $K_A$ and $K_B$ be the respective power–frequency constants of two separate power systems A and B, as shown in Figure 4.10. Let the change in the power transferred from A to B when a change resulting in an out-of-balance power $\Delta P$ occurs in system B, be $\Delta P_t$, where $\Delta P_t$ is positive when power is transferred from A to B. The change in frequency in system B, due to an extra load $\Delta P$ and an extra input of $\Delta P$, from A, is $-(\Delta P - \Delta P_t)/K_B$ (the negative sign indicates a fall in frequency). The drop in frequency in A due to the extra load

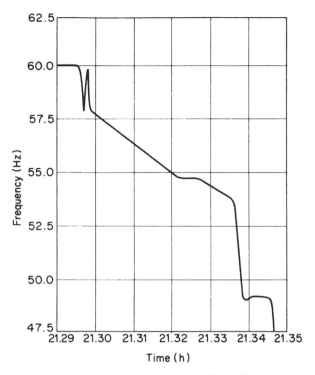

**Figure 4.9**   Decline of frequency with time of New York City system when isolated from external supplies (*Copyright © 1977 Institute of Electrical and Electronics Engineers, Inc. Reprinted by permission from I.E.E.E. Spectrum, Vol. 15, No. 2 (Feb. 1978) pp. 38–46*)

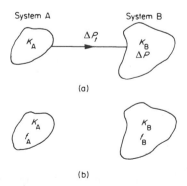

**Figure 4.10**   (a) Two interconnected power systems connected by a tie-line. (b) The two systems with the tie-line open

$\Delta P_t$ is $-\Delta P_t/K_A$, but the changes in frequency in each system must be equal, as they are electrically connected. Hence,

$$\frac{-(\Delta P - \Delta P_t)}{K_B} = \frac{-\Delta P_t}{K_A}$$

$$\therefore \Delta P_t = +\left(\frac{K_A}{K_A + K_B}\right)\Delta P \qquad (4.6)$$

Next, consider the two systems operating at a common frequency $f$ with A exporting $\Delta P_t$ to B. The connecting link is now opened and A is relieved of $\Delta P_t$ and assumes a freqency $f_A$, and B has $\Delta P_t$ more load and assumes $f_B$. Hence

$$f_A = f + \frac{\Delta P_t}{K_A} \quad \text{and} \quad f_B = f - \frac{\Delta P_t}{K_B}$$

from which

$$\frac{\Delta P_t}{f_A - f_B} = \frac{K_A K_B}{K_A + K_B} \qquad (4.7)$$

Hence, by opening the link and measuring the resultant change in $f_A$ and $f_B$ the values of $K_A$ and $K_B$ can be obtained.

When large interconnected systems are linked electrically to others by means of tie-lines the power transfers between them are usually decided by mutual agreement and the power is controlled by regulators. As the capacity of the tie-lines is small compared with the systems, care must be taken to avoid excessive transfers of power and corresponding cascade tripping.

### 4.6.1   Effect of governor characteristics

A fuller treatment of the performance of two interconnected systems in the steady state requires further consideration of the control aspects of the generation process.

A more complete block diagram for the steam turbine-generator connected to a power system is shown in Figure 4.11. For this system the following equation holds:

$$M s^2 \delta + K s \delta = \Delta P - \Delta P_L$$

where $M$ is a constant depending on inertia (see Chapter 8); $K$ is the stiffness or damping coefficient (i.e. change of power with speed) in MW/Hz or MW per rad/s

$$= \frac{\partial \, (\text{load power})}{\partial(s\delta)} - \frac{\partial \, (\text{turbine power})}{\partial(s\delta)}$$

where

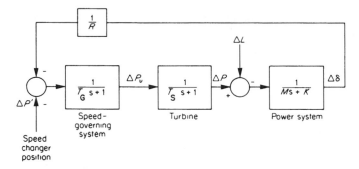

**Figure 4.11** Block diagram for turbine-generator connected to a power system (*From Economic Control of Interconnected Systems, by L. K. Kirchmayer. Copyright © 1959 John Wiley & Sons Ltd. By permission of John Wiley & Sons Inc.)*

$\Delta P$ and $\Delta P_L$ = change in prime mover and load powers;

$\quad\quad \Delta \delta$ = change from initial angular position;

$\quad\quad\quad R$ = present speed regulation (or governor droop), i.e. drop in speed or frequency when combined machines of an area change from no load to full load, expressed as p.u. or Hz or rad/s per MW;

$\quad\quad \Delta P'$ = change in speed-changer setting.

Therefore, change from normal speed or frequency,

$$s\delta = \frac{1}{Ms + K}(\Delta P - \Delta P_L)$$

This analysis holds for steam-turbine generation; for hydro-turbines, the large inertia of the water must be accounted for and the analysis is more complicated.

The representation of two systems connected by a tie-line is shown in Figure 4.12. The general analysis is as before except for the additional power terms due to the tie-line. The machines in the inidividual power systems are considered to be closely coupled and to possess one equivalent rotor.

For system (1),

$$M_1 s^2 \delta_1 + K_1 s \delta_1 + T_{12}(\delta_1 - \delta_2) = \Delta P_1 - \Delta P_{L1} \quad\quad (4.8)$$

where $T_{12}$ is the synchronizing torque coefficient of the tie-line.

For system (2),

$$M_2 s^2 \delta_2 + K_2 s \delta_2 + T_{12}(\delta_2 - \delta_1) = \Delta P_2 - \Delta P_{L2} \quad\quad (4.9)$$

The steady-state analysis of two interconnected systems may be obtained from the transfer functions given in the block diagram.

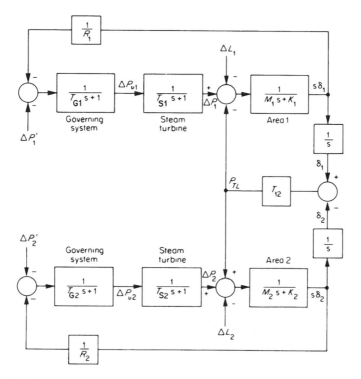

**Figure 4.12** Block control diagram of two power systems connected by a tie-line (*From Economic Control of Interconnected Systems by L. K. Kirchmayer. Copyright ©1959 John Wiley & Sons Ltd. By permission of John Wiley & Sons Inc.*)

The speed governor response is given by

$$\Delta P_V = \frac{-1}{(T_G s + 1)} \left( \frac{1}{R} s\delta + \Delta P' \right) \qquad (4.10)$$

In the steady state, from equation (4.10),

$$\Delta P_1 = \frac{1}{R_1} s\delta_1 \qquad (4.11)$$

and

$$\Delta P_2 = -\frac{1}{R_2} s\delta_2 \qquad (4.12)$$

Similarly, from equations (4.8) and (4.9), in the steady state,

$$\left( K_1 + \frac{1}{R_1} \right) s\delta_1 + T_{12}(\delta_1 - \delta_2) = -\Delta P_{L1} \qquad (4.13)$$

and

$$\left(K_2 + \frac{1}{R_2}\right)s\delta_2 + T_{12}(\delta_2 - \delta_1) = -\Delta P_{L2} \tag{4.14}$$

Adding equations (4.13) and (4.14) gives

$$\left(K_1 + \frac{1}{R_1}\right)s\delta_1 + \left(K_2 + \frac{1}{R_2}\right)s\delta_2 = -\Delta P_{L1} - \Delta P_{L2} \tag{4.15}$$

In a synchronous system, $s\delta_1 = s\delta_2 = s\delta = $ angular frequency and equation (4.15) becomes

$$\left[K_1 + K_2 + \left(\frac{1}{R_1} + \frac{1}{R_2}\right)\right]s\delta = -\Delta P_{L1} - \Delta P_{L2}$$

and

$$s\delta = \frac{-\Delta P_{L1} - \Delta P_{L2}}{(K_1 + K_2) + (1/R_1 + 1/R_2)} \tag{4.16}$$

From equations (4.13) and (4.14),

$$T_{12}(\delta_1 - \delta_2) = \frac{-\Delta P_{L1}[K_2 + (1/R_2)] + \Delta P_{L2}[K_1 + (1/R_1)]}{[K_2 + (1/R_2) + K_1 + (1/R_1)]} \tag{4.17}$$

It is normally required to keep the system frequency constant and to maintain the interchange through the tie-line at its scheduled value. To achieve this, additional controls are necessary to operate the speed-changer settings, as follows.

For area (1),

$$s\Delta P_1^1 \propto \gamma_{1t}T_{12}(\delta_1 - \delta_2) + \gamma_{1f}s\delta_1$$

$$\propto \gamma_{1t}\left[T_{12}(\delta_1 - \delta_2) + \frac{\gamma_{1f}}{\gamma_{1t}}s\delta_1\right] \tag{4.18}$$

Similarly, for area (2),

$$s\Delta P_2' \propto \gamma_{2t}\left[T_{12}(\delta_2 - \delta_1) + \frac{\gamma_{2f}}{\gamma_{2t}}s\delta_2\right] \tag{4.19}$$

where $\gamma_t$ and $\gamma_f$ refer to the control constants for power transfer and frequency, respectively. When a load change occurs in a given area the changes in tie-line power and frequency have opposite signs, i.e. the frequency falls for a load increase and the power transfer increases, and vice versa. In the interconnected area, however, the changes have the same sign. Typical system parameters have the following orders of magnitude:

$K = 0.75$ p.u. on system-capacity base

$T_{12} = 0.1$p.u. (10 per cent of system capacity results in 1 rad displacement between areas (1) and (2))

$R = 0.04$p.u. on a base of system capacity

$\gamma_{f}, = 0.005$

$\gamma_{t}, = 0.0009$

## *Example 4.4*

Two power systems, A and B, each have a regulation ($R$) of 0.1 p.u. (on respective capacity bases) and a stiffness $K$ of 1 p.u. The capacity of system A is 1500 MW and of B 1000 MW. The two systems are interconnected through a tie-line and are initially at 60 Hz. If there is a 100 MW load change in system A, calculate the change in the steady-state values of frequency and power transfer.

## *Solution*

$$K_A = 1 \times 1500 \text{ MW per Hz}$$
$$K_B = 1 \times 1000 \text{ MW per HZ}$$
$$R_A = \frac{\Delta f (\text{no load to full load})}{\text{full load capacity}} = \frac{0.1 \times 60}{1500}$$
$$= 6/1500 \text{ Hz per MW}$$
$$R_B = (6/1000) \text{ Hz per MW}$$

From equation (4.16),

$$\Delta f = \frac{-\Delta P_1}{(K_1 + 1/R_1) + (K_2 + 1/R_2)}$$
$$= \frac{-100}{1500 + \dfrac{1500}{6} + 1000 + \dfrac{1000}{6}}$$
$$= \frac{-600}{17500} = -0.034 \text{ Hz}$$
$$P_{12} = T_{12}(\delta_1 - \delta_2) = \frac{-\Delta P_1 (K_2 + 1)/R_2)}{(K_1 + 1/R_1) + (K_2 + 1/R_2)}$$
$$= \frac{-100 \left( \dfrac{7000}{6} \right)}{10500/6} = \frac{-7000}{105} = -6 \text{ MW}$$

Note that without the participation of governor control,

$$P_{12} = \left(\frac{-K_B}{K_A + K_B}\right) \qquad \Delta P = -\frac{1000}{2500} \times 100 = -40\,\text{MW}$$

### 4.6.2 Frequency-bias–tie-line control

Consider three interconnected power systems A, B and C, as shown in Figure 4.13, the systems being of similar size. Assume that initially A and B export to C, their previously agreed power transfers. If C has an increase in load the overall frequency tends to decrease and hence the generation in A, B, and C increases. This results in increased power transfers from A and B to C. These transfers, however, are limited by the tie-line power controller to the previously agreed values and therefore instructions are given for A and B to reduce generation and hence C is not helped. This is a severe drawback of what is known as straight tie-line control, which can be overcome if the systems are controlled by using consideration of both load transfer and frequency, such that the following equation holds:

$$\sum \Delta P + K\Delta f = 0 \qquad\qquad (4.20)$$

where $\sum \Delta P$ is the net transfer error and depends on the size of the system and the governor characteristic, and $\Delta f$ is the frequency error and is positive for high frequency. In the case above, after the load change in C, the frequency error is negative (i.e. low frequency) for A and B and the sum of $\Delta P$ for the lines AC and BC is positive. For correct control,

$$\sum \Delta P_A + K_A \Delta f = \sum \Delta P_B + K_B \Delta f = 0$$

Systems A and B take no regulating action despite their fall in frequency. In C, $\sum \Delta P_C$ is negative as it is importing from A and B and therefore the governor speeder motors in C operate to increase ouput and restore frequency. This system is known as frequency-bias–tie-line control and is often implemented automatically in interconnected systems.

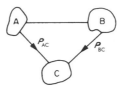

**Figure 4.13** Three power systems connected by tie-lines

# 4.7 Economic Power-System Operation

## 4.7.1 Introduction

The extensive interconnection of power sources has made the operation of a system in the most economical manner a complex subject (see also Chapter 12). Economy must be balanced against considerations such as security of supply. The use of the *merit order* ensures that as far as possible the most economical generating sets are used. A knowledge of the flows of real and reactive power and other parameters in the network, and effective means of dealing with the analysis of large systems, is required for the operation to attain an economic optimum, although experienced operators certainly approach this aim.

Nowadays, the use of digital computers for load flows and fault calculations and the development of optimization techniques in control theory have resulted in much attention being given to this topic.

Apart from financial considerations, it is becoming difficult for operators to cope with the information produced by large complex systems in times of emergency, such as with major faults. Computers with on-line facilities can more readily digest this information and take correcting action by instructing control gear and settings, or by the suitable display of relevant information to enable human operators to take appropriate action. In the attempt to obtain economic optimization the limitations of the system, such as plant ratings and stability limits, must be observed.

Optimization may be considered in a number of ways according to the time scale involved, namely: daily, yearly (especially with hydro stations), and over much longer periods when planning for future developments, although this latter is not strictly operational optimization. In an existing system the various factors involved are the fixed and variable costs. The former includes labour, administration, interest and depreciation, etc., and the latter, mainly fuel. A major problem is the effective prediction of the future load whether it occurs in 10 minutes, a few hours, or in several year' time.

For operational planning, daily operation, and the setting of economic schedules, the following data is normally required.

- *For each generator:*
  1. maximum and economic output capacities;
  2. fixed and incremental heat rates;
  3. minimum shut-down time;
  4. minimum stable output, maximum run-up and run-down rates.

- *For each station:*
    1. cost and calorific value of fuel (thermal stations);
    2. factors reflecting recent operational performance of the station;
    3. minimum time between loading and unloading successive generators;
    4. any constraints on station output.

- *For the system:*
    1. load demand at given intervals for the specified period;
    2. specified constraints imposed by transmission capability;
    3. running-spare requirements;
    4. transmission circuit parameters, including maximum capacities and reliability factors.

The input–output characteristic of a turbine is of great importance when economical operation is considered. A typical characteristic is shown in Figure 4.14. The *incremental heat rate* is defined as the slope of the input–output curve at any given output. The graph of the incremental heat rate against output is known as the Willans line. For large turbines with a single valve, and for gas turbines, the incremental heat rate is approximately constant over the operating range (most steam turbines in Britain are of this type); with multivalve turbines (as used in the U.S.A.) the Willans line is not horizontal but curves upwards and is often represented by the closest linear law. The value taken for the incremental heat rate of a generating set is sometimes complicated because if only one or two shifts are being operated (there are normally three shifts per day) heat has to be expended in banking boilers when the generator is not required to produce output.

Instead of plotting incremental heat rate or fuel consumption against power output for the turbine-generator, the incremental fuel cost may be used. This is

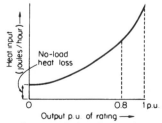

**Figure 4.14** Input–output characteristic of a turbogenerator set. Often the curve above an output of 0.8 p.u. is steeper than the remainder and the machine has a maximum rating at 1 p.u. and a maximum economic rating at, say 0.8 p.u.

advantageous when allocating load to generators for optimum economy as it incorporates differences in the fuel costs of the various generating stations. Usually, the graph of incremental fuel cost against power output can be approximated by a straight line (Figure 4.15). Consider two turbine-generator sets having the following different incremental fuel costs, $dC_1/dP_1$ and $dC_2/dP_2$, where $C_1$ is the cost of the fuel input to unit number 1 for a power output of $P_1$, and, similarly, $C_2$ and $P_2$ relate to unit number 2. It is required to load the generators to meet a given requirement in the most economical manner. Obviously the load on the machine with the higher $dC/dP$ will be reduced by increasing the load taken by the machine with the lower $dC/dP$. This transfer will be beneficial until the values of $dC/dP$ for both sets are equal, after which the machine with the previously higher $dC/dP$ now becomes the one with the lower value, and vice versa. There is no economic advantage in further transfer of load, and the condition when $dC_1/dP_1 = dC_2/dP_2$ therefore gives optimum economy; this can be seen by considering Figure 4.15. The above argument can be extended to several machines supplying a load. Generally, for optimum economy the *incremental fuel cost should be identical for all contributing turbine-generator sets* on free-governor action. In practice, most generators will be loaded to their maximum output.

The above reasoning must be modified when the distances of generating stations from the common loads are different; here the cost of transmission losses will affect the argument.

As important as the transmission losses is the optimum method of transporting fuel from the production centres to the generating stations. The transport of both electrical energy and fuel in the optimum manner forms *transportation problems*, which may be dealt with by special techniques or by the general method of linear programming. In a competitive situation the transportation costs will be included in the generator bid price.

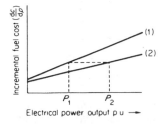

**Figure 4.15** Idealized graphs of incremental fuel cost against output for two machines sharing a load equal to $P_1$ and $P_2$

## *Example 4.5*

Four generators are available to supply a power system peak load of 472.5 MW. The cost of power $C(P_i)$ from each generator, and maximum output, is given in ($ U.S.) by:

$$C(P_1) = 200 + 15P_1 + 0.20P_1^2 \qquad \text{Max. output } 100 \text{ MW}$$
$$C(P_2) = 300 + 17P_2 + 0.10P_2^2 \qquad \text{Max. output } 120 \text{ MW}$$
$$C(P_3) = 150 + 12P_3 + 0.15P_3^2 \qquad \text{Max.output } 160 \text{ MW}$$
$$C(P_4) = 500 + 2P_4 + 0.07P_4^2 \qquad \text{Max. output } 200 \text{ MW}$$

The spinning reserve is to be 10 per cent of peak load and the transmission losses can be neglected.

Calculate the optimal loading of each generator and the cost of operating the system for 1 h at peak.

## *Solution*

The generators' combined output is to be 472.5 MW.

Marginal costs are given (in U.S.$/MW) by:

$$\frac{dC(P_1)}{dP_1} = 15 + 0.40P_1 \qquad P_1 \leqslant 100 \text{ MW}$$

$$\frac{dC(P_2)}{dP_2} = 17 + 0.20P_2 \qquad P_2 \leqslant 120 \text{ MW}$$

$$\frac{dC(P_3)}{dP_3} = 12 + 0.30P_3 \qquad P_3 \leqslant 160 \text{ MW}$$

$$\frac{dC(P_4)}{dP_4} = 2 + 0.14P_4 \qquad P_4 \leqslant 200 \text{ MW}$$

The marginal cost curves are plotted in Figure 4.16 for eachof the generators up to their maximum output.

From the curves, at 30 $/MW the outputs are

$$P_1 = 23$$
$$P_2 = 64$$
$$P_3 = 60$$
$$P_4 = 200$$
$$\text{Total} = 347 \text{ MW}$$

Hence, $P_4$ runs at full output and the remaining generators must sum to 272.5 MW.

At 40 $/MW the outputs sum to 446 MW. At 41 $/MW $P_2$ reaches its maximum of 120 MW, and $P_1 = 40$ MW and $P_3 = 98$ MW (total 458 MW).

This leaves 14.5 MW to be found from $P_1$ and $P_3$.

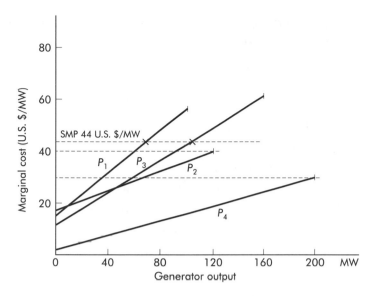

**Figure 4.16** Marginal cost curves for four generators

Adjusting the marginal cost line now to 44 $/MW provides 44 MW from $P_1$ and 108 MW from $P_3$, giving a total of 472 MW, which is near enough when using this graphical method. The spinning reserve on these generators is 108 MW, more than enough to cover 10 per cent of 472.5 MW.

The cost of operating the system for 1 h at peak is

$$C(P_1) = 200 + 15 \times 44 + 0.2 \times 44^2 = \$1247$$

Similarly

$$C(P_2) = \$3780$$
$$C(P_3) = \$3195$$
$$C(P_4) = \$3700$$
$$\text{Total} = \$11922$$

i.e.

$$\frac{11922}{472.5} = 25.2\,\$/MW$$

(Note: This is the *average* cost per hour whereas the *marginal cost* was calculated as 44 $/MW. It can be seen that the use of large units on base load (e.g. the 200 MW generator) reduces system marginal price (SMP) considerably (see Chapter 12).)

## 4.7.2  Basic formulation of the short-term optimization problem for thermal stations

Kirchamayer (1958) uses Lagrange multipliers in formulating equations, including transmission losses.

Let

$$P_i = \text{power output of } i \text{ (MW)}$$
$$P_R = \text{total load on system (MW)}$$
$$P_L = \text{transmission losses (MW)}$$
$$F_T = \text{total cost of generating units (money/h)}$$
$$\lambda = \text{Lagrange multiplier ((money/MWh)}$$
$$n = \text{number of generating units}$$

The total input to the system from all generators $P_T \sum_{i=1}^{n} P_i$, and

$$\left( \sum_{1}^{n} P_i \right) - P_L - P_R = 0$$

Using Lagrangian multipliers, the expression

$$\gamma = F_T - \lambda \left( \sum_{1}^{n} P_i - P_L - P_R \right)$$

is formulated, where for minimum cost ($F_T$), $\partial \gamma / \partial P_i = 0$ for all values of $i$. (Note the use of partial differentiation here.) This is given by

$$\frac{\partial \gamma}{\partial P_i} = \frac{\mathrm{d}Fi}{\mathrm{d}P_i} - \lambda + \lambda \frac{\partial P_L}{\partial P_i} = 0$$

since $P_R$ is assumed fixed. Hence

$$\frac{\mathrm{d}F_i}{\mathrm{d}P_i} + \lambda \frac{\partial P_L}{\partial P_i} = \lambda \tag{4.21}$$

In equation (4.21), $\partial P_L / \partial P_i$ is the incremental transmission loss. One way of solving the equations described by (4.21) is known as the penalty-factor method in which equation (4.21) is rewritten as

$$\frac{\mathrm{d}F_i}{\mathrm{d}P_i} L_i = \lambda \tag{4.22}$$

where

$$L_i = \left( \frac{1}{1 - \partial P_L / \partial P_i} \right) = \text{penalty factor of plant } i$$

where $i = 1 \ldots n$ (number of plants). In practice the determination of $\partial P_L / \partial P_i$ is difficult and the use of the so-called loss or '$B$' coefficients is made, i.e.

$$\frac{\partial P_{\mathrm{L}}}{\partial P_i} = \sum_i 2 B_{\mathrm{m}i} P_{\mathrm{m}} + B_{i0}$$

where the $B$ coefficients are determined from the network (see Kirchmayer, 1958).

There are many drawbacks to the above treatment e.g. limitations on power flows by equipment ratings transformer settings, and maximum phase angles allowable across transmission lines on stability grounds. Also, it is concerned only with active power, reactive power being neglected or taken into account by limiting MW flows across defined group boundaries.

The value of $\lambda$ at any particular period is known as 'system lambda' or 'system marginal price' (SMP).

## 4.8  Computer Control of Load and Frequency

### 4.8.1  Control of tie lines

Automatic control of area power systems connected by tie-lines has already been discussed. The methods used will now be extended to include optimum economy as well as power transfer and frequency control. The basic systems described are typical of U.S. and European practice and have been comprehensively discussed by many authors. In the previous section, methods for economic analysis and optimization as developed by Kirchmeyer (1958) have been summarized. The choice of generating units to be operated is largely decided by spinning reserve, voltage control, stability, and protection requirments. The methods discussed decide the allocation of load to particular machines.

If transmission losses are to be neglected, it has been shown that optimum economy results when $dF_i/dP_i = \lambda$. Control equipment to adjust the governor speed-change settings such that all units comply with the appropriate value of $dF_i/dP_i$ is incorporated in the control loops for frequency and power-transfer adjustment, as shown in Figure 4.17. The frequency and load-transfer control acts quickly, and once these quantities have been decided the slower acting economic controls then act. For example, if an increase in load occurs in the controlled area, a signal requiring increased generation transmits through the control system. These changes alter the value of $\lambda$ and cause the economic control apparatus to call for generation to be operated at the same incremental cost. Eventually the system is again in the steady state, the load change having been absorbed, and all units operate at an identical value of incremental loss.

If transmission losses are included, the basic economic requirement calls for $dF_i/dP_i = \lambda/L_i$ where $L_i$ is the penalty factor; $(1/L_i)$ signals are generated by a computer from a knowledge of system parameters in the form of the so-called $B$ coefficients. However, in practice, most generators are being run at their

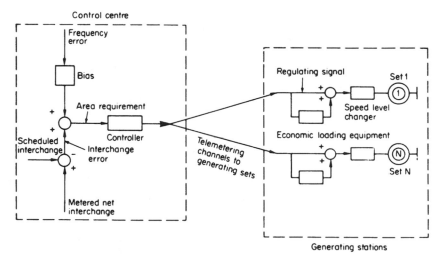

**Figure 4.17** Schematic diagram of automatic control arrangements covering frequency, tie-line power flows, and economic loading of generating sets (*From Economic Control of Interconnected Systems by L. K. Kirchmayer. Copyright © 1959 John Wiley & Sons Ltd. By permission of John Wiley & Sons Inc.*)

maximum rated output because their cost is below the system $\lambda$. Only a few units on the cost margin are setting the value of $\lambda$.

In the U.S.A. the many separate power companies are connected into power pools. The aim is optimal control of the participating machines in each pool, within security constraints. Figure 4.18 shows a block diagram of the New York Power Pool. Among the advantages of such pools are; the more economic use of large generators, emergency assistance to neighbouring utilities, reduced spinning reserves, and lower overall generation costs. Each pool is connected to others by tie-lines, e.g. the New York Pool to Ontario Hydro, New England Power Exchange and the Pennsylvania–New Jersey–Maryland interconnection. Control of the generators in the pool is achieved either centrally or from each of the constituent areas (i.e. area control centres) as indicated in Figure 4.19

The central control mechanisms for pools are basically the same as for areas and the systems of Figure 4.17 may be used with 'pool' substituted for 'area'. Allocation between generators in an area, or between areas in a pool, may be accomplished by the use of base points or loadings and participation factors. The former gives the economic allocation for a specified total generation and they are normally established every few minutes or when loadings change. When the generation allocations are established they are compared with the actual values being generated and a control error formed. The unit participation factor $(K_n)$ for any unit $(n)$ in the pool is given by

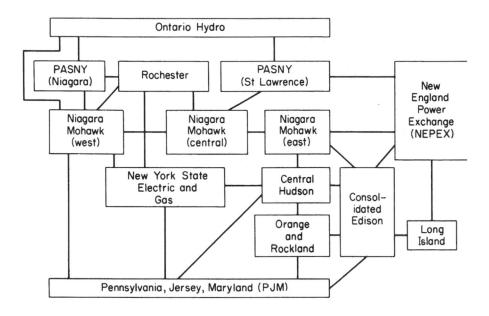

**Figure 4.18** Block diagram showing the eight power companies within the New York Power Pool, and three other pools that are interconnected with it (*Copyright © 1973 Institute of Electrical and Electronics Engineers, Inc. Reprinted by permission from I.E.E.E. Spectrum, Vol. 10, No. 3, March 1973, pp. 54–61*)

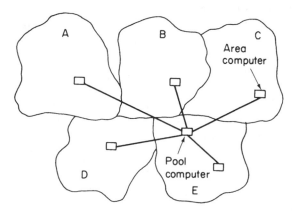

**Figure 4.19** Multicomputer configuration formed by a five-area pool (*Copyright © 1973 Institute of Electrical and Electronics Engineers, Inc. Reprinted by permission from I.E.E.E. Spectrum, Vol. 10, No. 3, March 1973, pp. 54–61*)

$$K_n = \left(\frac{1}{F_n L_n}\right) \Bigg/ \sum \frac{1}{F_n L_n} \qquad n = 1, 2, 3, \ldots$$

where $F_n$ is the slope of the incremental cost curve for unit $n$, and $L_n$ is the penalty factor for unit $n$. The above method may be extended with the control of areas from the pool centre replacing that of the individual units; the areas, in turn, control the units. Area base points are the sum of the unit base points in that area and represent the economic area allocation for a specified total pool generation. This is achieved by a multi-area dispatch program which is run every few minutes and accounts for limits on interchanges and tie-lines, as well as the usual parameters. In this case the additional pool generation ($\Delta G$) required between updating base points is allocated among areas according to the area participation factors defined as follows:

$$K_a = \frac{\Delta P_a}{\Delta G}$$

where $\Delta P_a$ is the power allocated to area $a$. The electric capacity of some pools approaches the capacities of state-run centrally controlled supply organizations in many countries outside the U.S.A.

In the U.K., following privatization in 1989/90, the England and Wales transmission network is run on a pool basis into which generators (including contributions from Scotland and France) bid a price for their electrical energy for the following day. The pool then computes an economic schedule of generation from a load forecast for the following day, predicting the system $\lambda$ price for each half-hour if the schedule is adhered to as planned. Although some producers may be required to provide a variable output to cover reserves and contingencies, consumers can vary their requirements depending on the SMP as displayed over an information network to (mainly) large users and other interested parties. Many generators, now belonging to a number of independent power producers (IPPs), can be directly instructed through tele-control by the system operator to start up, shut down, or run at a part load. Consequently, frequency can be controlled to within $\pm 0.05\,\text{Hz}$ throughout the day and night by the system operator, who has complete information at the control centre through the System Control and Data Acquisition (SCADA) system. Chapter 11 describes the use of SCADA and Chapter 12 outlines the economics of pool-type operation.

# Problems

**4.1**   Two identical 60 MW synchronous generators operate in parallel. The governor settings on the machines are such that they have 4 per cent and 3 per cent droops (no-load to full-load percentage speed drop). Determine (a) the load taken by each machine for a total of 100 MW; (b) the percentage adjustment in the no-load speed to be made by the speeder motor if the machines are to share the load equally.

(Answer: (a) 42.8 and 57.2 MW; (b) 0.83 per cent increase in no-load speed on the 4 per cent droop machine)

**4.2** (a) Explain how the output power of a turbine-generator operating in a constant frequency system is controlled by adjusting the setting of the governor. Show the effect on the generator power–frequency curve.

(b) Generator A of rating 200 MW and generator B of rating 350 MW have governor droops of 5 per cent and 8 per cent, respectively, from no-load to full-load. They are the only supply to an isolated system whose nominal frequency is 50 Hz. The corresponding generator speed is 3000 r.p.m. Initially, generator A is at 0.5 p.u. load and generator B is at 0.65 p.u. load, both running at 50 Hz. Find the no-load speed of each generator if it is disconnected from the system.

(c) Also determine the total output when the first generator reaches its rating.
(Answer: (b) Generator B 3156 r.p.m; generator A 3075 r.p.m; (c) 337 MW)
*(From E C Examination, 1996)*

**4.3** The incremental fuel costs of two units in a generating station are as follows:

$$\frac{dF_1}{dP_1} = 0.003P_1 + 0.7$$

$$\frac{dF_2}{dP_2} = 0.004P_2 + 0.5$$

where $F$ is in £/h and $P$ is in MW.

Assuming continuous running with a total load of 150 MW calculate the saving per hour obtained by using the most economical division of load between the units as compared with loading each equally. The maximum and minimum operational loadings are the same for each unit and are 125 MW and 20 MW.

(Answer: $P_1 = 57$ MW, $P_2 = 93$ MW; saving £1.14 per hour)

**4.4** What is the merit order used for when applied to generator scheduling?

A power system is supplied by three generators. The functions relating the cost (in £/h) to active power output (in MW) when operating each of these units are:

$$C_1(P_1) = 0.04P_1^2 + 2P_1 + 250$$

$$C_2(P_2) = 0.02P_2^2 + 3P_2 + 450$$

$$C_3(P_3) = 0.01P_3^2 + 5P_3 + 250$$

The system load is 525 MW. Assuming that all generators operate at the same marginal cost, calculate:

(a) the marginal cost;
(b) optimum output of each generator;
(c) the total hourly cost of this dispatch.

How is the economy of operation of the power system balanced against security requirements for all demands?
(Answer: (a) £10/MWh; (b) $P_1$ 100 MW, $P_2$, 175 MW, $P_3$, 250 MW; (c) £4562.5/h)
*(From E.C. Examination, 1997)*

**4.5** Two power systems A and B are interconnected by a tie-line and have power-frequency constants $K_A$ and $K_B$ MW/Hz. An increase in load of 500 MW on system A

causes a power transfer of 300 MW from B to A. When the tie-line is open the frequency of system A is 49 Hz and of system B 50 Hz. Determine the values of $K_A$ and $K_B$, deriving any formulae used.
(Answer: $K_A$ 500 MW/Hz; $K_B$ 750 MW/Hz)

**4.6** Two power systems, A and B, having capacities of 3000 and 2000 MW, respectively, are interconnected through a tie-line and both operate with frequency-bias–tie-line control. The frequency bias for each area is 1 per cent of the system capacity per 0.1 Hz frequency deviation. If the tie-line interchange for A is set at 100MW and for B is set (incorrectly) at 200 MW, calculate the steady-state change in frequency.
(Answer: 0.6 Hz; use $\Delta P_A + \sigma_A \Delta f = \Delta P_B + \sigma_B \Delta f$)

**4.7** (a) (i)   Why do power systems operate in an interconnected arrangement?
      (ii)   How is the frequency controlled in a power system?
      (iii) What is meant by the *stiffness* of a power system?

   (b) Two 50 Hz power systems are interconnected by a tie-line, which carries 1000 MW from system A to system B, as shown in Figure 4.20. After the outage of the line shown in the figure, the frequency in system A increases to 50.5 Hz, while the frequency in system B decreases to 49 Hz.

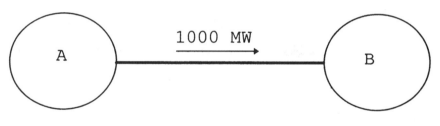

**Figure 4.20**   Interconnected systems of Problem 4.7 (b)

   (i)   Calculate the stiffness of each system.
   (ii) If the systems operate interconnected with 1000 MW being transferred from A to B, calculate the flow in the line after outage of a 600 MW generator in system B.
(Answer: (b) (i) $K_A$ 2000 MW/Hz, $K_B$ 1000 MW/Hz; (ii) 1400 MW)
  (*From, E.C. Examination, 1997*)

# 5

# Control of Voltage and Reactive Power

## 5.1  Introduction

The approximate relationship between the scalar voltage difference between two nodes in a network and the flow of reactive power was shown in Chapter 2 to be

$$\Delta V = \frac{RP + XQ}{V} \qquad (2.9)$$

Also it was shown that the transmission angle $\delta$ is proportional to

$$\frac{XP - RQ}{V} \qquad (2.10)$$

For networks where $X \gg R$, i.e. most power circuits, $\Delta V$, the voltage difference, determines $Q$. Consider the simple interconnector linking two generating stations A and B, as shown in Figure 5.1(a). The machine at A is in phase advance of that at B and $V_1$ is greater than $V_2$; hence there is a flow of power and reactive power from A to B. This can be seen from the phasor diagram shown in Figure 5.1(b). It is seen that $I_d$ and hence $P$ is determined by $\angle\delta$ and the value of $I_q$, and hence $Q$ mainly, by $V_1 - V_2$. In this case $V_1 > V_2$ and reactive power is transferred from A to B; by varying the generator excitations such that $V_2 > V_1$, the direction of the reactive power is reversed, as shown in Figure 5.1(c). Hence, power can be sent from A to B or B to A by suitably adjusting the amount of steam (or water) admitted to the turbine, and reactive power can be sent in either direction by adjusting the voltage magnitudes. These two operations are approximately independent of each other if $X \gg R$, and the flow of reactive power can be studied almost independently of the power flow. The phasor diagrams show that if a scalar voltage difference exists across a largely reactive link, the reactive power flows towards the node of lower voltage. From another point of view, if, in a network, there is

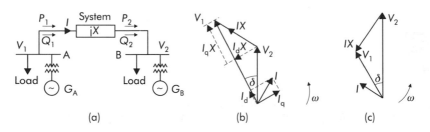

(a)                    (b)                    (c)

**Figure 5.1**   (a) System of two generators interconnected. (b) Phasor diagram when $V_1 > V_2$. $I_d$ and $I_q$ are components of $I$. (c) Phasor diagram when $V_2 > V_1$

a deficiency of reactive power at a point, this has to be supplied from the connecting lines and hence the voltage at that point falls. Conversely if there is a surplus of reactive power generated (e.g. lightly loaded cables generate positive vars), then the voltage will rise. This is a convenient way of expressing the effect of the power factor of the transferred current, and although it may seem unfamiliar initially, the ability to think in terms of var flows, instead of exclusively with power factors and phasor diagrams, will make the study of power networks much easier.

   If it can be arranged that $Q_2$ in the system in Figure 5.1(a) is zero, then there will be no voltage drop between A and B, a very satisfactory state of affairs. Assuming that $V_1$ is constant, consider the effect of keeping $V_2$, and hence the voltage drop $\Delta V$, constant. From equation (2.9)

$$Q_2 = \frac{V_2 \cdot \Delta V - R \cdot P_2}{X} = K - \frac{R}{X}P_2 \qquad (5.1)$$

where $K$ is a constant and $R$ is the resistance of the system.

   If this value of $Q_2$ does not exist naturally in the circuit then it will have to be obtained by aritificial means, such as the connection at B of capacitors or inductors. If the value of the power changes from $P_2$ to $P_2'$ and if $V_2$ remains constant, then the reactive power at B must change to $Q_2'$ such that

$$Q_2' - Q_2 = \frac{R}{X}(P_2' - P_2)$$

i.e. an increase in power causes an increase in reactive power. The change, however, is proportional to $(R/X)$, which is normally small. It is seen that voltage can be controlled by the injection into the network of reactive power of the correct sign. Other methods of a more obvious kind for controlling voltage are the use of tap-changing transformers and voltage boosters.

## 5.2 The Generation and Absorption of Reactive Power

In view of the findings in the previous section, a review of the characteristics of a power system from the viewpoint of reactive power is now appropriate.

### 5.2.1 Synchronous generators

These can be used to generate or absorb reactive power. The limits on the capability for this can be seen in Figure 3.14. The ability to supply reactive power is determined by the short-circuit ratio (1/synchronous reactance) as the distance between the power axis and the theoretical stability-limit line in Figure 3.14 is proportional to the short-circuit ratio. In modern machines the value of this ratio is made low for economic reasons, and hence the inherent ability to operate at leading power factors is not large. For example, a 200 MW 0.85 p.f. machine with a 10 per cent stability allowance has a capability of 45 MVAr at full power output. The var capacity can, however, be increased by the use of continuously acting voltage regulators, as explained in Chapter 3. An over-excited machine, i.e. one with greater than normal excitation, generates reactive power whilst an underexcited machine absorbs it. The generator is the main source of supply to the system of both positive and negative vars.

### 5.2.2 Overhead lines and transformers

When fully loaded, lines absorb reactive power. With a current $I$ amperes for a line of reactance per phase $X(\Omega)$ the vars absorbed are $I^2 X$ per phase. On light loads the shunt capacitances of longer lines may become predominant and the lines then become var generators.

Transformers always absorb reactive power. A useful expression for the quantity may be obtained for a transformer of reactance $X_T$ p.u. and a full-load rating of $3V \cdot I_{rated}$.

The ohmic reactance

$$= \frac{V \cdot X_T}{I_{rated}}$$

Therefore the vars absorbed

$$= 3 \cdot I^2 \cdot \frac{V \cdot X_T}{I_{rated}}$$

$$= 3 \cdot \frac{I^2 V^2}{(IV)_{rated}} \cdot X_T = \frac{(VA \text{ of load})^2}{\text{Rated } VA} \cdot X_T$$

### 5.2.3  Cables

Cables are generators of reactive power owing to their high shunt capacitance. A 275 kV, 240 MVA cable produces 6.25–7.5 MVAr per km; a 132 kV cable roughly 1.9 MVAr per km; and a 33kV cable, 0.125 MVAr per km.

### 5.2.4  Loads

A load at 0.95 power factor implies a reactive power demand of 0.33 kVAr per kW of power, which is more appreciable than the mere quoting of the power factor would suggest. In planning a network it is desirable to assess the reactive power requirements to ascertain whether the generators are able to operate at the required power factors for the extremes of load to be expected. An example of this is shown in Figure 5.2, where the reactive losses are added for each item until the generator power factor is obtained.

### *Example 5.1*

In the radial transmission system shown in Figure 5.2, all p.u. values are referred to the voltage bases shown and 100 MVA. Determine the power factor at which the generator must operate.

### *Solution*

Voltage drops in the circuits will be neglected and the nominal voltages assumed.

Busbar A,

$$P = 0.5\,\text{p.u.} \qquad Q = 0$$

$I^2 X$ loss in 132 kV lines and transformers

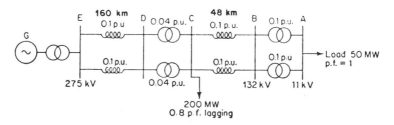

**Figure 5.2**  Radial transmission system with intermediate loads. Calculation of reactive-power requirement

$$= \frac{P^2 + Q^2}{V^2} X_{CA} = \frac{0.5^2}{1^2} \cdot 0.1$$

$$= 0.025 \, \text{p.u.}$$

Busbar C,

$$P = 2 + 0.5 = 2.5 \, \text{p.u.}$$
$$Q = 1.5 + 0.025 \, \text{p.u.}$$
$$= 1.525 \, \text{p.u.}$$

$I^2 X$ loss in 275 kV lines and transformers

$$= \frac{2.5^2 + 1.525^2}{1^2} 0.07$$

$$= 0.6 \, \text{p.u.}$$

The $I^2 X$ loss in the large generator-transformer will be negligible, so that the generator must deliver $P = 2.5$ and $Q = 2.125$ p.u. and operate at a power factor of 0.76 lagging. It is seen in this example that, starting with the consumer load, the vars for each circuit, in turn, are added to obtain the total.

## 5.3 Relation Between Voltage, Power, and Reactive Power at a Node

The phase voltage $V$ at a node is a function of $P$ and $Q$ at that node, i.e.

$$V = \phi(P, Q)$$

The voltage is also dependent on adjacent nodes and the present treatment assumes that these are infinite buses.

The total differential of $V$,

$$dV = \frac{\partial V}{\partial P} \cdot dP + \frac{\partial V}{\partial Q} \cdot dQ$$

and using

$$\frac{\partial P}{\partial V} \cdot \frac{\partial V}{\partial P} = 1 \quad \text{and} \quad \frac{\partial Q}{\partial V} \cdot \frac{\partial V}{\partial Q} = 1$$

$$dV = \frac{dP}{(\partial P / \partial V)} + \frac{dQ}{(\partial Q / \partial V)} \tag{5.2}$$

It can be seen from equation (5.2) that the change in voltage at a node is defined by the two quantities

$$\left( \frac{\partial P}{\partial V} \right) \quad \text{and} \quad \left( \frac{\partial Q}{\partial V} \right).$$

As an example, consider a line with series impedance $(R + jX)$ $\Omega$ and zero shunt admittance. From equation (2.9),

$$(V_1 - V)V - PR - XQ = 0 \qquad (5.3)$$

where $V_1$, the sending-end voltage, is constant, and $V$, the receiving-end voltage, depends on $P$ and $Q$ (Figure 5.3).

From equation (5.3)

$$\frac{\partial P}{\partial V} = \frac{V_1 - 2V}{R} \qquad (5.4)$$

Also,

$$\frac{\partial Q}{\partial V} = \frac{V_1 - 2V}{X} \qquad (5.5)$$

Hence,

$$\begin{aligned} dV &= \frac{dP}{\partial P/\partial V} + \frac{dQ}{\partial Q/\partial V} \\ &= \frac{dP \cdot R + dQ \cdot X}{V_2 - 2V} \end{aligned} \qquad (5.6)$$

For constant $V$ and $\Delta V$, $R\,dP + X\,dQ = 0$ and $dQ = -(R/X)dP$, which is obtainable directly from equation (5.1).

Normally, $\partial Q/\partial V$ is the quantity of greater interest. It can be found experimentally using a load-flow calculation (see Chapter 6) by the injection of a known quantity of vars at the node in question and calculating the difference in voltage produced. From the results obtained,

$$\frac{\Delta Q}{\Delta V} = \frac{Q_{\text{after}} - Q_{\text{before}}}{V_{\text{after}} - V_{\text{before}}}$$

$\Delta V$ should be small for this test, a few per cent of the normal voltage, thereby giving the *sensitivity* of the node to the var change.

From the expression,

$$\frac{\partial Q}{\partial V} = \frac{V_1 - 2V}{X}$$

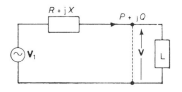

**Figure 5.3** Single-phase equivalent circuit of a line supplying a load of $P + jQ$ from an infinite busbar of voltage $V_1$

proved for the line, it is evident that the smaller the reactance associated with a node, the larger the value of $\partial Q/\partial V$ for a given voltage drop, i.e. the voltage drop is inherently small. The greater the number of lines meeting at a node, the smaller the resultant reactance and the larger the value of $\partial Q/\partial V$. Obviously, $\partial Q/\partial V$ depends on the network configuration, but a high value would lie in the range 10–15 MVAr/kV. If the natural voltage drop at a point without the artificial injection of vars is, say, 5 kV, and the value of $\partial Q/\partial V$ at this point is 10 MVAr/kV, then to maintain the voltage at its no-load level would require $5 \times 10$ or 50 MVAr. Obviously, the greater the value of $\partial Q/\partial V$, the more expensive it becomes to maintain voltage levels by injection of reactive power.

### 5.3.1  $\partial Q/\partial V$ *and the short-circuit current at a node*

It has been shown that for a connector of reactance $X(\Omega)$ with a sending-end voltage $V_1$ and a received voltage $V$ per phase

$$\frac{\partial Q}{\partial V} = \frac{V_1 - 2V}{X} \tag{5.5}$$

If the three-phases of the connector are now short-circuited at the receiving end (i.e. three-phase symmetrical short circuit applied), the current flowing in the lines

$$= \frac{V_1}{X} \text{amperes}, \qquad \text{assuming } R \ll X$$

With the system on no-load

$$V = V_1 \qquad \text{and} \qquad \frac{\partial Q}{\partial V} = \frac{V_1}{X}$$

Hence the magnitude of $(\partial Q/\partial V)$ is equal to the short-circuit current; the sign decides the nature of the reactive power. With normal operation, $V$ is within a few per cent of $V_1$ and hence the value of $\partial V/\partial Q$ at $V = V_1$ gives useful information regarding reactive power/voltage characteristics for small excursions from the nominal voltage. This relationship is especially useful as the short-circuit current will normally be known at all substations.

### *Example 5.2*

Three supply points A, B, and C are connected to a common busbar M. Supply point A is maintained at a nominal 275 kV and is connected to M through a 275/132 kV transformer (0.1 p.u. reactance) and a 132 kV line of reactance 50 Ω. Supply point B is nominally at 132 kV and is connected to M through a 132 kV line of 50 Ω reactance. Supply point C is nominally at 275 kV and is connected to M by a 275/132 kV transformer (0.1 p.u. reactance) and a 132 kV line of 50 Ω reactance.

If, at a particular system load, the line voltage of M falls below its nominal value by 5 kV, calculate the magnitude of the reactive volt-ampere injection required at M to re-establish the original voltage.

The p.u. values are expressed on a 500 MVA base and resistance may be neglected throughout.

## Solution

The line diagram and equivalent single-phase circuit are shown in Figures 5.4 and 5.5.

It is necessary to determine the value of $\partial Q/\partial V$ at the node or busbar M; hence the current flowing into a three-phase short-circuit at M is required.

The base value of reactance in the 132 kV circuit is

$$\frac{132^2 \times 1000}{500\,000} = 35\,\Omega$$

Therefore the line reactances

$$= \frac{j50}{35} = j1.43\,\text{p.u.}$$

The equivalent reactance from M to N $= j0.5\,\text{p.u.}$

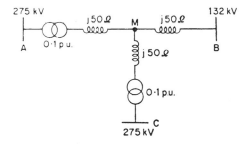

Figure 5.4   Schematic diagram of the system for Example 5.2

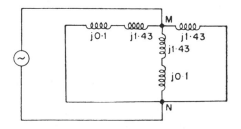

**Figure 5.5**   Equivalent single-phase network with the node M short-circuited to neutral (refer to Chapter 7 for full explanation of the derivation of this circuit)

Hence the fault MVA at M

$$= \frac{500}{0.5} = 1000 \, \text{MVA}$$

and the fault current

$$= \frac{1000 \times 10^6}{\sqrt{(3)} \times 132\,000}$$

$$= 4380 \, \text{A} \qquad \text{at zero power factor lagging}$$

It has been shown that $\partial Q_M / \sqrt{(3)} \partial V_M =$ three-phase short-circuit current when $Q_M$ and $V_M$ are three-phase and line values

$$\therefore \frac{\partial Q_M}{\partial V_M} = 7.6 \, \text{MVAr/kV}$$

The natural voltage drop at M $= 5 \, \text{kV}$. Therefore the value of the injected vars required, $\Delta Q_M$, to offset this drop

$$= 7.6 \times 5 = 38 \, \text{MVAr}$$

## 5.4 Methods of Voltage Control: (i) Injection of Reactive Power

The background to this method has been given in the previous sections. This is the most fundamental method, but in transmission systems it lacks the flexibility and economy of transformer tap-changing. Hence it is only used in schemes when transformers alone will not suffice. The provision of static capacitors to improve the power factors of factory loads has been long established. The capacitance required for the power-factor improvement of loads for optimum economy is determined as follows.

Let the tariff of a consumer be

$$£A \times \text{kVA} + B \times \text{kWh}$$

A load of $P_1$ kilowatts at power factor $\phi_1$ lagging has a kVA of $P_1 / \cos \phi_1$. If this power factor is improved to $\cos \phi_2$, the new kVA is $P_1 / \cos \phi_2$. The saving is therefore

$$£P_1 A \left( \frac{1}{\cos \phi_1} - \frac{1}{\cos \phi_2} \right)$$

The reactive power required from the correcting capacitors

$$= (P_1 \tan \phi_1 - P_1 \tan \phi_2) \, \text{kVAr}$$

Let the cost per annum in interest and depreciation on the capacitor installation be $£C$ per kVAr or

$$£C(P_1 \tan \phi_1 - P_1 \tan \phi_2)$$

The net saving

$$= £\left[AP_1\left(\frac{1}{\cos \phi_1} - \frac{1}{\cos \phi_2}\right) - CP_1(\tan \phi_1 - \tan \phi_2)\right]$$

This saving is a maximum when

$$\frac{d(\text{saving})}{d\phi_2} = 0$$

i.e. when $\sin \phi_2 = C/A$.

It is interesting to note that the optimum power factor is independent of the original one. The improvement of load power factors in such a manner will help to alleviate the whole problem of var flow in the transmission system.

The main effect of transmitting power at non-unity power factors is as follows. It is evident from equation (2.9) that the voltage drop is largely determined by the reactive power ($Q$). The line currents are larger, giving increased $I^2R$ losses and hence reduced thermal capability. One of the obvious places for the artificial injection of reactive power is at the loads themselves. In general, four methods of injection are available, involving the use of:

1. static shunt capacitors;

2. static series capacitors;

3. synchronous compensators;

4. current compensation by series injection.

### 5.4.1 Shunt capacitors and reactors

Shunt capacitors are used for lagging power-factor circuits, whereas reactors are used on those with leading power factors such as created by lightly loaded cables. In both cases the effect is to supply the requisite reactive power to maintain the values of the voltage. Capacitors are connected either directly to a busbar or to the tertiary winding of a main transformer and are disposed along the route to minimize the losses and voltage drops. Unfortunately, as the voltage falls, the vars produced by a shunt capacitor or reactor fall; thus, when needed most, their effectiveness falls. Also, on light loads when the voltage is high, the capacitor output is large and the voltage tends to rise to excessive levels, requiring some capacitors or cable circuits to be switched out by local overvoltage relays. Switched capacitors in parallel with semiconductor-controlled reactors provides modern variable var compensators for fast control of voltages in power systems.

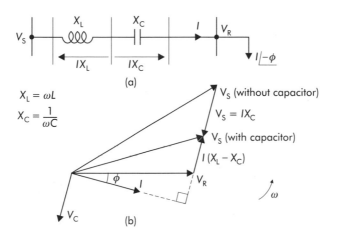

**Figure 5.6** (a) Line with series capacitor, $C$ load. (b) Phasor diagram for fixed $V_R$

### 5.4.2 Series capacitors

These are connected in series with the line conductors and are used to reduce the inductive reactance between the supply point and the load. One major drawback is the high overvoltage produced when a short-circuit current flows through the capacitor, and special protective devices need to be incorporated (e.g. spark gaps) and non-linear resistors. The phasor diagram for a line with a series capacitor is shown in Figure 5.6(b).

The relative merits between shunt and series capacitors may be summarized as follows:

1. If the load var requirement is small, series capacitors are of little use.

2. With series capacitors the reduction on line current is small; hence if thermal considerations limit the current, little advantage is obtained and shunt compensation should be used

3. If voltage drop is the limiting factor, series capacitors are effective; also, voltage fluctuations due to arc furnaces, etc., are evened out.

4. If the total line reactance is high, series capacitors are very effective and stability is improved.

### 5.4.3 Synchronous compensators

A synchronous compensator is a synchronous motor running without a mechanical load and, depending on the value of excitation, it can absorb or

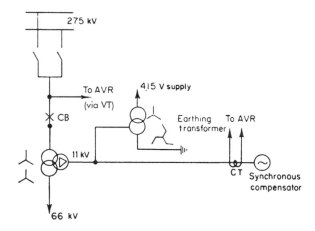

**Figure 5.7**  Typical installation with synchronous compensator connected to tertiary (delta) winding of main transformer. A neutral point is provided by the earthing transformer shown. The automatic voltage regulator on the compensator is controlled by a combination of the voltage on the 275 kV system and the current output; this gives a droop to the voltage-var output curve which may be varied as required

generate reactive power. As the losses are considerable compared with static capacitors, the power factor is not zero. When used with a voltage regulator the compensator can automatically run overexcited at times of high load and underexcited at light load. A typical connection of a synchronous compensator is shown in Figure 5.7 and the associated voltage–var output characteristic in Figure 5.8. The compensator is run up as an induction motor in 2.5 min and then sychronized.

A great advantage is the flexibility of operation for all load conditions. Although the cost of such installations is high, in some circumstances it is justified, e.g. at the receiving-end busbar of a long high-voltage line where transmission at power factors less than unity cannot be tolerated. Being a rotating machine, its stored energy is useful for riding through transient disturbances, including voltage sags.

### 5.4.4  Series injection

With the development of high-power, high-voltage semiconductor-controlled devices, including pulse turn-on and turn-off (e.g. IGBT, as discussed in Chapter 9), inverters are now being designed and constructed that can inject a voltage in series with a line whose angle can have any desired relation with the phase voltage. This is equivalent to the series capacitor case of Figure 5.6, except that $V_C$ is not confined to being only 90° out of phase with the current

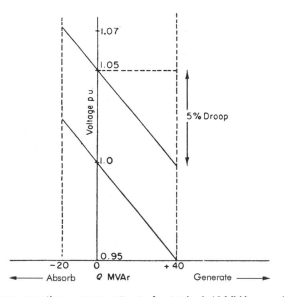

**Figure 5.8** Voltage-reactive power output of a typical 40 MVAr synchronous compensator

and dependent upon the $IX_C$ voltage rise. Such a regulator, known as a Universal Power Controller (UPC) is shown, in principle, in Figure 5.9.

It can be realised that if $V_i$ is at $90°$ to the current $I$, then no energy is required from the source. At any other angle, energy is either drawn from the system or required from the source. Most conveniently, the source is a transformer connected to the system busbars feeding a rectifier, from which a sinusoidal injected voltage at the desired magnitude and angle is synthesized. Alternatively, the source could be a storage device (battery, capacitor, superconducting energy store, etc.), in which case some auxiliary charging may be necessary, but peak lopping of even uninterruptible power may also be provided.

## 5.5 Methods of Voltage Control: (ii) Tap-Changing Transformers

The basic operation of the tap-changing transformer has been discussed in Chapter 3; by changing the transformation ratio, the voltage in the secondary circuit is varied and voltage control is obtained. This constitutes the most popular and widespread form of voltage control at all voltage levels.

Consider the operation of a radial transmission system with two tap-changing transformers, as shown in the equivalent single-phase circuit of

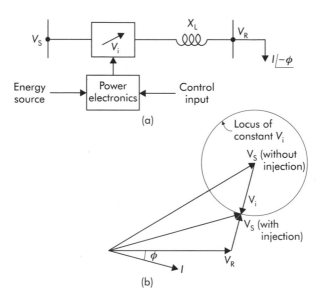

**Figure 5.9** Principle of Universal Power Controller for series injection. (a) System diagram. (b) Phasor diagram

Figure 5.10. Here, $t_s$ and $t_r$ are fractions of the nominal transformation ratios, i.e. the tap ratio/nominal ratio. For example, a transformer of nominal ratio 6.6 to 33 kV when tapped to give 6.6 to 36 kV has a $t_s - 36/33 = 1.09$. $V_1$ and $V_2$ are the nominal voltages; at the ends of the line the actual voltages are $t_s V_1$ and $t_r V_2$. It is required to determine the tap-changing ratios required to completely compensate for the voltage drop in the line. The product $t_s t_r$ will be made unity; this ensures that the overall voltage level remains in the same order and that the minimum range of taps on both transformers is used.

(Note that all values are in per unit; $t$ is the off-nominal tap ratio.)

Transfer all quantitites to the load circuit. The line impedance becomes $(R + jX)/t_r^2$; $V_s = V_1 t_s$ and, as the impedance has been transferred, $V_r = V_1 t_s$. The input voltage to the load circuit becomes $V_1 t_s/t_r$ and the equivalent circuit is as shown in Figure 5.10(c). The arithmetic voltage drop

$$= (V_1 t_s/t_r) - V_2 \approx \frac{RP + XQ}{t_r^2 V_2}$$

When $t_r = 1/t_s$,

$$t_s^2 V_1 V_2 - V_2^2 = (RP + XQ)t_s^2$$

and

$$V_2 = \tfrac{1}{2}[t_s^2 V_1 \pm t_s(t_s^2 V_1^2 - 4(RP + XQ))^{\frac{1}{2}} \tag{5.7}$$

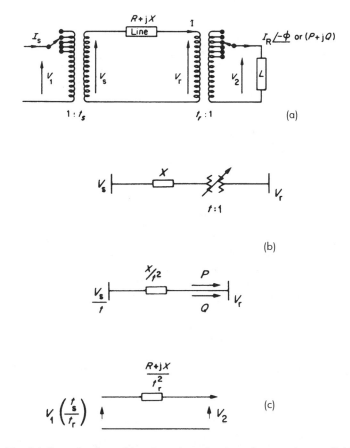

**Figure 5.10** (a) Coordination of two tap-changing transformers in a radial transmission link (b) and (c) Equivalent circuits for dealing with off-nominal tap ratio. (b) Single transformer. (c) Two transformers

Hence if $t_s$ is specified, there are two values of $V_2$ for a given $V_1$.

## *Example 5.3*

A 13 kV line is fed through an 11/132 kV transformer from a constant 11 kV supply. At the load end of the line the voltage is reduced by another transformer of nominal ratio 132/11 kV. The total impedance of the line and transformers at 132 kV is $(25 + j66)$ $\Omega$. Both transformers are equipped with tap-changing facilities which are arranged so that the product of the two off-nominal settings is unity. If the load on the system is 100 MW at 0.9 p.f. lagging, calculate the settings of the tap-changers required to maintain the voltage of the load busbar at 11 kV. Use a base of 100 MVA.

*Solution*

The line diagram is shown in Figure 5.11. As the line voltage drop is to be completely compensated, $V_1 = V_2 = 132\,\text{kV} = 1\,\text{p.u.}$ Also, $t_s \times t_r = 1$. The load is $100\,\text{MW}$, $48.3\,\text{MVAr.}$, i.e. $1 + \text{j}0.483\,\text{p.u.}$

Using equation (5.7)

$$1 = \tfrac{1}{2}[(t_s^2 \cdot 1) \pm t_s(t_s^2 - 4(0.14 \times 1 + 0.38 \times 0.48))^{\frac{1}{2}}]$$

$$\therefore 2 = t_s^2 \pm t_s(t_s^2 - 1.28)^{\frac{1}{2}}$$

$$\therefore (2 - t_s^2)^2 = t_s^2(t_s^2 - 1.28)$$

Hence,

$$t_s = 1.21 \qquad \text{and} \qquad t_r = 1/1.21 = 0.83$$

These settings are large for the normal range of tap-changing transformers (usually not more than $\pm 20$ per cent tap range). It would be necessary, in this system, to inject vars at the load end of the line to maintain the voltage at the required value.

It is important to note that the transformer does not improve the var-flow position and also that the current in the supplying line is increased if the ratio is increased. In countries with long and inadequate distribution circuits, it is often the practice to boost the received voltage by a variable ratio transformer so as to maintain rated voltage as the power required increases. Unfortunately, this has the effect of increasing the primary supply circuit current by the transformer ratio, thereby decreasing the primary voltage still further until voltage collapse occurs. This will be considered in more detail later in this chapter.

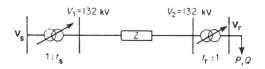

**Figure 5.11**    Schematic diagram of system for Example 5.3

# 5.6 Combined Use of Tap-Changing Transformers and Reactive-Power Injection

The usual practical arrangement is shown in Figure 5.12, where the tertiary winding of a three-winding transformer is connected to a compensator. For given load conditions it is proposed to determine the necessary transformation ratios with certain outputs of the compensator.

The transformer is represented by the equivalent star connection and any line impedance from $V_1$ or $V_2$ to the transformer can be lumped together with the transformer branch impedances. Here, $V_n$ is the phase voltage at the star point of the equivalent circuit in which the secondary impedance ($X_s$) is usually

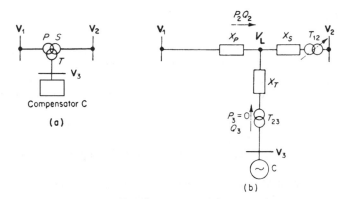

**Figure 5.12** (a) Schematic diagram with combined tap-changing and synchronous compensation. (b) Equivalent network

approaching zero and hence is neglected. Resistance and losses are neglected. The allowable ranges of voltage for $V_1$ and $V_2$ are specified and the values of $P_2, Q_2, P_3,$ and $Q_3$ are given; $P_3$ is usually taken as zero.

The volt drop $V_1$ to $V_L$

$$= \Delta V_p \approx X_p \frac{Q_2/3}{V_n}$$

or

$$\Delta V_p = X_p \frac{Q_2}{V_L \sqrt{3}},$$

where $V_L$ is the line voltage $= \sqrt{(3)} V_n$, and $Q_2$ is the total vars. Also,

$$\Delta V_q = X_p \frac{P_2}{V_L \sqrt{3}}$$

$$\therefore (V_n + \Delta V_p)^2 + (\Delta V_q)^2 = V_1^2$$

(see phasor diagram of Figure 2.22; phase values used) and

$$\left( V_n + X_p \frac{Q_2}{V_L \sqrt{3}} \right)^2 + X_p^2 \left( \frac{P_2^2}{3 V_L^2} \right) = V_1^2$$

$$\therefore (V_L^2 + X_p Q_2)^2 + X_p^2 P_2^2 = V_{1L}^2 V_L^2$$

where $V_{1L}$ is the line voltage $= \sqrt{(3)} V_1$

$$\therefore V_L^2 = \frac{V_{1L}^2 - 2X_p Q_2}{2} \pm \frac{1}{2} \sqrt{[V_{1L}^2 (V_{1L}^2 - 4X_p Q_2) - 4 \cdot X_p^2 P_2^2]}$$

Once $V_L$ is obtained, the transformation ratio is easily found. The procedure is best illustrated by an example.

## *Example 5.4*

A three-winding grid transformer has windings rated as follows: 132 kV (line), 75 MVA, star connected; 33 kV (line), 60 MVA, star connected; 11 kV (line), 45 MVA, delta connected. A compensator is available for connection to the 11 kV winding.

The equivalent circuit of the transformer may be expressed in the form of three windings, star connected, with an equivalent 132 kV primary reactance of 0.12 p.u., negligible secondary reactance, and an 11 kV tertiary reactance of 0.08 p.u. (both values expressed on a 75 MVA base).

In operation, the transformer must deal with the following extremes of loading:

1. Load of 60 MW, 30 MVAr with primary and secondary voltages governed by the limits 120 kV and 34 kV; compensator disconnected.

2. No load, Primary and secondary voltage limits 143 kV and 30 kV; compensator in operation and absorbing 20 MVAr.

Calculate the range of tap-changing required. Ignore all losses.

## *Solution*

The value of $X_p$, the primary reactance (in ohms)

$$= 0.12 \times 132^2 \times 1000/75 \times 1000 = 27.8 \ \Omega$$

Similarly, the effective reactance o the tertiary winding is $18.5 \ \Omega$. The equivalent star circuit is shown in Figure 5.13.

The first operating conditions are as follows:

$$P_1 = 60 \, \text{MW} \qquad Q_1 = 30 \, \text{MVAr} \qquad V_{1L} = 120 \, \text{kV}$$

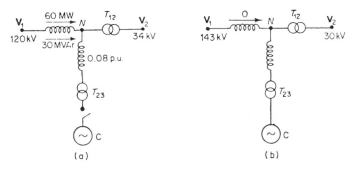

(a)                                    (b)

**Figure 5.13** Systems for Example 5.4. (a) System with loading condition 1. (b) System with loading condition 2

Hence,

$$V_L^2 = \tfrac{1}{2}(120\,000^2 - 2 \times 27.8 \times 30 \times 10^6)$$
$$\pm \tfrac{1}{2}\sqrt{[120\,000^2(120\,000^2 - 4 \times 27.8 \times 30 \times 10^6) - 4 \times 27.8^2 \times 60^2 \times 10^{12}]}$$
$$= \left(63.61 \pm \frac{122}{2}\right)10^8 = 124.4 \times 10^8$$

$$\therefore V_L = 111\,\text{kV}$$

The second set of conditions are:

$$V_{1L} = 143\,\text{kV} \qquad P_2 = 0 \qquad Q_2 = 20\,\text{MVAr}$$

Again, using the formula for $V_L$,

$$V_L = 138.5\,\text{kV}$$

The transformation ratio under the first condition

$$= 111/34 = 3.27$$

and, for the second condition

$$\frac{138.5}{30} = 4.61$$

The actual raio will be taken as the mean of these extremes, i.e. 3.94, varying by $\pm 0.67$ or $3.94 \pm 17$ per cent. Hence the range of tap-changing required is $\pm 17$ per cent.

A further method of var production is the use of adjustment of tap settings on transformers connecting large interconnected systems. Consider the situation in Figure 5.14(a), in which $V_s$ and $V_r$ are constant voltages representing the two connected systems. The circuit may be rearranged as shown in Figure 5.14(b), where $t$ is the off-nominal (per unit) tap setting; resistance is zero. The voltage drop between busbars

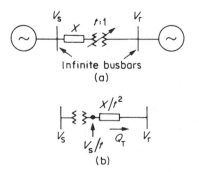

**Figure 5.14** (a) Two power systems connected via a tap-change transformer. (b) Equivalent circuit with impedance transferred to receiver side

$$= \left(\frac{V_{s}}{t}\right) - V_{r} = \frac{X}{t^{2}} \cdot \frac{Q_{T}}{V_{r}}$$

Hence,

$$(V_{s}V_{r}t - V_{r}^{2}t^{2})1/X = Q_{T}$$

and

$$t(1-t)V^{2}/X = Q_{T} \qquad \text{when} \qquad V_{s} = V_{r} = V$$

Also, $Q_{T} = t(1-t)S$, where $S =$ short-circuit level, i.e. $V^{2}/X$. When

$t < 1$,      $Q_{T}$ is positive, i.e. a flow of vars into $V_{r}$

$t > 1$,      $Q_{T}$ is negative, a flow of vars out of $V_{r}$

Thus, by suitable adjustment of the tap setting, an appropriate injection of reactive power is obtained.

The idea can be extended to two transformers in parallel between networks. If one transformer is set to an off-nominal ratio of, say, 1:1.1 and the other to 1:0.8 (i.e. in opposite directions), then a circulation of reactive power occurs round the loop, resulting in a net absorption of vars. This is known as 'tap-stagger' and is a comparatively inexpensive method of var absorption.

## 5.7 Booster Transformers

It may be desirable, on technical or economic grounds, to increase the voltage at an intermediate point in a line rather than at the ends as with tap-changing transformers, or the system may not warrant the expense of tap-changing. Here, booster transformers are used as shown in Figure 5.15. The booster can be brought into the circuit by the closure of relay B and the opening of A, and vice versa. The mechanism by which the relays are operated can be controlled from a change in either the voltage or the current. The latter method is the more sensitive, as from no-load to full-load represents a 100 per cent change in current, but only in the order of a 10 per cent change in voltage. This booster gives an in-phase boost, as does a tap-changing transformer. An economic advantage is that the rating of the booster is the product of the current

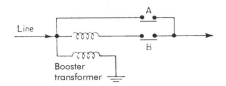

**Figure 5.15**  Connection of in-phase booster transformer. One phase only shown

and the injected voltage, and is hence only about 10 per cent of that of a main transformer. Boosters are often used in distribution feeders where the cost of tap-changing transformers is not warranted.

## Example 5.5

In the system shown by the line diagram in Figure 5.16, each of transformers $T_A$ and $T_B$ have tap ranges of $\pm$ 10 per cent in 10 steps of 1.0 per cent. It is required to find the voltage boost needed on transformer $T_A$ to share the power flow equally between the two lines.

The system data is as follows:

$$\text{All transformers: } X_T = 0.1 \text{ p.u.}$$

$$\left.\begin{array}{l} \text{Transmission lines: } R = R' = 0 \\ \qquad\qquad\qquad X = 0.20 \text{ p.u.} \\ \qquad\qquad\qquad X' = 0.15 \text{ p.u.} \end{array}\right\} \text{ All to same base}$$

$$V_A = 1.1\underline{/5^\circ} \qquad V_B = 1.0\underline{/0^\circ}$$

## Solution

We must first calculate the current sharing in the two parallel lines:

$$I_1 = \frac{1.1\underline{/5^\circ} - 1.0\underline{/0}}{\text{j}0.40} = 0.2397 - \text{j}0.2397$$

$$I_2 = \frac{1.1\underline{/5^\circ} - 1.0\underline{/0}}{\text{j}0.35} = 0.2740 - \text{j}0.2740$$

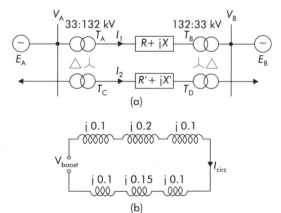

(a)

(b)

**Figure 5.16**   (a) Line diagram of system for Example 5.5. (b) Equivalent network with voltage boost $V_{\text{boost}}$ acting

Any boost by transformer $T_A$ will cause a current to circulate between the two busbars because the voltages $V_A$ and $V_B$ are assumed to be held constant by the voltage regulators on the generators.

To equalize the currents, a circulating current

$$I_{circ} = \frac{I_2 - I_1}{2}$$

is required, as in Figure 5.16(b), giving

$$I_{circ} = \frac{0.0343 - j0.0343}{2} = 0.0241 \underline{/-45°}$$

$$\therefore V_{Boost} = 0.0241\underline{/-45} \times j0.75$$

$$= 0.0180\underline{/45°}\ V$$

To achieve this boost, ideally $T_A$ should be equipped with a phase changer of 45° and taps to give 1.8 per cent boost. In practice, a tap of 2 per cent would be used in either an in-phase boost (such as obtainable from a normal tapped transformer) or a quadrature boost (obtainable from a phase-shift transformer—see next section). In transmission networks it should be noted that because of the generally high $X/R$ ratio, an in-phase boost gives rise to a quadrative current whereas a quadrature boost produces an in-phase circulating current, thereby adding to or subtracting from the real power flow. This technique has also been used to de-ice lines in winter by producing extra $I^2R$ losses for heating. Alternatively, two transformers in parallel can be tap-staggered to produce $I^2X$ absorption under light-load, high-voltage conditions.

### 5.7.1   Phase-shift transformer

A quadrature phase shift can be achieved by the connections shown in Figure 5.17(a). The booster arrangement shows the injection of voltage into one phase only; it is repeated for the other two phases. In Figure 5.17(b), the corresponding phasor diagram is shown and the nature of the angular shift of the voltage boost $V'_{YB}$ indicated. By the use of tappings on the energizing transformer, several values of phase shift may be obtained.

### *Example 5.6*

In the system shown in Figure 5.18, it is required to keep the nominally 11 kV busbar at constant voltage. The range of taps is not sufficient and it is proposed to use shunt capacitors connected to the tertiary winding. The data are as follows, per unit quantities being referred to a 15 MVA base: line 16 km, 115 mm², OHL (overhead line), 33 kV, $Z_L = (0.0304 = j0.0702)$; $Z_L$ referred to 33 kV side $= (2.2 + j5.22)\,\Omega$.

For the three-winding transformer the measured impedances between the windings and the resulting equivalent star impedances $Z_1, Z_2$ and $Z_3$ are given in Table 5.1. The equivalent circuit referred to 33 kV is shown in Figure 5.18(b).

The voltage across the receiving-end load

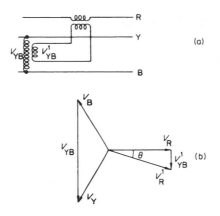

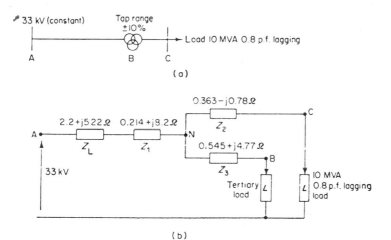

**Figure 5.17** (a) Connections for one phase of a phase-shifting transformer. Similar connections for the other two phases. (b) Corresponding phasor diagram

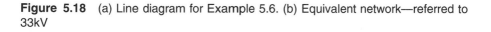

**Figure 5.18** (a) Line diagram for Example 5.6. (b) Equivalent network—referred to 33kV

$$= \frac{33\,000}{\sqrt{3}} - \Delta V_\text{p}$$

where

$$\Delta V_\text{p} \approx \frac{RP + XQ}{V_\text{C}}$$

$$\therefore \Delta V_\text{p} \approx \frac{2.77 \times 8/3 \times 10^6 + 12.64 \times 6/3 \times 10^6}{33\,000/\sqrt{3}}$$

**Table 5.1**  Data for three-winding transformer

| Winding | MVA | Voltage (kV) | p.u. Z referred to nameplate MVA | p.u. Z on 15 MVA base | $(Z(\Omega))$ referred to 33 kV | Equivalent $(Z(\Omega))$ referred to 33 kV |
|---|---|---|---|---|---|---|
| P–S | 15 | 33/11 | 0.008 + j0.1 | 0.008 + j0.1 | 0.57 + j7.3 | $Z_1 =$ 0.214 + j8.2 |
| P–T | 5 | 33/1.5 | 0.0035 + j0.0595 | 0.0105 + j.179 | 0.76 + j4.32 | $Z_2 =$ 0.363 − j0.78 |
| S–T | 5 | 11/1.5 | 0.0042 + j0.0175 | 0.0126 + j0.0525 | 0.915 + j1.27 | $Z_3 =$ 0.545 + j4.77 |

As $V_C$ referred to the 33 kV base is not known because of the system volt drop, 33 kV is assumed initially. The revised value is then used and the process is repeated.

$$\Delta V_p = \frac{7.4 + 25.28}{19} kV = 1.715\,kV \qquad \text{and} \qquad V_C = 17.285\,kV$$

Repeating the calculation for $\Delta V_p$ with the new $V_C$,

$$\Delta' V_p = \frac{2.77 \times 8/3 \times 10^6 + 12.64 \times 6/3 \times 10^6}{17.285} = 1.89\,kV$$

Hence

$$V'_C = 19 - 1.89 = 17.11\,kV$$
$$\Delta' V = 1.9 \qquad \text{and} \qquad V'_C = 17.1\,kV$$

This will be the final value of $V_C$.

$V_C$ referred to 11 kV = 17.1/3 = 5.7 kV (phase) or 9.9 kV (line). In order to maintain 11 kV at C, the voltage is raised by tapping down on the transformer. Using the full range of 10 per cent, i.e. $t_r = 0.9$, the voltage at C is

$$\frac{29.7}{(33 \times 0.9)/11} = 11\,kV$$

The true voltage will be less than this as the primary current will have increased by $(1/0.9)$ because of the change in transformer ratio. It is evident that the tap-changing transformer is not able to maintain 11 kV at C and the use of a static capacitor connected to the tertiary will be investigated.

Consider a shunt capacitor of capacity 5 MVAr (the capacity of the tertiary). Assume the transformer to be at its nominal ratio 33/11 kV. The voltage drop

$$= \frac{2.414 \times 8/3 \times 10^6 + 13.42 \times 1/3 \times 10^6}{V_N(\approx 19\,kV)}$$
$$= 0.587\,kV$$
$$V'_N = 19 - 0.587 = 18.413\,kV \text{ (phase)}$$
$$\therefore \Delta' V_N = 0.606 \qquad \text{and} \qquad V'_N = 18.394\,kV$$

Therefore the volt drop N to C

$$\Delta V_C = \frac{0.363 \times 8/3 - 0.78 \times 6/3}{18.394} \, kV$$

$$= -0.032 \, kV$$

$$\therefore V_C = 18.394 + 0.032$$

$$= 18.426 \, kV \text{ (phase)}$$

As $\Delta V_C$ is so small there is no need to iterate further.

Referred to 11 kV, $V_C = 10.55$ kV (line). Hence, to have 11 kV the transformer will tap such that $t_r = (1 - 0.35/11) = 0.97$, i.e. a 3 per cent tap change, which is well within the range and leaves room for load increases. On *no-load*

$$\Delta V_p = \frac{2.959 \times 0 + 18.19 \times (-5.3)}{19} \, kV$$

$$= \frac{30.3}{19} = -1.595 \, kV \text{ (phase)}$$

The shunt capacitor is a constant impedance load and hence as $V_N$ rises the current taken increases, causing further volt increase.

Ignoring this effect initially,

$$V_C = 19 + 1.6 = 20.6 \, kV \text{ (phase)}$$

On the 11 kV side

$$V_C = 11.85 \, kV \text{ (line)}$$

therefore the tap change will have to be at least 7.15 per cent, which is well within the range. Further refinement in the value of $V_C$ will be unnecessary. If an accurate value of $V_C$ is required, then the reactance of the capacitor must be found and the current evaluated.

## 5.8  Voltage Collapse

Voltage collapse is essentially an aspect of system stability to be discussed in Chapter 8. As the voltages to be maintained in a system are influenced by system stability it is appropriate to discuss the subject here.

Consider the circuit shown in Figure 5.19(a). If $V_s$ is fixed (i.e. an infinite busbar supply), the graph of $V_R$ against $P$ for given power factors is as shown in Figure 5.19(b). In Figure 5.19(b), $Z$ represents the series impedance of a 160 km long, double-circuit, 400 kV, 260 mm$^2$ conductor overhead line. The fact that two values of voltage exist for each value of power is easily demonstrated by considering the analytical solution of this circuit. At the lower voltage a very high current is taken to produce the power. The seasonal thermal ratings of the line are also shown, and it is apparent that for loads of power factor less than unity the possibility exists that, before the thermal rating is reached, the operating power may be on that part of the characteristic where

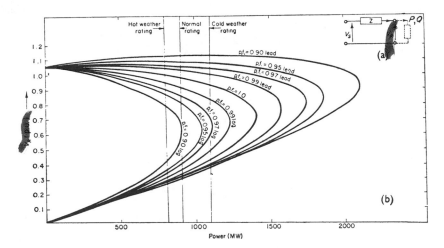

**Figure 5.19** (a) Equivalent circuit of a line supplying a load $P + jQ$. (b) Relation between load voltages and received power at constant power factor for a 400 kV, $2 \times 260$ mm$^2$ conductor line, 160 km in length. Thermal ratings of the line are indicated

small changes in load cause large voltage changes and *voltage instability* will have occurred. In this condition the action of tap-changing transformers is interesting. If the receiving-end transformers 'tap up' to maintain the load voltage, the line current increases, thereby causing further increase in the voltage drop. It would, in fact, be more profitable to 'tap down', thereby reducing the current and voltage drop. It is feasible therefore for a 'tapping-down' operation to result in increased secondary voltage, and vice versa.

The possibility of an actual voltage collapse depends upon the nature of the load. If this is stiff (constant power), e.g. induction motors, the collapse is aggravated. If the load is soft, e.g. heating, the power falls off rapidly with voltage and the situation is alleviated. Referring to Figure 5.19 it is evident that a critical quantity is the power factor; at full load a change in lagging power factor from 0.99 to 0.90 will precipitate voltage collapse. On long lines, therefore, for reasonable power transfers it is necessary to keep the power factor of transmission approaching unity, certainly above 0.97 lagging, and it is economically justifiable to employ var injection by static capacitors, synchronous compensators or UPC close by the load.

A problem arises with the operation of two or more lines in parallel, e.g. the system shown in Figure 5.20, in which the shunt capacitance has been represented as in a $\pi$ section. If one of the three lines is removed from the circuit because of a fault, the total system reactance will increase from $X/3$ to $X/2$, and the capacitance, which normally improves the power factor, decreases to $2C$ from $3C$. Thus the overall voltage drop is greatly increased and, owing to the increased $I^2X$ loss of the lines and the decreased generation of vars by the

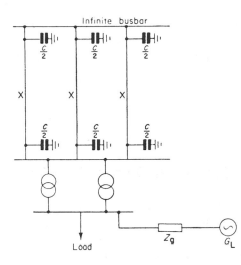

**Figure 5.20** Line diagram of three long lines in parallel—effect of the loss of one line. $G_L$ = local generators

shunt capacitances, the power factor decreases; hence the possibility of voltage instability. The same argument will, of course, apply to two lines in parallel.

Usually, there will be local generation or compensation feeding the receiving-end busbars at the end of long lines. If this generation is electrically close to the load busbars, i.e. low connecting impedance $Z_g$, a fall in voltage will automatically increase the local var generation, and this may be sufficient to keep the reactive power transmitted low enough to avoid large voltage drops in the long lines. Often, however, the local generators supply lower voltage networks and are electrically remote from the high-voltage busbar of Figure 5.20, and $Z_g$ is high. The fall in voltage now causes little change in the local generator var output and the use of static-controlled capacitors at the load may be required. As $Z_g$ is inversely proportional to the three-phase short-circuit level at the load busbar because of the local generation, the reactive-power contribution of the local machines is proportional to this fault level. When a static or synchronous compensator reaches its rated limit, voltage can no longer be controlled and rapid collapse of voltage can follow because any vars demanded by the load must now be supplied from sources further away electrically over the high-voltage system.

In the U.K. and some other countries where 'unbundled' systems are now extant, many generators are some distance from the load centres. Consequently, the transmission system operator is required to install local flexible var controllers or compensators to maintain a satisfactory voltage at the delivery substations supplying the local distribution systems. Such flexible controllers, based on semiconductor devices which can vary the var absorption

~~in a reactor or generation in a capacitor, are called FACTS (Flexible a.c.~~ ~~Transmission System).~~ These are discussed further in Chapter 9.

Typical values of compensation required for a 400 kV or 500 kV network are:

$$\text{Peak load} = 0.3 \, \text{kVAr/kW absorb}$$
$$\text{Light load} = 0.25 \, \text{kVAr/kW generate}$$

## 5.9 Voltage Control in Distribution Networks

Single-phase supplies to houses and other small consumers are tapped off from three-phase feeders. Although efforts are made to allocate equal loads to each phase the loads are not applied at the same time and some unbalance occurs.

In the distribution network (British practice) shown in Figure 5.21 an 11 kV distributor supplies a number of lateral feeders in which the voltage is 420 V, and then each phase, loaded separately.

The object of design is to keep the consumers' nominal 415 V supply within the statutory ±6 per cent of the declared voltage. The main 33/11 kV transformer gives a 5 per cent rise in voltage between no-load and full-load. The distribution transformers have taps that are only adjustable off-load, and a

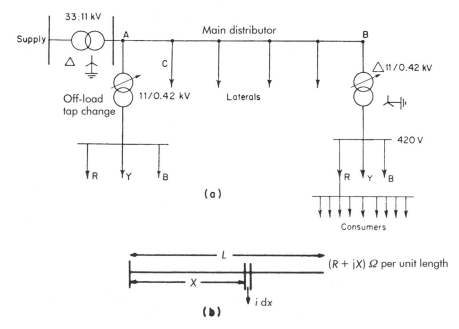

**Figure 5.21** Line diagrams of typical radial distribution schemes

secondary phase voltage of 250 V which is 4 per cent high on the nominal value of 240 V. A typical distribution of voltage drops would be as follows: main distributor, 6 per cent; 11/0.42 kV transformer, 3 per cent; 420 V circuit, 7 per cent; consumer circuit, 1.5 per cent; giving a total of 17.5 per cent. On very light load (10 per cent of full load) the corresponding drop may be 1.5 per cent. To offset these drops, various voltage boosts are employed as follows: main trans-former, +5 per cent (zero on light load); distribution transformer, inherent boost of +4 per cent (i.e. 250 V secondary) plus a 2.5 per cent boost. These add to give a total boost on full load of 11.5 per cent and on light load of 6.5 per cent. Hence the consumers' voltage varies between (−17.5 + 11.5), i.e. −6 per cent and (6.5 − 1.5) i.e. +5 per cent, which is just permissible. There will also be a difference in consumer voltage depending upon the position of the lateral feeder on the main distributor; obviously, a consumer supplied from C will have a higher voltage than one supplied from B.

### 5.9.1  Uniformly loaded feeder from one end

In areas with high load densities a large number of tappings are made from feeders and a uniform load along the length of a feeder may be considered to exist. Consider the voltage drop over a length $dx$ of the feeder distant $x$ metres from the supply end. Let $iA$ be the current tapped per unit length and $R$ and $X$ be the resistance and reactance per phase per metre, respectively. The length of the feeder is $L$ (m) (see Figure 5.21(b)).

The voltage across $dx = Rix\, dx \cos\phi + Xix\, dx \sin\phi$, where $\cos\phi$ is the power factor (assumed constant) of the uniformly distributed load.

The total voltage drop

$$= R \int_0^L ix\, dx \cos\phi + X \int_0^L ix\, dx \sin\phi$$

$$= R_i \frac{L^2}{2} \cos\phi + Xi \frac{L^2}{2} \sin\phi$$

$$= \frac{LR}{2} I \cos\phi + \frac{LX}{2} I \sin\phi$$

where $I = Li$, the total current load. Hence the uniformly distributed load may be represented by the total load tapped at the centre of the feeder length.

## 5.10  Long Lines

On light loads the charging volt-amperes of a line exceed the inductive vars consumed and the voltage rises, causing problems for generators. With very long lines the voltage drop can be massive. A length of 1500 km at 50 Hz corresponds to a quarter-wavelength line. Series capacitors would normally be installed to improve the power capacity and these effectively shorten the line electrically. In addition, shunt reactors are switched in circuit at times of light load to absorb the generated vars.

A 500 km line can operate within ±10 per cent voltage variation without shunt reactors. However, with, say, a 800 km line, shunt reactors are essential and the effects of these are shown in Figure 5.22. For long lines in general, it is usual to divide the system into sections with compensation at the ends of each section. This controls the voltage profile, helps switching, and reduces short-circuit currents. Shunt compensation can be varied by switching discrete amounts of inductance. A typical 500 kV, 1000 km scheme uses compensation totalling 1200 MVAr.

Improvement in voltage profile may be obtained by compensation, using FACTS devices, at intermediate points, as well as at the ends of the line, as shown in Figure 5.22. If the natural load is transmitted there is, of course, constant voltage along the line with no compensation. If the various busbars of a sectioned line can be maintained at *constant voltage* regardless of load, each section has a theoretical maximum transmission angle of 90°. Thus, for a three-section line a total angle of much greater than 90° would be possible. This is illustrated in Figure 5.23 for a three-section, 1500 km line with a unity power-factor load.

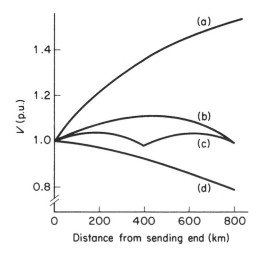

**Figure 5.22**  Voltage variation along a long line; (a) on no load with no compensation; (b) on no load with compensation at ends; (c) on no load with compensation at ends and at centre. (d) Transmitting natural load, compensation at ends and centre

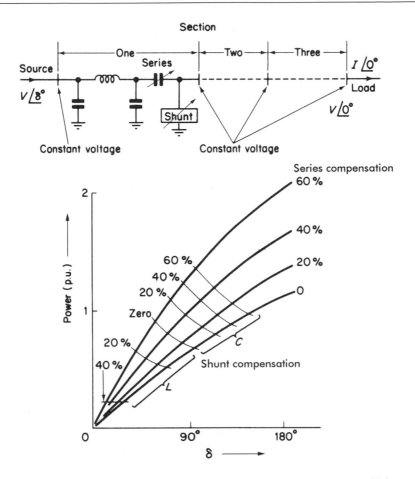

**Figure 5.23**  Power-angle curves for 1500 km line in three sections. Voltages at section-busbars maintained constant by variable compensation. Percentage of series and shunt compensation indicated (Reproduced by permission of the *Institution of Electrical Engineers*)

## 5.10.1 Reactive-power requirements for the voltage control of long lines

It is advantageous to use the generalized line equation,

$$\mathbf{V_S} = \mathbf{AV_r} + \mathbf{BI_r}$$

which takes into account the presence of transformers.

Let the received load current $\mathbf{I_r}$ lag $\mathbf{V_r}$ by $\phi_r$ where $\mathbf{V_r}$ is the reference phasor

$$\mathbf{A} = A\angle\alpha \quad \text{and} \quad \mathbf{B} = B\angle\beta$$

$$V_s\angle\delta_s = AV_r\angle\alpha + BI_r\angle(\beta - \phi_r)$$

$$\therefore V_s = AV_r\cos\alpha + jAV_r\sin\alpha + BI_r\cos(\beta - \phi_r) + jBI_r\sin(\beta - \phi_r)$$

As the modulus of the left-hand side is equal to that of the right

$$V_s^2 = A^2 V_r^2 + B^2 I_r^2 + 2ABV_r I_r \cos(\alpha = \beta + \phi_r)$$
$$= A^2 V_r^2 + B^2 I_r^2 + 2ABV_r I_r[\cos(\alpha - \beta)\cos\phi_r - \sin(\alpha - \beta)\sin\phi_r] \quad (5.8)$$
$$P_r = V_r I_r \cos\phi_r \quad \text{and} \quad Q_r = V_r I_r \sin\phi_r$$

Hence equation (5.8) becomes

$$V_s^2 = A^2 V_r^2 + B^2 I_r^2 + 2ABP_r\cos(\alpha - \beta) - 2ABQ_r\sin(\alpha - \beta)$$

Also

$$\mathbf{I_r} = I_p - jI_q \qquad I_r^2 = I_p^2 + I_q^2 \qquad I_p = \frac{P_r}{V_r} \qquad \text{and} \qquad I_q = \frac{Q_r}{V_r}$$

$$\therefore V_s^2 = A^2 V_r^2 + B^2\left(\frac{P_r^2}{V_r^2} + \frac{Q_r^2}{V_r^2}\right) + 2ABP_r\cos(\alpha - \beta) - 2ABQ_r\sin(\alpha - \beta) \quad (5.9)$$

For a given network, $P_r$, $\mathbf{A}$, and $\mathbf{B}$ will be known. The magnitude of $Q_r$ such that $V_r$ is equal to, or a specified ratio of, $V_S$ can be determined by the use of equation (5.9).

## 5.10.2 Subsynchronous oscillation

The combination of series capacitors and the natural inductance of the line (plus that of the connected systems) creates a resonant circuit of subsynchronous resonant frequency. This resonance can interact with the generator-shaft critical-torsional frequency, and a mechanical oscillation is superimposed on the rotating generator shaft that may have sufficient magnitude to cause mechanical failure.

Subsynchronous resonance has been reported, caused by line-switching in a situation where trouble-free switching was normally carried out with all capacitors in service, but trouble occurred when one capacitor bank was out of service. Although this phenomena may be a rare occurrence, the damage resulting is such that, at the design stage, an analysis of possible resonance effects is required.

## 5.11   General System Considerations

Because of increasing voltages and line lengths, and also the wider use of underground circuits, the light-load reactive-power problem for an interconnected system becomes substantial, particularly with modern generators of limited var absorption capability. At peak load, transmission systems need to increase their var generation, and as the load reduces to a minimum (usually during the night) they need to reduce the generated vars by the following methods, given in order of economic viability:

1.  switch out shunt capacitors;

2.  switch in shunt inductors;

3.  run hydro plant on maximum var absorption;

4.  switch out one cable in a double-circuit link;

5.  tap-stagger transformers;

6.  run base-load generators at maximum var absorption.

## Problems

**5.1.**   An 11 kV supply busbar is connected to an 11/132 kV, 100 MVA, 10 per cent reactance transformer. The transformer feeds a 132 kV transmission link consisting of an overhead line of impedance $(0.014 + j0.04)$ p.u. and a cable of impedance $(0.03 + j0.01)$ p.u. in parallel. If the receiving end is to be maintained at 132 kV when delivering 80 MW, 0.9 p.f. lagging, calculate the power and reactive power carried by the cable and the line. All p.u. values relate to 100 MVA and 132 kV bases.
(Answer: Line $(23 + j38)$ MVA; cable $(57 + j3.8)$ MVA)

**5.2**   A three-phase induction motor delivers 500 hp at an efficiency of 0.91, the operating power factor being 0.76 lagging. A loaded synchronous motor with a power consumption of 100 kW is connected in parallel with the induction motor. Calculate the necessary kVA and the operating power factor of the synchronous motor if the overall power factor is to be unity.
(Answer: 365 kVA, 0.274)

**5.3**   The load at the receiving end of a three-phase overhead line is 25 MW, power factor 0.8 lagging, at a line voltage of 33 kV. A synchronous compensator is situated at the receiving end and the voltage at both ends of the line is maintained at 33 kV. Calculate the MVAr of the compensator. The line has resistance of 5 Ω per phase and inductive reactance of 20 Ω per phase.
(Answer: 25 MVAr)

**5.4**   A transformer connects two infinite busbars of equal voltage. The transformer is rated at 500 MVA and has a reactance of 0.15 p.u. Calculate the var flqw for a tap setting of (a) 0.85:1; (b) 1.1:1.

(Answer: (a) 427 MVAr; (b) −367 MVAr)

**5.5**   A three-phase transmission line has resistance and inductive reactance of 25 Ω and 90 Ω, respectively. With no load at the receiving end, a synchronous compensator there takes a current lagging by 90°; the voltage is 145 kV at the sending end and 132 kV at the receiving end. Calculate the value of the current taken by the compensator.

When the load at the receiving end is 50 MW, it is found that the line can operate with unchanged voltages at the sending and receiving ends, provided that the compensator takes the same current as before, but now leading by 90°.

Calculate the reactive power of the load.
(Answer: 83.5 A; $Q_L$ 24.2 MVAr)

**5.6**   Repeat Problem 5.3 making use of $\partial Q/\partial V$ at the receiving end.

**5.7**   In Example 5.3, determine the tap ratios if the receiving-end voltage is to be maintained at 0.9 p.u. of the sending-end voltage.
(Answer: $1.19t_s$, $0.84t_r$)

**5.8**   In the system shown in Figure 5.24, determine the supply voltage necessary at D to maintain a phase voltage of 240 V at the consumer's terminals at C. The data in Table 5.2 apply.
(Answer: 33 kV)

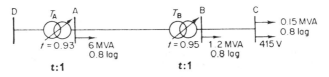

**Figure 5.24**   Line diagram for system in Problem 5.8

**Table 5.2**   Data for Problem 5.8

| Line or/ transformers | Rated voltage (kV) | Transmission rating | Nominal tap ratio | Impedance (Ω) |
|---|---|---|---|---|
| BC | 0.415 | | | 0.0127 + j0.00909 |
| AB | 11 | | | 1.475 + j2.75 |
| DA | 33 | | | 1.475 + j2.75 |
| $T_A$ | 33/11 | 10 MVA | 30.69/11 | 1.09 + j9.8 referred to 33 kV |
| $T_B$ | 11/0.415 | 2.5 MVA | 10.450/0.415 | 0.24 + j1.95 referred to 33 kV |

**5.9**   A load is supplied through a 275 kV link of total reactance 50 Ω from an infinite busbar at 275 kV. Plot the receiving-end voltage against power graph for a constant load power factor of 0.95 lagging. The system resistance may be neglected.

**5.10** (a) Describe two methods of controlling voltage in a power system.

(b) Show how the scalar voltage difference between two nodes in a network is given approximately by:

$$\Delta V = \frac{RP + XQ}{V}$$

(c) Each phase of a 50 km, 132 km overhead line has a series resistance of $0.156\,\Omega/\text{km}$ and an inductive reactance of $0.4125\,\Omega/\text{km}$. At the receiving end the voltage is 132 kV with a load of 100 MVA at a power factor of 0.9 lagging. Calculate the magnitude of the sending-end voltage.

(d) Calculate also the approximate angular difference between the sending-end and receiving-end voltages.

(Answer: (c) 144.5 kV; (d) 4.55°)

(*From E.C. Examination 1997*)

**5.11** Explain the limitations of tap-changing transformers. A transmission link (Figure 5.25(a) connects an infinite busbar supply of 400 kV to a load busbar supplying 1000 MW, 400 MVAr. The link consists of lines of effective impedance $(7+\text{j}70)\,\Omega$ feeding the load busbar via a transformer with a maximum tap ratio of 0.9:1. Connected to the load busbar is a compensator. If the maximum overall voltage drop is to be 10 per cent with the transformer taps fully utilized, calculate the reactive power requirement from the compensator.

(Answer: 148 MVAr)

Note: Refer voltage and line $Z$ to load side of transformer in Figure 5.25(b).

$$V_R = \frac{V_s}{t} - \left( \frac{\frac{RP}{t^2} + \frac{X}{t^2}Q}{V_R} \right)$$

**5.12** A generating station consists of four 500 MW, 20 kV, 0.95 p.f. (generate) generators, each feeding through a 525 MVA, 0.1 p.u. reactance transformer onto a common busbar. It is required to transmit 2000 MW at 0.95 p.f. lagging to a substation

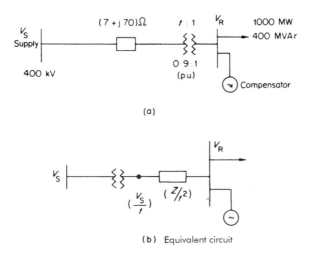

(a)

(b) Equivalent circuit

**Figure 5.25**  Circuits for Problem 5.11

maintained at 500 kV in a power system at a distance of 500 km from the generating station. Design a suitable transmission link of nominal voltage 500 kV to achieve this, allowing for a reasonable margin of stability and a maximum voltage drop of 10 per cent. Each generator has synchronous and transient reactances of 2 p.u. and 0.3 p.u. respectively, and incorporates a fast-acting automatic voltage regulator. The 500 kV transmission lines have an inductive reactance per phase of $0.4 \, \Omega/\text{km}$ and a shunt capacitive reactance per phase of $0.3 \times 10^6 \, \Omega/\text{km}$. Both series and shunt capacitors may be used if desired and the number of three-phase lines used should be not more than three—fewer if feasible. Use approximate methods of calculation, ignore resistance, and state clearly any assumption made. Assume shunt capacitance to be lumped at the receiving end only. (Use two 500 kV lines with series capacitors compensating to 70 per cent of series inductance.)

**5.13** It is required to transmit power from a hydroelectric station to a load centre 480 km away using two lines in parallel for security reasons.

Assume sufficient bundle conductors are used such that there are no thermal limitations, and the effective reactance per phase per km is $0.44 \, \Omega$ and that the resistance is negligible. The shunt capactive reactance of each line is $0.44 \, \text{M}\Omega$ per phase per km, and each line may be represented by the nominal $\pi$-circuit with half the capacitance at each end. The load is 2000 MW at 0.95 lagging and is independent of voltage over the permissible range.

Investigate, from the point of view of stability and voltage drop, the feasibility and performance of the link if the sending-end voltage is 345, 500, and 765 kV assuming the transmission angle is not to exceed $30°$.

The lines may be compensated up to 70 per cent by series capacitors and the load-end compensators of 120 MVAr capacity are available. The maximum permissible voltage drop is 10 per cent. As two lines are provided for security reasons, your studies should include the worst-operating case of only one line in use.

**5.14** Explain the action of a variable-tap transformer, showing, with a phasor diagram, how reactive power may be despatched from a generator down a mainly reactive line by use of the taps. How is the level of real power despatch controlled?

Power flows down an H.V. line of impedance $0 + j0.15$ p.u. from a generator whose output passes through a variable-ratio transformer to a large power system. The voltage of the generator and the distant large system are both kept at 1.0 p.u. Determine the tap setting if 0.8 p.u. power and 0.3 p.u. VAr are delivered to a lagging load at the power system busbar. Assume the reactance of the transformer is negligible.
(Answer: $t = 1.052$)
(*From E.C. Examination, 1995*)

**5.15** Two substations are connected by two lines in parallel, of negligible impedance, each containing a transformer of reactance 0.18 p.u. and rated at 120 MVA. Calculate the net absorption of reactive power when the transformer taps are set to 1:1.15 and 1:0.85, respectively (i.e. tap-stagger is used). The p.u. voltages are equal at the two ends and are constant in magnitude.
(Answer: 30 MVAr)

# 6
# *Load Flows*

## 6.1  Introduction

A *load flow* is power-system jargon for the steady-state solution of a network. This does not essentially differ from the solution of any other type of network except that certain constraints are peculiar to power supply. In previous chapters the manner in which the various components of a power system may be represented by equivalent circuits has been demonstrated. It should be stressed that the simplest representation should always be used, consistent with the accuracy of the information available. There is no merit in using very complicated machine and line models when the load and other data are known only to a limited accuracy, e.g. the long-line representation should only be used where absolutely necessary. Similarly, synchronous-machine models of more sophistication than given in this text are needed only for very specialized purposes, e.g. in some stability studies. Usually, the size and complexity of the network itself provides more than sufficient intellectual stimulus without undue refinement of the components. Often, resistance may be neglected with little loss of accuracy and an immense saving in computation.

The following combinations of quantities are usually specified at the system busbars for load-flow studies.

*Slack, swing, or floating busbar*   One node is always specified by a voltage, constant in magnitude and phase. The effective generator at this node supplies the losses to the network; this is necessary because the magnitude of the losses will not be known until the calculation of currents is complete, and this cannot be achieved unless one busbar has no power constraint and can feed the required losses into the system. The location of the slack node can influence the complexity of the calculations; the node most closely approaching an infinite busbar should be used.

***Load nodes*** The complex power $S = P \pm jQ$ is specified. Designated $P, Q$ node.

***Generation nodes*** The voltage magnitude and power are usually specified. Often, limits to the value of the reactive power are given depending upon the characteristics of individual machines. Designated $P, V$ node. Load flow studies are performed to investigate the following features of a power system network:

1.  Flow of MW and MVAr in the branches  of the network;
2.  Busbar (node) voltages;
3.  Effect of rearranging circuits and incorporating new circuits on system loading;
4.  Effect of temporary loss of generation and transmission circuits on system loading (mainly for security studies);
5.  Effect of injecting in-phase and quadrature boost voltages on system loading;
6.  Optimum system running conditions and load distribution;
7.  Minimizing system losses;
8.  Optimum rating and tap-range of transformers;
9.  Improvements from change of conductor size and system voltage.

Studies will normally be performed for minimum-load conditions (possibility of instability due to high voltage levels and self-excitation of induction machines) and maximum-load conditions (possibility of instability). Having ascertained that a network behaves reasonably under these conditions, further load flows will be performed to attempt to optimize various quantities. The design and operation of a power network to obtain optimum economy is of paramount importance and the furtherance of this ideal is achieved by the use of centralized automatic control of generating stations through system control centres.

Although the same approach can be used to solve all problems, e.g. the nodal voltage method, the object should be to use the quickest and most efficient method for the particular type of problem. Radial networks will require less sophisticated methods than closed loops. In very large networks the problem of organizing the data is almost as important as the method of solution, and the calculation must be carried out on a systematic basis where the nodal-voltage method is the most advantageous. Such methods as network reduction combined with the Thevenin or superposition theorems are at their best with smaller networks. In the nodal method, great numerical accuracy is required in the computation as the currents in the branches are derived from the voltage differences between the ends. These differences are small in well-designed networks so the method is ideally suited for computation using digital computers.

## 6.2 Radial and Simple Loop Networks

In radial networks the phase shifts due to transformer connections along the circuit are not usually important. The following examples illustrate the solution of this type of network.

### *Example 6.1*

Distribution feeders with several tapped loads. A distribution feeder with several tapped inductive loads (or laterals), and fed at one end only, is shown in Figure 6.1(a). Determine the total voltage drop.

### *Solution*

The current in AB

$$
\begin{aligned}
=&(I_1 \cos \theta_1 + I_2 \cos \theta_2 + I_3 \cos \theta_3 + I_4 \cos \theta_4) \\
&- j(I_1 \sin \theta_1 + I_2 \sin \theta_2 + I_3 \sin \theta_3 + I_4 \sin \theta_4)
\end{aligned}
$$

Similarly, the currents in the other sections of the feeder are obtained. The approximate voltage drop is obtained from $\Delta V = RI \cos \theta + XI \sin \theta$ for each section. That is

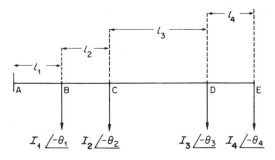

**Figure 6.1**(a)   Feeder with several load tappings along its length

$$\Delta V_{AE} = R_1(I_1 \cos\theta_1 + I_2 \cos\theta_2 + I_3 \cos\theta_3 + I_4 \cos\theta_4)$$
$$+ R_2(I_2 \cos\theta_2 + I_3 \cos\theta_3 + I_4 \cos\theta_4)$$
$$+ R_3(I_3 \cos\theta_3 + I_4 \cos\theta_4) + R_4(I_4 \cos\theta_4)$$
$$+ X_1(I_1 \sin\theta_1 + I_2 \sin\theta_2 + I_3 \sin\theta_3 + I_4 \sin\theta_4), \text{ and so on}$$

Rearranging and letting the resistance per loop-metre be $r\,\Omega$ and the reactance per loop-metre be $x\,\Omega$ (the term loop-metre refers to single-phase circuits and includes the go and return conductors),

$$\Delta V_{AE} = r[I_1 \cdot l_1 \cos\theta_1 + I_2 \cos\theta_2 \cdot (l_1 + l_2) + I_3 \cos\theta_3(l_1 + l_2 + l_3)$$
$$+ I_4 \cos\theta_4(l_1 + l_2 + l_3 + l_4)] + x[I_1 l_1 \sin\theta_1 + I_2 \sin\theta_2(l_1 + l_2)$$
$$+ I_3 \sin\theta_3(l_1 + l_2 + l_3) + I_4 \sin\theta_4(l_1 + l_2 + l_3 + l_4)]$$

In the system shown in Figure 6.1(b), calculate the size of cable required if the voltage drop at the end load on the feeder must not exceed 12 V (line value). Let the resistance and reactance per metre per phase be $r(\Omega)$ and $x(\Omega)$, respectively. Then, referring to Figure 6.1(c), voltage drop is given by

$$r[100 \times 40 + 250 \times 20 + 330 \times 25] + x[100 \times 30 + 250 \times 0 + 330 \times 25] = 12/\sqrt{3}$$

i.e.

$$r(4000 + 5000 + 8250) + x(3000 + 0 + 8250) = 12/\sqrt{3}$$
$$17250r + 11250x = 12/\sqrt{3}$$

The procedure now is to consult the appropriate overhead line or cable specifications to select a cross-section that gives values or $r$ and $x$ which best fit the above equation. Usually, if underground cables are used the size of cross-section is selected on thermal considerations and then the voltage drop is calculated. With overhead lines the volt drop is the prime consideration and the conductor size will be determined accordingly.

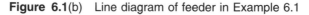

**Figure 6.1**(b)   Line diagram of feeder in Example 6.1

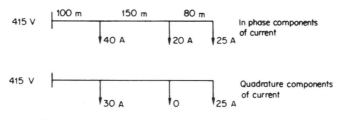

**Figure 6.1**(c)   In-phase and quadrature quantities

## Example 6.2

The system shown in figure 6.2(a) feeds two distinct loads, a group of domestic consumers, and a group of induction motors which, on starting, take $5\times$ full-load current at zero power-factor lagging. This is a 'worst-case' calculation. The induction motor voltage is nominally 6 kV. Calculate the dip in voltage on the domestic-load busbar when the induction motor group is started.

## Solution

The problem of the drop in the voltage to other consumers when an abnormally large current is taken for a brief period from an interconnected busbar is often serious. The present problem poses a typical situation. The 132 kV system is not an infinite busbar and is represented by a voltage source in series with a 0.04 p.u. reactance on a 50MVA base. Using a 50 MVA base the equivalent single-phase circuit is as shown in Figure 6.2(b).

Starting current of the induction motors

$$= -j\frac{15000 \times 10^3}{\sqrt{3} \times 6000} \times 5 \,\text{A}$$

$$= -j7217 \,\text{A}$$

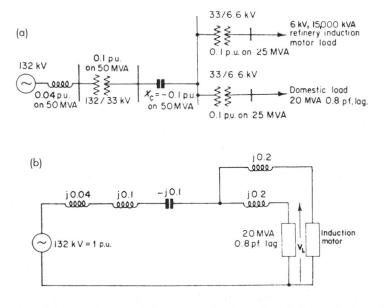

**Figure 6.2** (a) Line diagram of the system in Example 6.2; (b) equivalent circuit of the system in Example 6.2

Domestic load voltage before the induction-motor load is switched on

$$\mathbf{V}_L = 1.0 - \mathbf{I}\mathbf{Z}$$

where

$$\mathbf{Z} = j(0.04 + 0.1 - 0.1 + 0.2) = j0.24 \text{ p.u.}$$

and

$$\mathbf{I} = \frac{20 \times 10^6}{\sqrt{3} \times 6000} = 1925 \text{ A}, 0.8 \text{ p.f. lagging}$$

But base current at 6.6 kV

$$= \frac{50 \times 10^6}{\sqrt{3} \times 6.6 \times 10^3}$$

$$= 4374 \text{ A}$$

$$\therefore \mathbf{I}_{\text{p.u.}} = \frac{1925}{4374} = 0.440 \text{ p.u. at } 0.8 \text{ lag}$$

Hence

$$\mathbf{V}_L = 1.0 - 0.440(0.8 - j0.6)(j0.24) \text{ p.u.}$$

$$= 0.9366 - j0.0845 \text{ p.u.}$$

$$\therefore V_L = 0.9404 \text{ p.u.}$$

Starting current of the induction motors

$$= \frac{-j7217}{4374} = -j1.640$$

Voltage $\mathbf{V}_L'$ when the motors start is given by

$$= 1 - j0.04[0.440(0.8 - j0.6) - j1.640] - j0.2[0.440(0.8 - j0.6)]$$

$$= 0.871 - j0.084 \text{ p.u.}$$

and

$$V_L' = 0.875 \text{ p.u.}$$

Hence the voltage dips from 0.940 to 0.87 p.u. or from 6.204 kV to 5.74 kV.

It will be noticed that a series capacitor has been installed partly to neutralize the network reactance. Without this capacitor the dip will be much more serious; it is left to the reader to determine by how much. Often, in steel mills, large pulses of power are taken at regular intervals and the result is a regular fluctuation on the consumer busbars. This is known as *voltage flicker* owing to the effect on electric lights.

## Load flows in closed loops

In radial networks any phase shifts due to transformer connections are not usually of importance as the currents and voltages are shifted by the same amount. In a closed loop, to avoid circulating currents, the product of the transformer transformation ratios (magnitudes) round the loop should be unity and the sum of the phase shifts in a common direction round the loop should be zero. This will be illustrated by the system shown in Figure 6.3.

Neglecting phase shifts due to the impedance of components, for the closed loop formed by the two lines in parallel, the total shift and transformation ratio is

$$\left(\frac{33}{13.8}\angle 30°\right)\left(\frac{132}{33}\angle -30°\right)\left(\frac{13.8}{132}\angle 0°\right) = 1\angle 0°$$

In practice, the transformation ratios of transformers are frequently changed by the provision of tap-changing equipment. This results in the product of the ratios round a loop being no longer unity, although the phase shifts are still equal to zero. To represent this condition a fictitious autotransformer is connected as shown in Figure 6.4. The increase in voltage on the 132 kV line by tap-changing has the effect, in a largely reactive path, of changing the flow of reactive power and thus the power factor of the current. An undesirable effect is the circulating current set up around the loop, unless load sharing is required (see Section 5.7).

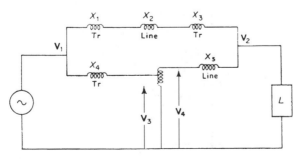

**Figure 6.3**  Loop with transformer phase shifts

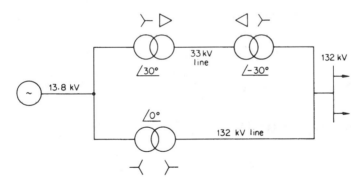

**Figure 6.4**  Equivalent circuit of network in Figure 6.3 showing autotransformer

Often, in lower voltage systems the out-of-balance voltage represented by the autotransformer can be neglected. If this is not the case, the best method of calculation is to determine the circulating current and consequent voltages due to the out-of-balance voltage acting alone, and then superpose these values on those obtained for operation with completely nominal voltage ratios.

## 6.3  Computation of Power Flows in a Network

Consider the three-node system of Figure 6.5.

With the voltages, currents, and line impedances as shown, we can formulate the following equations for each node using Kirchoff's laws:

$$
\left.
\begin{array}{ll}
\text{Node 1:} & I_1 - I_{12} + I_{31} = 0 \\
\text{Node 2:} & I_2 - I_{23} + I_{12} = 0 \\
\text{Node 3:} & -I_3 + I_{23} - I_{31} = 0
\end{array}
\right\}
\tag{6.1}
$$

(Note that currents flowing *into* each node are positive.)

Replacing the network currents by voltages and impedances, we obtain

$$\text{Node 1:} \quad I_1 - (\mathbf{V}_1 - \mathbf{V}_2)\left(\frac{1}{jX_{12}}\right) + (\mathbf{V}_3 - \mathbf{V}_1)\left(\frac{1}{jX_{31}}\right) = 0$$

$$\text{Node 2:} \quad I_2 - (\mathbf{V}_2 - \mathbf{V}_3)\left(\frac{1}{jX_{23}}\right) + (\mathbf{V}_1 - \mathbf{V}_2)\left(\frac{1}{jX_{12}}\right) = 0$$

$$\text{Node 3:} \quad -I_3 + (\mathbf{V}_2 - \mathbf{V}_3)\left(\frac{1}{jX_{23}}\right) - (\mathbf{V}_3 - \mathbf{V}_1)\frac{1}{(jX_{32})} = 0$$

Rearranging these equations and putting $\mathbf{Y} = \dfrac{1}{jX}$ gives:

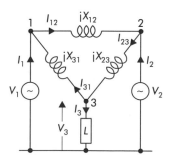

**Figure 6.5**   Three-node network for power-flow formulation

$$\mathbf{I}_1 = (\mathbf{Y}_{12} + \mathbf{Y}_{31})\mathbf{V}_1 + (-\mathbf{Y}_{12})\mathbf{V}_2 + (-\mathbf{Y}_{31})\mathbf{V}_3$$
$$\mathbf{I}_2 = (-\mathbf{Y}_{12})\mathbf{V}_1 + (\mathbf{Y}_{12} + \mathbf{Y}_{23})\mathbf{V}_2 + (-\mathbf{Y}_{23})\mathbf{V}_3$$
$$-\mathbf{I}_3 = (-\mathbf{Y}_{31})\mathbf{V}_1 + (-\mathbf{Y}_{23})\mathbf{V}_2 + (\mathbf{Y}_{31}\mathbf{Y}_{23})\mathbf{V}_3$$

These equations can be put into matrix form as:

$$\begin{bmatrix} \mathbf{I}_1 \\ \mathbf{I}_2 \\ -\mathbf{I}_3 \end{bmatrix} = \begin{bmatrix} (\mathbf{Y}_{12} + \mathbf{Y}_{31}) & -\mathbf{Y}_{12} & -\mathbf{Y}_{31} \\ -\mathbf{Y}_{12} & (\mathbf{Y}_{12} + \mathbf{Y}_{23}) & -\mathbf{Y}_{23} \\ -\mathbf{Y}_{31} & -\mathbf{Y}_{23} & (\mathbf{Y}_{31} + \mathbf{Y}_{23}) \end{bmatrix} \begin{bmatrix} \mathbf{V}_1 \\ \mathbf{V}_2 \\ \mathbf{V}_3 \end{bmatrix} \qquad (6.2)$$

It will be seen that the diagonal elements are the *sum* of the admittances connected to the node (known as self-admittances) whilst the off-diagonal elements are the admittances connected between the relevant nodes, and are negative. Consequently, if the currents on the left-hand side are known, the voltages can be calculated by inverting the **Y** matrix.

## *Example 6.3*

Perform a load flow on the interconnected busbars shown in Figure 6.6.

## *Solution*

The nodal equations may be written directly as follows:

$$\begin{bmatrix} 2 & -0.5 & -1.5 \\ -0.5 & 1.25 & -0.75 \\ -1.5 & -0.75 & +2.25 \end{bmatrix} \begin{bmatrix} V_1 \\ V_2 \\ V_3 \end{bmatrix} = \begin{bmatrix} I_1 \\ I_2 \\ I_3 \end{bmatrix}$$

It is usual to write the admittance matrix direct. If the values in the current column vector are specified, it is required to determine the values of $V_1$, $V_2$, and $V_3$. [**Y**] may be inverted or the equations solved by elimination.

As    $[\mathbf{Y}][\mathbf{V}] = [\mathbf{I}]$    then    $[\mathbf{Y}]^{-1}[\mathbf{I}] = [\mathbf{V}]$

Hence it is required to invert the admittance matrix. A simple method will be used in this case:

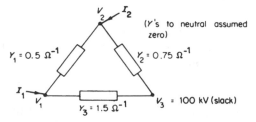

**Figure 6.6**  Three-busbar system–load flow using matrix inversion [note: $\Omega^{-1}$ is an alternative to $S$ (siemen)]

$$V_1 = 0.5I_1 + 0.25V_2 + 0.75V_3$$

Substitute for $V_1$ in equations for $I_2$ and $I_3$:

$$-0.25I_1 + 1.125V_2 - 1.125V_3 = I_2$$
$$-0.75I_1 - 1.125V_2 + 1.125V_3 = I_3$$

Also, from the equation for $I_2$

$$V_2 = 0.222I_1 + 0.888I_2 + V_3$$

Substitute for $V_2$ in the other equations:

$$0.555I_1 + 0.222I_2 + V_3 = V_1$$
$$0.222I_1 + 0.888I_2 + V_3 = V_2$$
$$-I_1 - I_2 - 0V_3 = I_3$$

i.e.

$$\begin{bmatrix} 0.555 + 0.222 + 1.0 \\ 0.222 + 0.888 + 1.0 \\ -1 \quad\quad -1 \quad\quad 0.0 \end{bmatrix} \begin{bmatrix} I_1 \\ I_2 \\ V_3 \end{bmatrix} = \begin{bmatrix} V_1 \\ V_2 \\ I_3 \end{bmatrix}$$

The node at which the voltage is specified, the slack busbar, is node (3); hence the manipulation of the matrix into the form shown. Next, some loadings at the nodes will be specified:

$$I_1 = 1000 \text{ A (generator node)}$$
$$I_2 = 1500 \text{ A (load node)}$$
$$V_3 = 100 \text{ kV (phase value), slack busbar}$$

Dealing in kiloamperes and kilovolts,

$$V_1 = 0.555(1) + 0.222(-1.5) + (+1)(100)$$
$$= 100.222 \text{ kV}$$
$$V_2 = 0.222(1) + 0.888(-1.5) + (1)(100)$$
$$= 98.89 \text{ kV}$$
$$I_3 = -(1)(1) + (1)(1.5) + 0(100)$$
$$= +0.5 \text{ kA i.e. a generator node}$$

(Note that $I_1 + I_2 + I_3 = 0$.)

Although Example 6.3 shows how a computed solution can be readily obtained, in practice it is the power inputs and outputs of the network that are specified, not the nodal injected currents.

A common calculation, particularly for quick planning studies and security calculations, is known as the 'real power', megawatt, or d.c. load flow. Here, the injected MW value is specified for all the nodes except the slack, where the voltage magnitude is given. It is required to find the MW flows in the network to satisfy the specified conditions.

## Example 6.4

Perform a real-power load flow for the network shown in Figure 6.7.

## Solution

An estimate may be obtained by assuming zero resistance in the network and using the fact that in a well-designed and well-operated system, all the voltages will be close to 1 p.u. such that $P$ in p.u. is the same as $I$ in p.u. Consequently, voltage drops given by $PX$ around the network loops must sum to zero. Hence $\Sigma P_{ij} X_{ij} = 0$ where $i$ and $j$ refer to the branch between nodes $i$ and $j$ of the network. This follows from the fact that $\Delta V \propto \angle \delta \approx (XP/V)$ and the sum of the angular shifts round the loop is zero. Labelling the powers as in Figure 6.7, the following equations hold:

$$13.2P_1 + 2.06P_3 - 19.8P_4 - 6.6P_2 = 0$$
$$P_4 + P_3 = 100 \quad \text{or} \quad P_4 = 100 - P_3$$
$$P_2 - P_4 = 50 \quad P_1 = 60 + P_3$$
$$P_2 + P_1 = 210 \quad P_2 = 210 - 60 - P_3 = 150 - P_3$$
$$\therefore 13.2(60 + P_3) + 2.06(P_3) - 19.8(100 - P_3) - 6.6(150 - P_3) = 0$$

and

$$792 + 13.2P_3 + 2.06P_3 - 1980 + 19.8P_3 - 990 + 6.6P_3 = 0$$

From which $P_3 = 52.2\,\text{MW}$, which is a reasonably accurate estimate.
Once $P$ in BC is known, the flows in the other branches follow readily, i.e.

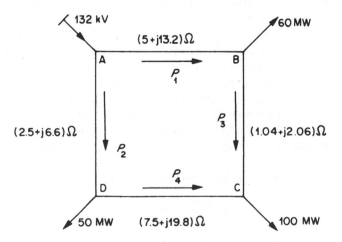

**Figure 6.7**  Line diagram of loop network for Example 6.4

$$P_4 = P_{DC} = 100 - 52.2 = 47.8\,\text{MW}$$
$$P_1 = P_{AB} = 60 + 52.2 = 112.2\,\text{MW}$$
$$P_2 = P_{AD} = 47.8 + 50 = 97.8\,\text{MW}$$

Quite often, for networks where $X/R > 10$, small changes to network flows, e.g. less than 10 per cent of existing flows, can be estimated by using the principle of superposition. For example, in the network of Figure 6.7, if node C supplied an extra 10 MW to the load, this would need to come from node A and flow over AD and DC in series, but in parallel with AB and BC in series. This 10 MW would divide in the ratio of $(\text{j}13.2 + \text{j}2.06)$ to $(\text{j}6.6 + \text{j}19.8)$ giving

$$10 \times \left(\frac{15.26}{15.26 + 26.4}\right) \text{MW} = 3.66\,\text{MW}$$

to be added to the path ADC. Hence $P_2$ would now become $97.8 + 3.66 = 101.46\,\text{MW}$ and $P_4 = 51.46\,\text{MW}$. The remaining flow of $10 - 3.66 = 6.33\text{MW}$ would be added to $P_1$ and $P_3$.

Using the same superposition principle it is possible to estimate the change in flows due to the removal of a circuit by injecting into the simplified network a flow equal and opposite to that previously in the removed circuit, and calculating the additional flows due to this negative injection. Such estimates are useful to assess the system security (see later in this chapter).

## 6.4  Complex Flows in Large Systems

In large practical power systems, for both planning and operational purposes it is necessary to carry out many load flows taking into account the complex impedances of the circuits, the limits caused by circuit capacities, and the voltages that can be satisfactorily provided at all nodes in the system. Systems consisting of up to 3000 busbars, 6000 circuits, and 500 generators may have to be solved in reasonable time scales (e.g. 1–2 min) with accuracies requiring 32 or 64 bits for numerical stability. Many system states possible in a day's operation may have to be considered. Consequently, a systematic way of computing the desired states of the system is required and an organized method for handling the data is necessary. Data base organization and rapid retrieval and modification are just as important as having an efficient and fast computation technique.

Basically, the equations to be solved are given by

$$\left.\begin{array}{c} [\mathbf{Y}][\mathbf{V}] = [\mathbf{I}] \\[2mm] [\mathbf{I}] = \left[\dfrac{\mathbf{S}^*}{\mathbf{V}^*}\right] \end{array}\right\} \tag{6.3}$$

where * denotes conjugate. This solution can only be obtained by iteration as the problem is non-linear, but iterative matrix inversion using Gaussian elimination can be employed as indicated in the next section.

## 6.4.1 Direct methods

### Triangulation and partitioning

Direct methods solve only linear systems, i.e. $[\mathbf{Y}][\mathbf{V}] = [\mathbf{I}]$ where $[\mathbf{I}]$ is specified. The fact that powers are specified in practice makes the problem non-linear. A new value for $\mathbf{I}$ must be obtained from $\mathbf{S} = \mathbf{VI}^*$ after each direct solution and this value used to obtain a new one.

Partitioning is useful for handling large problems on small computers, for network reduction, and the elimination of unwanted nodes. A matrix

$$[\mathbf{Y}] = \begin{bmatrix} Y_{11} & Y_{12} & Y_{13} \\ Y_{21} & Y_{22} & Y_{23} \\ Y_{31} & Y_{32} & Y_{33} \end{bmatrix} \tag{6.4}$$

can be partitioned into four submatrices by the dividing lines shown in equation (6.4). Hence, in this typical admittance matrix for a network,

$$[\mathbf{Y}] = \begin{bmatrix} \mathbf{B} & \mathbf{C} \\ \mathbf{C}^t & \mathbf{D} \end{bmatrix}$$

where $\mathbf{C}^t$ is the transpose of $\mathbf{C}$,

$$\mathbf{B} = \begin{bmatrix} Y_{11} & Y_{12} \\ Y_{21} & Y_{22} \end{bmatrix} \quad \text{and so on}$$

A saving in computation time and storage is often obtained by manipulating the submatrices instead of the main one, and in the equation $[\mathbf{I}] = [\mathbf{Y}][\mathbf{V}]$ it is possible to eliminate nodes at which $\mathbf{I}$ is not injected. The matrix $[\mathbf{Y}]$ is partitioned so that the nodes to be eliminated are grouped together in a submatrix, e.g.

$$\begin{bmatrix} \mathbf{I}_w \\ \mathbf{I}_u \end{bmatrix} = \begin{bmatrix} \mathbf{B} & \mathbf{C} \\ \mathbf{C}^t & \mathbf{D} \end{bmatrix} \begin{bmatrix} \mathbf{V}_w \\ \mathbf{V}_u \end{bmatrix} \tag{6.5}$$

where the subscript u indicates unwanted, and subscript w wanted, nodes. For example, if the currents at the unwanted nodes are zero,

$$\mathbf{I}_u = 0 \qquad -\mathbf{C}^t\mathbf{V}_w = \mathbf{D}\mathbf{V}_u \qquad \text{and} \qquad -\mathbf{D}^{-1}\mathbf{C}^t\mathbf{V}_w = \mathbf{V}_u$$

Substituting this value of $\mathbf{V}_u$ in the expression for $\mathbf{I}_w$ produces

$$\mathbf{I}_w = \mathbf{B}\mathbf{V}_w - \mathbf{C}\mathbf{D}^{-1}\mathbf{C}^t\mathbf{V}_w$$

which gives a new admittance matrix containing only wanted nodes, i.e.

$$[\mathbf{Y}] = \mathbf{B} - \mathbf{CD}^{-1}\mathbf{C}^{t} \qquad (6.6)$$

Methods are available for modifying the admittance matrix to allow for changes in the configuration of the network, say, due to line outages. These modifications can be performed such that a completely fresh inversion of the new admittance matrix is not necessary.

Most power-system matrices are symmetric, i.e. $\mathbf{Y}_{ij} = \mathbf{Y}_{ji}$, and it is only required to store elements on and above the diagonal. Frequently, most of the non-zero elements lie within a narrow band about the diagonal and, again, considerable savings in computer storage may be attained.

*Accuracy*   When solving the set of equations $[\mathbf{A}][\mathbf{x}] = [\mathbf{b}]$, the residual vector $\mathbf{r} = [\mathbf{b}] - [\mathbf{A}][\mathbf{x}]$ is not zero because of rounding errors. Troubles arise when ill-conditioned equations are obtained in which, although the residual is small, the solution may be inaccurate. Such matrices are often large and sparse and for most rows the diagonal element is equal to the sum of the non-diagonal elements and of opposite sign.

### 6.4.2   Iterative methods

The two sets of equations of (6.3) are solved simultaneously as the iteration proceeds. A number of methods are available, and as three of these have gained wide adoption they will be described in detail. The first is the Gauss–Seidel method which has been widely used for many years and is simple in approach; the second is the Newton–Raphson method which, although more complex, has certain advantages. A third, which is almost an industry standard, is the fast decoupled load flow. The speed of convergence of these methods is of extreme importance as, apart from the cost of computer time, the use of these methods in schemes for the automatic control of power systems requires very fast load-flow solutions.

### (i) The Gauss–Seidel method

In this method the unknown quantities are initially assumed and the value obtained from the first equation for, say $V_1$ is then used when obtaining $V_2$ from the second equation, and so on. Each equation is considered in turn and then the complete set is solved again until the values obtained for the unknowns converge to within required limits.

Application of the method to the simple network of Example 6.3 gives the nodal equations as

$$2V_1 - 0.5V_2 - 1.5V_3 = I_1 = 1$$
$$-0.5V_1 + 1.25V_2 - 0.75V_3 = I_2 = -1.5$$
$$-1.5V_1 - 0.75V_2 + 2.25V_3 = I_3$$

where the voltages are in kV and the currents in kA.

As $V_3$ is known, i.e. 100 kV (slack busbar voltage), it is necessary only to solve the first two equations. Initially, make $V_2 = 100$ kV.

$$\therefore {}^1V_1 = 0.5(1 + 150 + 50) = 100.5000$$

Using this value to evaluate $V_2$,

$${}^1V_2 = \frac{1}{1.25}(-1.5 + 75 + 50.25) = 99.0000$$

Using this value of $V_2$ to evaluate $V_1$,

$${}^2V_1 = 0.5(1 + 150 + 49.5) = 100.2500$$

$${}^2V_2 = \frac{1}{1.25}(-1.5 + 75 + 50.125) = 98.9000$$

$${}^3V_1 = 100.2250 \qquad {}^2V_2 = 98.8900$$

$${}^4V_1 = 100.2225 \qquad {}^4V_2 = 98.8890$$

$${}^5V_1 = 100.22225 \qquad {}^5V_2 = 98.8888$$

Iterations are now producing changes only in the fourth place of decimals and the process may be stopped; hence

$$I_3 = -1.5 \times 100.22225 - 0.75 \times 98.8888 + 2.25 \times 100 = 0.5 \text{ kA}$$

The solution has been obtained with much less computation than the more direct method used previously.

In the three-node system of Example 6.3, the iterative form of the three nodal equations with the nodal constraints that $\mathbf{S} = \mathbf{VI}^*$ is obtained as follows ($p$ indicates the iteration number and is *not* a power). For node 1:

$$\mathbf{I}_1 = \mathbf{V}_1\mathbf{Y}_{11} + \mathbf{V}_2\mathbf{Y}_{12} + \mathbf{V}_3\mathbf{Y}_{13}$$

$$\therefore \mathbf{V}_1 = -\mathbf{V}_2\frac{\mathbf{Y}_{12}}{\mathbf{Y}_{11}} - \mathbf{V}_3\frac{\mathbf{Y}_{13}}{\mathbf{Y}_{11}} + \frac{\mathbf{I}_1}{\mathbf{Y}_{11}}$$

$$\therefore \mathbf{V}_1^* = -\frac{\mathbf{Y}_{12}^*}{\mathbf{Y}_{11}^*}\mathbf{V}_2^* - \frac{\mathbf{Y}_{13}^*}{\mathbf{Y}_{11}^*}\mathbf{V}_3^* + \frac{\mathbf{I}_1^*}{\mathbf{Y}_{11}^*}$$

Substituting $\mathbf{I}_1^* = \mathbf{S}_1/\mathbf{V}_1$ and writing in the iterative form,

$$\mathbf{V}_1^{p+1} = -\frac{\mathbf{Y}_{12}}{\mathbf{Y}_{11}}\mathbf{V}_2^p - \frac{\mathbf{Y}_{13}}{\mathbf{Y}_{11}}\mathbf{V}_3^p + \frac{\mathbf{S}_1^*}{\mathbf{Y}_{11}}\frac{1}{\mathbf{V}_1^{*P}}$$

Similarly,

$$V_2^{p+1} = -\frac{Y_{21}}{Y_{22}} V_1^{p+1} - \frac{Y_{23}}{Y_{22}} V_3^p + \frac{S_2^*}{Y_{22}} \frac{1}{V_2^{*p}}$$

and

$$V_3^{p+1} = -\frac{Y_{31}}{Y_{33}} V_1^{p+1} - \frac{Y_{32}}{Y_{33}} V_2^{p+1} + \frac{S_3^*}{Y_{33}} \frac{1}{V_3^{*p}}$$

At any node $i$ the already scanned nodes up to $i$ will have new values appropriate to the $(p + 1)$ iteration and the nodes yet to be scanned ($j > i$) are appropriate to iteration $p$. Generally,

$$(V_i^{p+1}) = +\frac{S_i^*}{Y_{ii}} \cdot \frac{1}{V_i^{*p}} - \sum_{\substack{j>1 \\ j \neq i}} \frac{Y_{ij}}{Y_{ii}} V_j^p - \sum_{\substack{j>i \\ j \neq i}} \frac{Y_{ij}}{Y_{ii}} V_i^{p+1} \qquad (6.7)$$

In this particular case, node 3 is the slack-bus, and as $V_i$ is known, the equation for it is not required. It is seen in the above equations that the new value of $V_i$ in the preceding equation is immediately used in the next equation, i.e. $V_1^{p+1}$ is used in the $V_2$ equation. Each node is scanned, in turn, over a complete iteration.

Equation (6.7) refers to a busbar with $P$ and $Q$ specified. At a generator node ($i$), usually $V_i$ and $P_i$ are specified with perhaps upper and lower limits to $Q_i$.

The magnitude of $V_i$ is fixed, but its phase depends on $Q_i$. The values of $V_i$ and $Q_i$ from the previous iteration are not related by equation (6.7) because $V_i$ has been modified to give a constant magnitude. It is necessary at the next iteration to calculate the value of $Q_i$ corresponding to $V_i$ from equation (6.7), i.e. $Q_i =$ imaginary part of $S_i$, i.e. of

$$Y_{ii}^* V_i^{p-1} \left[ V_i^{*p} + \sum_{\substack{j \neq i \\ j>i}} \frac{Y_{ij}^*}{Y_{ii}^*} V_j^{*p-1} + \sum_{\substack{j \neq i \\ j<i}} \frac{Y_{ij}^*}{Y_{ii}^*} V_j^{*p} \right]$$

This value of $Q_i$ holds for existing value of $V_i$ and is then substituted into

$$-\sum_{\substack{j \neq i \\ j>i}} \frac{Y_{ij}^*}{Y_{ii}^*} V_j^{*p} + \frac{P_i + jQ_i}{Y_{ii}^* V_i^p} \sum_{\substack{j \neq i \\ j<i}} \frac{Y_{ij}^*}{Y_{ii}^*} V_j^{*p+1}$$

to obtain $(V_i^{p+1})^*$.

The real and imaginary components of $V_i^{p+1}$ are then multiplied by the ratio $V_i/V_i^{p+1}$, thus complying with the constant $V_i$ constraint. This final step is a slight approximation, but the error involved is small and the saving in computation is large. The phase of $V_i$ is thus found and the iteration can proceed to the next node. The process continues until the value of $V^{p+1}$ at any node differs from $V^p$ by a specified amount, a common figure being 0.0001 p.u. The study is commenced by assuming 1 p.u. voltage at all nodes except one; this exception is necessary in order that current flows may be obtained in the first iteration.

***Acceleration factors***   The number of iterations required to reach the specified convergence can be greatly reduced by the use of acceleration factors. The correction in voltage from $V^p$ to $V^{p+1}$ is multiplied by such a factor so that the new voltage is brought closer to its final value, i.e.

$$^1V^{p+1} = V^p + \alpha(V^{p+1} - V^p)$$
$$= V^p + \alpha\Delta V^p$$

where $^1V^{p+1}$ is the accelerated new voltage. It can be shown that a complex value of $\alpha$ reduces the number of iterations more than a real value. However, up to the present, only real values seem to be used; the actual value depends on the nature of the system under study but a value of 1.6 is widely used. The Gauss–Seidel method with the use of acceleration factors is known as the method of successive overrelaxation.

## Transformer tap-changing

Further changes which must be accommodated in the admittance matrix are those due to transformer tap-changing. When the ratio is at the nominal value the transformer is represented by a single series impedance, but when off-nominal, adjustments have to be made as follows.

Consider a transformer of ratio $t : 1$ connected between two nodes $i$ and $j$; the series admittance of the transformer is $Y_t$. Referring to Figure 6.8(a), the following nodal-voltage equation holds:

$$I_j = V_j(Y_{jr} + Y_{j0} + Y_{js} + Y_t) - (V_r Y_{jr} + V_x Y_t + V_s Y_{js} + V_0 Y_{j0})$$

As $V_0 = 0$ and $V_x = V_i/t$,

$$I_j + V_s Y_{js} = V_j(Y_{jr} + Y_{j0} + Y_{js} + Y_t) - \left(Y_{jr}V_r + \frac{Y_t V_i}{t}\right)$$

where $x$ is an artificial node between the voltage transforming element and the transformer admittance. From this last equation it is seen that for the node on

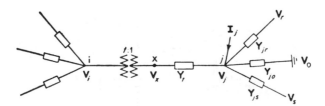

**Figure 6.8**(a)   Equivalent circuit of transfofrmer with off-nominal tap ratio. Transformer series admittance on non-tap side

the off-tap side of the transformer (i.e. the more remote of the two nodes $i$ and $j$), the following conditions apply:

when forming $\mathbf{Y}_{jj}$ use $\mathbf{Y}_t$ for the transformer, and when forming $\mathbf{Y}_{ij}$ use $\mathbf{Y}_t/t$ for the transformer.

It can similarly be shown for the tap-side node that the following conditions apply:

when forming $\mathbf{Y}_{ii}$ use $\mathbf{Y}_t/t^2$, and when forming $\mathbf{Y}_{ij}$ use $\mathbf{Y}_t/t$.

These conditions may be represented by the $\pi$ section shown in Figure 6.8(b), although it is probably easier to modify the mutual and self-admittances directly.

## (ii) The Newton–Raphson method

Although the Gauss–Seidel was the first popular method, the Newton–Raphson method was subsequently used increasingly. With some systems the latter gives a greater assurance of convergence and is at the same time economical in computer time.

The basic iterative procedure is as follows:

value at new iteration,

$$x^{p+1} = x^p - \frac{f(x^p)}{f'(x^p)}$$

extending this to a multi-equation system,

$$\mathbf{x}^{p+1} = \mathbf{x}^{(p)} - J^{-1}(\mathbf{x}^p)f(\mathbf{x}^p)$$

where the $\mathbf{x}$'s in $f$ are column vectors and $J(\mathbf{x}^p)$ is a matrix known as the Jacobian matrix, of the form

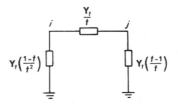

**Figure 6.8(b)**   The $\pi$ section to represent transformer with off-nominal tap ratio

$$
\begin{bmatrix}
\dfrac{\partial f_1}{\partial x_1} & \cdot & \cdot & \cdot & \dfrac{\partial f_1}{\partial x_n} \\
\cdot & & & & \cdot \\
\cdot & \cdot & \dfrac{\partial f_k}{\partial x_k} & & \cdot \\
\cdot & \cdot & \cdot & \cdot & \cdot \\
\dfrac{\partial f_n}{\partial x_1} & \cdot & \cdot & \cdot & \dfrac{\partial f_n}{\partial x_n}
\end{bmatrix}
p\text{th iteration}
$$

Consider now the application to an $n$-node power system, for a link connecting nodes $k$ and $j$ of admittance $\mathbf{Y}_{kj}$,

$$
P_k + jQ_k = \mathbf{V}_k\mathbf{I}_k^* = \mathbf{V}_k \sum_{j=1}^{n-1}(\mathbf{Y}_{kj}\mathbf{V}_j)^*
$$

Let

$$
\mathbf{V}_k = a_k + jb_k \qquad \text{and} \qquad \mathbf{Y}_{kj} = G_{kj} - jB_{kj}
$$

Then,

$$
P_k + jQ_k = (a_k + jb_k) \sum_{1}^{n-1}[(G_{kj} - jB_{kj})(a_j + jb_i)]^* \tag{6.8}
$$

from which,

$$
P_k = \sum_{j=1}^{n-1}[a_k(a_jG_{kj} + b_jB_{kj}) + b_k(b_jG_{kj} - a_jB_{kj})] \tag{6.9}
$$

$$
Q_k = \sum_{j=1}^{n-1}[b_k(a_jG_{kj} + b_jB_{kj}) - a_k(b_jG_{kj} - a_jB_{kj})] \tag{6.10}
$$

Hence, there are two non-linear simultaneous equations for each node. Note that $(n-1)$ nodes are considered because the slack node $n$ is completely specified.

Changes in $P$ and $Q$ are related to changes in $a$ and $b$ by equations (6.9) and (6.10), e.g.

$$
\Delta P_1 = \frac{\partial P_1}{\partial a_1}\Delta a_1 + \frac{\partial P_1}{\partial a_2}\Delta a_2 + \cdots\frac{\partial P_1}{\partial a_{n-1}}\Delta a_{n-1}
$$

Similar equations hold in terms of $\Delta P$ and $\Delta b$, and $\Delta Q$ in terms of $\Delta a$ and $\Delta b$. The equations may be expressed generally in the following manner:

$$\begin{bmatrix} \Delta P_1 \\ \cdot \\ \cdot \\ \cdot \\ \Delta P_{n-1} \\ \Delta Q_1 \\ \cdot \\ \cdot \\ \Delta Q_{n-1} \end{bmatrix} = \begin{bmatrix} \dfrac{\partial P_1}{\partial a_1} & \cdots & \dfrac{\partial P_1}{\partial a_{n-1}} & \dfrac{\partial P_1}{\partial b_1} & \cdots & \dfrac{\partial P_1}{\partial b_{n-1}} \\ & \cdots & & & \cdots & \\ \dfrac{\partial P_{n-1}}{\partial a_1} & \cdots & \dfrac{\partial P_{n-1}}{\partial a_{n-1}} & \dfrac{\partial P_{n-1}}{\partial b_1} & \cdots & \dfrac{\partial P_{n-1}}{\partial b_{n-1}} \\ \dfrac{\partial Q_1}{\partial a_1} & \cdots & \dfrac{\partial Q_1}{\partial a_{n-1}} & \dfrac{\partial Q_1}{\partial b_1} & \cdots & \dfrac{\partial Q_1}{\partial b_{n-1}} \\ & \cdots & & & \cdots & \\ \dfrac{\partial Q_{n-1}}{\partial a_1} & \cdots & \dfrac{\partial Q_{n-1}}{\partial a_{n-1}} & \dfrac{\partial Q_{n-1}}{\partial b_1} & \cdots & \dfrac{\partial Q_{n-1}}{\partial b_{n-1}} \end{bmatrix} \begin{bmatrix} \Delta a_1 \\ \cdot \\ \cdot \\ \cdot \\ \Delta a_{n-1} \\ \Delta b_1 \\ \cdot \\ \cdot \\ \Delta b_{n-1} \end{bmatrix} \qquad (6.11)$$

For convenience, denote the Jacobian matrix by

$$\begin{bmatrix} J_A & J_B \\ J_C & J_D \end{bmatrix}$$

The elements of the matrix are evaluated for the values of $P, Q,$ and $V$ at each iteration as follows.

For the submatrix $J_A$ and from equation (6.9),

$$\frac{\partial P_k}{\partial a_j} = a_k G_{kj} - b_k B_{kj} \qquad (6.12)$$

where $k \neq j$, i.e. off-diagonal elements.

Diagonal elements,

$$\frac{\partial P_k}{\partial a_k} = 2a_k G_{kk} + b_k B_{kk} - b_k B_{kk} + \sum_{\substack{j=1 \\ j \neq k}}^{n-1}(a_j G_{kj} + b_j B_{kj}) \qquad (6.13)$$

This element may be more readily obtained by expressing some of the quantities in terms of the current at node $k, \mathbf{I}_k$, which can be determined separately at each iteration

Let

$$\mathbf{I}_k = c_k + jd_k = (G_{kk} - jB_{kk})(a_k + jb_k) + \sum_{\substack{j=1 \\ j \neq k}}^{n-1}(G_{kj} - jB_{jk})(a_j + jb_j)$$

from which,

$$c_k = a_k G_{kk} + b_k B_{kk} + \sum_{\substack{j=1 \\ j \neq k}}^{n-1}(a_j G_{kj} + b_j B_{kj})$$

and

$$d_k = b_k G_{kk} - a_k B_{kk} + \sum_{\substack{i=1 \\ j \neq k}}^{n-1} (b_j G_{kj} - a_j B_{jk})$$

So that,

$$\frac{\partial P_k}{\partial a_k} = a_k G_{kk} - b_k B_{kk} + c_k$$

For $J_\mathrm{B}$,

$$\frac{\partial P_k}{\partial b_k} = a_k B_{kk} + b_k G_{kk} + d_k$$

and

$$\frac{\partial P_k}{\partial b_j} = a_k B_{kj} + b_k G_{kj} \qquad (k \neq j)$$

For $J_\mathrm{C}$,

$$\frac{\partial Q_k}{\partial a_k} = a_k B_{kk} + b_k G_{kk} - d_k$$

and

$$\frac{\partial Q_k}{\partial a_j} = a_k B_{kj} + b_k G_{kj} \qquad (k \neq j)$$

For $J_\mathrm{D}$,

$$\frac{\partial Q_k}{\partial b_j} = -a_k G_{kj} + b_k B_{kj} \qquad (k \neq j)$$

and

$$\frac{\partial Q_k}{\partial b_k} = -a_k G_{kk} + b_k B_{kk} + c_k$$

The process commences with the iteration counter 'p' set to zero and all the nodes except the slack-bus being assigned voltages, usually 1 p.u.

From these voltages, $P$ and $Q$ are calculated from equations (6.9) and (6.10). The changes are then calculated:

$$\Delta P_k^p = P_k \text{ (specified)} - P_k^p \text{ and } \Delta Q_k^p = Q_k \text{ (specified)} - Q_k^p$$

where $p$ is the iteration number.

Next the node currents are computed as

$$I_k^p = \left(\frac{P_k^p + jQ_k^p}{V^p}\right)^* = c_k^p + jd_k^p$$

The elements of the Jacobian matrix are than formed, and from equation (6.11),

$$
\begin{array}{|c|} \hline \Delta a \\ \hline \Delta b \\ \hline \end{array} = \begin{array}{|c|c|} \hline J_A & J_B \\ \hline J_C & J_D \\ \hline \end{array}^{-1} \begin{array}{|c|} \hline \Delta P \\ \hline \Delta Q \\ \hline \end{array} \tag{6.14}
$$

Hence, $a$ and $b$ are determined and the new values, $a_k^{p+1} = a_k^p + \Delta a_k^p$ and $b_k^{p+1} = b_k^p + \Delta b_k^p$, are obtained. The process is repeated $(p = p + 1)$ until $\Delta P$ and $\Delta Q$ are less than a prescribed tolerance.

The Newton–Raphson method has better convergence characteristics and for many systems is faster than the Gauss–Seidel method; the former has a much larger time per iteration but requires very few iterations (four is general), whereas the Gauss–Siedel requires at least 30 iterations, the number increasing with the size of system.

Acceleration factors may be used for the Newton–Raphson method. The quantities are frequently expressed in polar form.

The polar form of the equations has advantages and the equations are:

$$
\begin{aligned}
P_k &= P(V, \theta) \\
Q_k &= Q(V, \theta)
\end{aligned} \tag{6.15}
$$

The power at a bus is

$$
\begin{aligned}
\mathbf{S}_k &= P_k + jQ_k = \mathbf{V}_k \mathbf{I}_k^* \\
&= V_k \sum_{m \neq k} Y_{km}^* V_m^*
\end{aligned} \tag{6.16}
$$

$$
P_k = \sum_{m \neq k} V_k V_m (G_{km} \cos \theta_{km} + B_{km} \sin \theta_{km}) \tag{6.17}
$$

$$
Q_k = \sum_{m \neq k} V_k V_m (G_{km} \sin \theta_{km} - B_{km} \cos \theta_{km}) \tag{6.18}
$$

where $\theta_{km} = \theta_k - \theta_m$.

For a load bus,

$$
\Delta P_k = \sum_{m \neq k} \frac{\partial P_k}{\partial \theta_m} \Delta \theta_m + \sum_{m \neq k} \frac{\partial P_k}{\partial V_m} \Delta V_m \tag{6.19}
$$

$$
\Delta Q_k = \sum \frac{\partial Q_k}{\partial \theta_m} \Delta \theta_m + \sum \frac{\partial Q_k}{\partial V_m} \Delta V_m
$$

For a generator $(P, V)$ busbar, only the $\Delta P_k$ equation is used as $Q_k$ is not specified. The mismatch equation is

$$
\begin{array}{|c|} \hline \Delta \mathbf{P}^{P-1} \\ \hline \Delta \mathbf{Q}^{P-1} \\ \hline \end{array} = \begin{array}{|c|c|} \hline \mathbf{H}^{P-1} & \mathbf{N}^{P-1} \\ \hline \mathbf{J}^{P-1} & \mathbf{L}^{P-1} \\ \hline \end{array} \begin{array}{|c|} \hline \Delta \theta^P \\ \hline \left(\dfrac{\Delta \mathbf{V}^P}{\mathbf{V}^{P-1}}\right) \\ \hline \end{array} \tag{6.20}
$$

$\Delta\theta^P$ is the correction to $P, Q$, and $PV$ buses and $\Delta V^P/V^{P-1}$ is the correction to $P, Q$ buses.

For buses $k$ and $m$,

$$H_{km} = \frac{\partial P_k}{\partial\theta_m} = V_k V_m (G_{km} \sin\theta_{km} - B_{km} \cos\theta_{km})$$

$$N_{km} = V_m \frac{\partial P_k}{\partial V_m} = V_k V_m (G_{km} \cos\theta_{km} + B_{km} \sin\theta_{km})$$

$$J_{km} = \frac{\partial Q_K}{\partial\theta_m} = -V_k V_m (G_{km} \cos\theta_{km} + B_{km} \sin\theta_{km})$$

$$L_{km} = V_m \frac{\partial Q_m}{\partial V_m} = V_k V_m (G_{km} \sin\theta_{km} - B_{km} \cos\theta_{km})$$

Also,

$$H_{kk} = -Q_k - B_{kk} V_k^2$$
$$N_{kk} = P_k + G_{kk} V_k^2$$
$$J_{kk} = P_k - G_{kk} V_k^2$$
$$L_{kk} = Q_k - B_{kk} V_k^2$$

In the above, admittance of the link $km$ is $Y_{km} = G_{km} + jB_{km}$. The computational process can be enhanced by pre-ordering and dynamic ordering, defined as

- *Preordering*, in which nodes are numbered in sequence of increasing number of connections.

- *Dynamic ordering*, in which at each step in the elimination the next row to be operated on has the fewest non-zero terms.

### (iii) Decoupled load flow

The coupling between $P$–$\theta$ an $Q$–$V$ components is weak. Hence the equations can be reduced to

$$[P] = [T][\theta]$$
$$[Q] = [U][V - V_0] \tag{6.21}$$

At reference node $\theta_0 = 0$ and $V_k = V_0$, so elements of $T$ and $U$ are given by

$$T_{km} = -\frac{V_k V_m}{Z_{km}^2/X_{km}} \qquad U_{km} = -\frac{1}{Z_{km}^2/X_{km}}$$

$$T_{kk} = -\sum_{m\neq k} T_{km} \qquad U_{kk} = -\sum_{m\neq k} U_{km}$$

where $Z_{km}$ and $X_{km}$ are the branch impedance and reactance and $[U]$ is constant valued. Then

$$[\Delta \mathbf{P}] = [\mathbf{T}][\Delta \theta]$$
$$[\Delta \mathbf{Q/V}] = [\mathbf{U}][\Delta \mathbf{V}]$$

These are solved alternatively. Advantage is obtained by using the following:

$$[\Delta \mathbf{P/V}] = [\mathbf{A}][\Delta \theta]$$
$$[\Delta \mathbf{Q/V}] = [\mathbf{C}][\Delta \mathbf{V}]$$
(6.22)

***Fast decoupled load flow***   This makes the Jacobians of the decoupled method constant in value throughout the iteration. The following assumptions are made:

$$E_k = E_m = 1 \text{ p.u.}$$

$G_{km} \ll B_{km}$ and can be ignored; this is reasonable for lines and cables,

$$\cos (\theta_k - \theta_m) \approx 1$$
$$\sin (\theta_k - \theta_m) \approx 0$$

Hence,

$$[\Delta \mathbf{P}] = [\mathbf{B}][\Delta \theta]$$
$$[\Delta \mathbf{Q}] = [\mathbf{B}][\Delta \mathbf{V}]$$

where $B_{km} = -B_{km}$ for $m \neq k$, and

$$B_{kk} = \sum_{m \neq k} B_{km}$$

Further assumptions yield:

$$\left[\frac{\Delta P}{V}\right] = [B'][\Delta \theta]$$
$$\left[\frac{\Delta Q}{V}\right] = [B''][V]$$
(6.23)

where

$$B'_{km} = -\frac{1}{X_{km}}(m \neq k)$$

$$B'_{kk} = \sum_{m \neq k}\frac{1}{X_{km}}$$

$$B''_{km} = -B_{km}(m \neq k)$$

$$B''_{kk} = \sum_{m \neq k} B_{km}$$

Matrices $B'$ and $B''$ are real and constant in value and need to be triangulated only once.

## Example 6.5

Using the fast decoupled method calculate the angles and voltages after the first iteration for the three-node network described by the following admittance matrix (from Figure 6.9). $V_1$ is 230 kV, initially $V_2 = 220$ kV, $\theta_2 = 0$, and $V_3$ is 228 kV, $\theta_3 = 0$. Node 2 is a load consuming 200 MW, 120 MVAr; node 3 is a generator node set at 70 MW and 228 kV. $V_1$ is an infinite busbar.

| | 1 | 2 | 3 |
|---|---|---|---|
| 1 | $0.00819 - j0.049099$ | $-0.003196 + j0.019272$ | $-0.004994 + j0.030112$ |
| $Y =$ 2 | $-0.003196 + j0.019272$ | $0.007191 - j0.043099$ | $-0.003995 + j0.02409$ |
| 3 | $-0.004994 + j0.030112$ | $-0.003995 + j0.02409$ | $0.008989 - j0.053952$ |

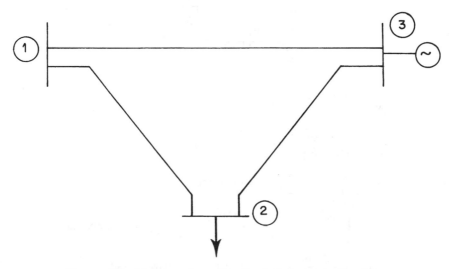

**Figure 6.9**  Three-node system for fast decoupled load flow

$$\begin{bmatrix} \dfrac{\Delta \mathbf{P}_2}{\mathbf{V}_2} \\[2ex] \dfrac{\Delta \mathbf{P}_3}{\mathbf{V}_3} \end{bmatrix} = \begin{bmatrix} \mathbf{B}'_{22} & \mathbf{B}'_{23} \\ \mathbf{B}'_{32} & \mathbf{B}'_{33} \end{bmatrix} \begin{bmatrix} \Delta\theta_2 \\ \Delta\theta_3 \end{bmatrix}$$

$V_1 = 230\,\text{kV}, \qquad V_2 = 220\,\text{kV}, \theta_2 = 0, \qquad V_3 = 228\,\text{kV}, \theta_3 = 0$

$B'_{22} = \dfrac{V_1}{X_{12}} + \dfrac{V_3}{X_{23}} = \dfrac{220}{50.5} + \dfrac{220}{40.4} = 9.80198$

$B'_{23} = -\dfrac{V_3}{X_{23}} = -\dfrac{220}{40.4} = -5.4455$

$B'_{32} = -\dfrac{V_2}{X_{23}} = -\dfrac{220}{40.4} = -5.4455$

$B'_{33} = \dfrac{V_1}{X_{13}} + \dfrac{V_2}{X_{23}} = \dfrac{220}{32.32} + \dfrac{220}{40.4} = 12.25247$

$C_{32} = L_{32} = \dfrac{B'_{32}}{B'_{22}} = -\dfrac{5.4455}{9.80198} = -0.5555$

$D_{32} = -B'_{22} = 9.80198$

$D_{33} = B'_{33} - \dfrac{B'_{32} B'_{23}}{B'_{22}} = 12.25247 - \dfrac{5.4455^2}{9.80198} = 9.2272$

$\Delta P_2^0 = -200 - 220(-0.003196 \times 230 + 0.00719 \times 220 - 0.003995 \times 228)$
$\qquad = -185.937\,\text{MW}$

$\Delta P_3^0 = 70 - 228(-0.004994 \times 230 - 0.003995 \times 220 + 0.008989 \times 228)$
$\qquad = 64.99\,\text{MW}$

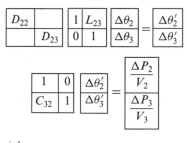

$\Delta\theta'_i = \dfrac{\Delta P_i}{V_i} - \sum_{k=2}^{i-1} C_{ik}\Delta\theta'_k \qquad (i = 2, 3)$

$\Delta\theta_i = \dfrac{\Delta\theta'_i}{D_{ii}} - \sum_{k=3}^{i+1} L_{ik}\Delta\theta_k \qquad (i = 3, 2)$

$\Delta\theta'_2 = \dfrac{\Delta P_2^{(0)}}{V_2} = \dfrac{-185.937}{220} = -0.845168$

$\Delta'\theta_3 = \left( \dfrac{\dfrac{64.99}{228} - 0.5555 \times 0.84516}{9.2272} \right) = \dfrac{-0.1844}{9.2272} = -0.1998\,\text{rad}$

$\Delta\theta'_2 = \dfrac{-0.845168}{9.80198} - 0.5555 \times 0.01988 = -0.09732\,\text{rad}$

Hence,

$$\theta_2' = \theta_2^{(0)} + \Delta\theta_2^{(0)} = -0.09732 \text{ rad} = -5.576°$$
$$\theta_3' = \theta_3^0 + \Delta\theta_3' = -0.012 \text{ rad} = -1.14477°$$

Considering reactive power at node 2,

$$
\begin{aligned}
\Delta Q_2^0 &= Q_2 - V_2^0\{V_1[G_{21}\sin(\theta_2'-\theta_1) - B_{21}\cos(\theta_2'-\theta_1)] - V_2^0 B_{22} \\
&\quad + V_3^0[G_{23}\sin(\theta_2'-\theta_3') - B_{23}\cos(\theta_2'-\theta_3')]\} \\
&= -120 - 220\{230[-0.003196\sin(-5.576-0) \\
&\quad - 0.019272\cos(-5.576-0)] - 220 \times 0.043099 \\
&\quad + 228[-0.003995\sin(-5.576+1.14477) \\
&\quad - 0.02409\cos(-5.576+1.14477)]\} = -59.476 \text{ MVAr}
\end{aligned}
$$

Hence,

$$\Delta V_2' = \frac{\Delta Q_2^0}{B_{22}'' V_2^0} = \frac{-59.476}{0.043049 \times 220} = -6.28 \text{ kV}$$

and

$$V_1 = 230 \text{ kV}$$
$$V_2' = V_2^0 + \Delta V_2' = 220 - 6.28 = 213.72 \text{ kV}$$
$$V_3' = V_3^0 = 228 \text{ kV}$$

The process is repeated for the next iteration, and so on, until convergence is reached, when,

$$P_1 = 134.389 \text{ MW}$$
$$Q_1 = 56.77 \text{ MVAr}$$

## 6.5  Example of a Complex Load Flow

The line diagram of a system is shown in Figure 6.10, along with details of line and transformer impedances and loads. This represents a slightly simplified arrangement of the network used by Ward and Hale (1956). All quantities are expressed as per unit and the nodes are numbered as indicated. They are not ordered consecutively to show, in the solution, that the numbering system is not important, although in more sophisticated studies on larger systems it has been shown that certain orderings of nodes can produce faster convergence and solutions.

The solution of this problem has been carried out by digital computer; a brief description of the arrangement of essential data will be givn as well as the first iteration performed on a hand calculator. This program was developed for instructional purposes and is not as refined or sophisticated as commercial

## Table 6.1

| | | | |
|---|---|---|---|
| +5 — — — — — — — — — — — — — — — −n | | | |
| +8 — — — — — — — — — — — — — — −l | | | |

| | | | |
|---|---|---|---|
| +1 | | | |
| +2 | | | |
| +2 | | | Off-diagonal elements per row |
| +1 | | | |
| +2 | | | |
| +2 | | | |
| +3 | +1 | | |
| +5 | +2 | | Column numbers |
| +5 | | | |
| +4 | +3 | | |
| +0.708238 | −0.033932 | $Y_{12}/Y_{11}$ | |
| +0.081853 | +0.050025 | $Y_{23}/Y_{22}$ | |
| +1.010063 | −0/055033 | $Y_{21}/Y_{22}$ | |
| +0.646931 | −0.043097 | $Y_{35}/Y_{33}$ | Off-diagonal elements $Y_{ij}/\mathbf{Y}_{ii}$ |
| +0.353069 | +0.043097 | $Y_{32}/Y_{33}$ | |
| +0.640699 | −0.052397 | $Y_{45}/Y_{44}$ | |
| +0.726081 | −0.090184 | $Y_{54}/Y_{55}$ | |
| +0.292797 | +0.087839 | $Y_{53}/Y_{55}$ | |

Generation Table 1

| | | |
|---|---|---|
| +1.000000 | +0.000000 | |
| +1.000000 | +0.000000 | |
| +1.000000 | +0.000000 | Voltage estimates, $v_i$ |
| +1.000000 | +0.000000 | |
| +1.000000 | +0.000000 | |
| −0.000369 | +0.007706 | |
| −0.019534 | +0.066298 | |
| +0.179357 | −0.162088 | $P + jQ/Y_{ii}^{*}$ values |
| −0.014702 | +0.076068 | |
| −0.046094 | +0.058905 | |
| +0.234539 | +0.039069 | |
| +0 | +0 | |
| +0 | +0 | $V_S Y_{is}/Y_{ii}$ values  slack-bus connections |
| +0.366206 | +0.055921 | |
| +0 | +0 | |
| +0.558269 | −11.652234 | |
| +0.444860 | −8.164860 | |
| +1.021401 | −1.954524 | $Y_{ii}$ values |
| +0.433934 | −5.306045 | |
| +0.576541 | −4.641795 | |

| | | | |
|---|---|---|---|
| +0.223371 | +0.037209 | +101.000000 | |
| +0.000000 | +0.000000 | +0.000000 | |
| + 0.000000 | + 0.000000 | + 0.000000 | $Y_{is}/Y_{ii}$ values + slacknumbers |
| +0.348767 | +0.053259 | +101.000000 | |
| +0.000000 | +0.000000 | +0.000000 | |

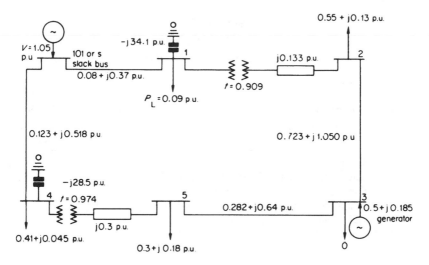

**Figure 6.10** System for complex load flow using iterative method

programs; it is hoped, however, that the beginner will find it easier to understand than the latter. In view of the nature of equation (6.7), the basic system data will be modified to allow less computation during the actual solution of the equations; this will be obvious as the method is described. The system admittance-matrix [**Y**] is stored by specifying the following: order of matrix ($n$), number of off-diagonal elements, number of off-diagonal elements per row, column numbers of the off-diagonal elements in each row, a list of off-diagonal elements ($Y_{ij}/Y_{ii}$). Tables 6.1A and 6.1B show the tabular arrangement for input data and Table 6.1 shows the actual data.

In the tables, $P$ and $Q$ are considered positive for watts and lagging vars generated or supplied, and negative when received. It should be noted that the $Y_{ij}$ values are, in fact, negative although they are shown as positive in the tables; this is useful as the $Y_{ij}$ terms appear as negative in the nodal equations. The connections to the slack-busbar ($V_s$) are shown separately, no equation is necessary for the slack-nodes as the voltages are fully specified. The input data are shown in Table 6.1 and identification of the various quantities will be made easier by reference to the following system matrix:

$$\begin{bmatrix} Y_{11} & Y_{12} & Y_{13} & Y_{14} & Y_{15} & Y_{1s} \\ Y_{21} & Y_{22} & Y_{23} & Y_{24} & Y_{25} & Y_{2s} \\ Y_{31} & Y_{32} & Y_{33} & Y_{34} & Y_{35} & Y_{3s} \\ Y_{41} & Y_{42} & Y_{43} & Y_{44} & Y_{45} & Y_{4s} \\ Y_{51} & Y_{52} & Y_{53} & Y_{54} & Y_{55} & Y_{5s} \end{bmatrix}$$

**Table 6.1A**

$l$ = number of links
$n$ = number of non-slack nodes

| $i$<br>Sending<br>end note | $j$<br>Receiving<br>end note | $R$<br>Resistance<br>per unit | $X$<br>Reactance<br>per unit | $B$<br>Susceptance<br>per unit | $t$<br>Off-nominal<br>transformer<br>ratio |
|---|---|---|---|---|---|
| | | | | | |

$n = 1$–99
Slack nodes are numbered starting at 101
Earth connections are numbered 0
When representing transformers, node $i$ must be the tap-side node and $R$
and $X$ refer to the non-tap side

**Table 6.1B**

$l$ = number of links
$n$ = number of non-slack nodes
$s$ = number of slack nodes
(subscripts **G** = generation, subscript **L** = loads)

| Node<br>number | $V_{\text{real}}$<br>per unit | $V_{\text{imag}}$<br>per unit | $P_{\text{G}}$<br>per unit | $Q_{\text{G}}$<br>per unit | $P_{\text{L}}$<br>per unit | $Q_{\text{L}}$<br>per unit |
|---|---|---|---|---|---|---|
| | | | | | | |

The initial part of the computer flow-diagram is shown in Figure 6.11.

Before proceeding with the first iteration, the calculation of the admittances associated with one of the transformers will be shown. Consider the transformer 1–2 of impedance j0.133 per unit and having $t = 0.909$. For the off-tap-side node 2, $\mathbf{Y}_{22}$ is formed using $\mathbf{Y}_t$ only, i.e.

$$\mathbf{Y}_{22} = \frac{1}{0.723 + j1.05} + \frac{1}{j0.133}$$
$$= 0.444860 - j8.164860$$

$$\mathbf{Y}_{21} = \frac{1}{j0.133} \cdot \frac{1}{0.909} = -j8.278000$$

The formation of the admittance matrix is necessary for both the Gauss–Seidel and Newton–Raphson methods. At this point, the methods differ, and the application of the Gauss–Seidel method will be given first with a complete computer solution, and then an outline will be given of the Newton–Raphson method applied to the problem. For node 1:

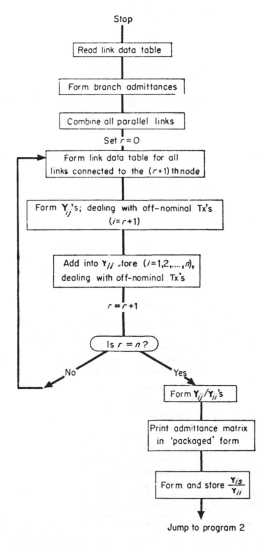

**Figure 6.11**(a)   Initial part of the flow diagram for data processing of the load-flow program

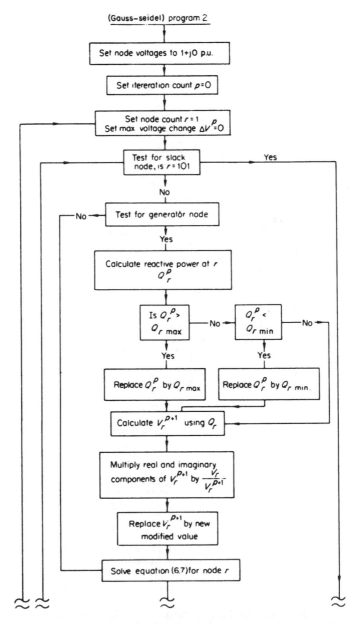

**Figure 6.11**(b)   Flow diagram for Gauss–Seidel method *(continued opposite)*

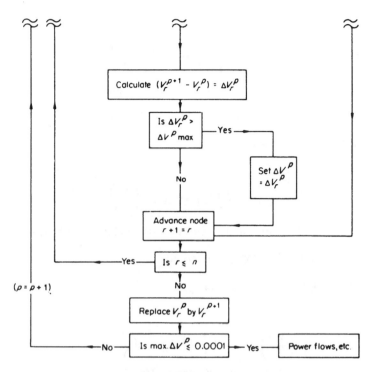

**Figure 6.11**(b)  (*continued*)

$$Y_{11} = \frac{1}{0.08 + j0.37} + \frac{1}{-j34.1} + \frac{1}{j0.133} \cdot \frac{1}{0.909^2} = 0.558269 - j11.652234$$

$$Y_{12} = Y_{21}$$

First iteration. Node 1:

$$V_1^{1*} = -\left(\frac{Y_{12}}{Y_{11}}\right)^* V_2^* - \left(\frac{Y_{1s}}{Y_{11}}\right)^* V_s^* + \frac{S_1}{Y_{11}^*} \frac{1}{V_1}$$

$$= (0.708238 - j0.033932)^* 1^* + (0.234539 + j0.039069)^* 1.05^*$$

$$+ (-0.000369 + j0.007706)\frac{1}{1}$$

$$\therefore V_1^{1*} = 0.942409 + j0.002569 \text{ p.u.}$$

Node 2:

$$V_2^{1*} = -\left(\frac{Y_{21}}{Y_{22}}\right)^* V_1^1 - \left(\frac{Y_{23}}{Y_{22}}\right)^* V_3^* + \left(\frac{S_2}{Y_{22}^*}\right)\frac{1}{V_2}$$

$$= (1.010063 + j0.055033)(0.942409 + j0.002569)$$

$$+ (0.081853 - j0.050025)1 + (-0.019534 + j0.066298)\frac{1}{1}$$

$$\therefore V_2^{1*} = 1.014073 + j0.070671 \text{ p.u.}$$

Node 3:

$$V_3^{1*} = -\left(\frac{Y_{32}}{Y_{33}}\right)^* V_2^{1*} - \left(\frac{Y_{35}}{Y_{33}}\right)^* V_5^* + \left(\frac{S_3}{Y_{33}^*}\right)\frac{1}{V_3}$$
$$= (0.353069 - j0.043097)(1.014073 + j0.070671)$$
$$+ (0.646931 + j0.043097)1 + (0.179357 - j0.162088)\frac{1}{1}$$
$$V_3^{1*} = 1.187371 - j0.137743 \text{ p.u.}$$

Node 4:

$$V_4^{1*} = -\left(\frac{Y_{45}}{Y_{44}}\right)^* V_5^* - \left(\frac{Y_{4s}}{Y_{44}}\right)^* V_s^* + \left(\frac{S}{Y_{44}^*}\right)\frac{1}{V_4}$$
$$= (0.640699 - j0.052397)(1) + (0.366206 - j0.055921)$$
$$+ (-0.014702 + j0.076068)\frac{1}{1}$$
$$V_4^{1*} = 0.992202 + j0.072543 \text{ p.u.}$$

Node 5:

$$V_5^{1*} = -\left(\frac{Y_{54}}{Y_{55}}\right)^* V_4^{1*} - \left(\frac{Y_{53}}{Y_{55}}\right)^* V_3^{1*} + \left(\frac{S}{Y_{55}^*}\right)\frac{1}{V_5}$$
$$= (0.726081 + j0.090184)(1.010513 + j0.069748)$$
$$+ (0.292797 - j0.087839)(1.187371 - j0.137743)$$
$$+ (-0.046094 + j0.058905)\frac{1}{1}$$

$$\therefore V_5^{1*} = 1.003342 - j0.056430 \text{ p.u.}$$
$$V_s \text{ or } V_{101} = 1.05000 + j0.000000 \text{ p.u.}$$

It will be noticed that the latest value of each nodal voltage is used. In this iteration no acceleration factor has been used (i.e. the factor = 1); if a factor of 1.6 is used, $V_1^1$ becomes

$$V_1 + 1.6(V_1^1 - V_1)$$

i.e.

$$1 + 1.6(-0.057591 - j0.002569)$$

or

$$0.907854 - j0.004110 \text{ p.u.}$$

This modified value of $V_1^1$ should then be used to evaluate $V_2^1$ when $V_2^1$ is modified to $V_2 + 1.6(V_2^1 - V_2)$, and so on. The busbar voltages after 30 iterations are

$$\mathbf{V}_1 = 0.918345 - j0.159312$$
$$\mathbf{V}_2 = 0.978674 - j0.221811$$
$$\mathbf{V}_3 = 1.101718 - j0.065242$$
$$\mathbf{V}_4 = 0.901468 - j0.194617$$
$$\mathbf{V}_5 = 0.903003 - j0.196604$$

## 6.5.1 Evaluation of line currents and power flows

Knowing the nodal voltages and admittances, the current, power, and var flows between nodes are readily obtained. Links with transformers, however, need special attention. Consider a transformer with its impedance referred to the non-tap side (Figure 6.8):

$$\mathbf{I}_i = \frac{\mathbf{I}_j}{t} \quad \text{and} \quad \mathbf{V}_i = t\mathbf{V}_x$$
$$\therefore \mathbf{I}_j = (\mathbf{V}_x - \mathbf{V}_j)\mathbf{Y}_t$$
$$= \left(\frac{\mathbf{V}_i}{t} - \mathbf{V}_j\right)\mathbf{Y}_t$$

The power at $j$ is

$$P_j = \mathbf{V}_j\left[\left(\frac{\mathbf{V}_i}{t} - \mathbf{V}_j\right)\mathbf{Y}_t\right]^*$$

Also,

$$\mathbf{I}_i = \left(\frac{\mathbf{V}_i}{t^2} - \frac{\mathbf{V}_j}{t}\right)\mathbf{Y}_t$$
$$\therefore P_i = \mathbf{V}_i\left[\left(\frac{\mathbf{V}_i}{t^2} - \frac{\mathbf{V}_j}{t}\right)\mathbf{Y}_t\right]^*$$

Therefore power transferred $i$ to $j$

$$= P_i - P_j$$

## 6.5.2 Application of the Newton–Raphson method to the system in Figure 6.10

The elements of the admittance matrix are determined as in the previous method, and it should be noted that for off-diagonal elements both $G$ and $B$ will be $(-1)$ times the values derived from the network.

Let all the busbar voltages be assigned a voltage of $(1 + j0)$ p.u.

At node 3:

$$P_3 = 0.5 \qquad Q_3 = 0.185 \text{(generated)}$$

$$\mathbf{Y}_{32} = 0.444860 - j0.646063 \text{ p.u.}$$
$$\mathbf{Y}_{35} = 0.576541 - j1.308461 \text{ p.u.}$$
$$\mathbf{Y}_{33} = 1.021401 - j1.954524 \text{ p.u.}$$

$$a_3 = 1, b_3 = 0 \qquad a_2 = 1, b_2 = 0 \qquad a_5 = 1, b_5 = 0$$

First iteration $(p = 0)$ :

$$
\begin{aligned}
P_3^0 = {} & 1 \times 1 \times (-0.444860) + 1 \times 0 \times (-0.646063) \\
& + 0 \times 0 \times (-0.444860) - 0 \times 1 \times (-0.646063) \\
& + 1 \times 1 \times (+1.021401) + 1 \times 0 \times (1.954524) \\
& + 0 \times 0 \times (1.021401) - 0 \times 1 \times (1.954524) \\
& + 1 \times 1 \times (-0.576541) + 1 \times 0 \times (-1.308461) \\
& + 0 \times 0 \times (-0.576541) - 0 \times 1 \times (-1.308461) \\
= {} & 0.0 \text{ p.u.}
\end{aligned}
$$

Note that $\mathbf{Y}_{31}, \mathbf{Y}_{34}$ do not exist. (The above result would be expected in the initial case as all the involved voltages are equal to 1 p.u.) Similarly,

$$Q_3^0 = 0.0 \text{ p.u.}$$

Therefore,

$$\Delta P_3^0 = 0.5 - 0 = 0.5 \text{ p.u.}$$

and

$$\Delta Q_3^0 = 0.185 - 0 = 0.185 \text{ p.u.}$$

$\Delta P$ and $\Delta Q$ for the remaining non-slack nodes are similarly obtained.

$$\mathbf{I}_k^p = \frac{P_k^p - jQ_k^p}{(V_k^p)^*} \qquad \text{and} \qquad \mathbf{I}_3^0 = \frac{0.5 - j0.185}{1 - j0}$$

hence

$$c_3^0 = 0.5 \qquad \text{and} \qquad d_3^0 = -0.185$$

The elements of the Jacobian are determined next:

$$\frac{\partial P_3}{\partial a_3} = a_3^0 G_{33} - b_3^0 B_{33} + C_3^0$$

$$= 1(1.021401) - 0.0 + 0.5 = 1.521401$$

$$\frac{\partial P_3}{\partial a_2} = a_3^0 G_{32} - b_3^0 B_{32}$$

$$= 1(-0.444860) + 0.0$$

$$\frac{\partial P_3}{\partial a_5} = -0.576541$$

$$\frac{\partial P_3}{\partial b_3} = a_3^0 B_{33} + b_3^0 G_{33} + d_3^0$$

$$= 1(1.954520) + 0.0 + (-0.185) = 1.76952$$

$$\frac{\partial P_3}{\partial b_2} = a_3^0 B_{32} + b_3^0 G_{32} = 1(-0.646063) + 0.0$$

$$\frac{\partial P_3}{\partial b_5} = 1(-1.308461) + 0.0$$

Similarly, we obtain

$$\frac{\partial Q_3}{\partial a_3}, \quad \frac{\partial Q_3}{\partial a_2}, \quad \frac{\partial Q_3}{\partial a_5}, \quad \frac{\partial Q_3}{\partial b_3}, \quad \frac{\partial Q_3}{\partial b_2}, \quad \frac{\partial Q_3}{\partial b_5}$$

The Jacobian matrix for the first iteration is thus formed and inverted, and then $\Delta a_k^0$ and $\Delta b_k^0$ ($k = 1 \ldots 5$) are evaluated. hence, $V_k^{0+1} = V_k^0 + \Delta a^0 + j\Delta b^0$. The process is repeated until changes in real and reactive power at each bus are less than a prescribed amount, say 0.01 p.u.

### 6.5.3 Summary

The direct method involving matrix inverstion and a final iterative procedure to deal with the restraints at the nodes has advantages for smaller networks. For large networks the fast decoupled load flow (FDLF) method is preferable. System information (such as specified generation and loads, line and transformer series, and shunt admittances) is fed into the computer. The data are expressed in per unit on arbitrary MVA and voltage bases. The computer formulates the self- and mutual-admittances for each node.

If the system is very sensitive to reactive power flows, i.e. the voltages change considerably with change in load and network configuration, the computer program may diverge. It is preferable to allow the reactive-power outputs of generators to be initially without limits to ensure an initial convergence. Convergence having been attained, the computer evaluates the real and reactive power flows in each branch of the system, along with losses, absorption of vars, and any other information that may be required. Programs are available which automatically adjust the tap settings of transformers to optimum values. Also, facilities exist for outputting only information regarding overloaded and underloaded lines; this is very useful when carrying out a series of load flows investigating the outages of plant and lines for security assessment purposes.

## 6.6 Optimal Power Flows

In Chapter 4, economic loading of generating plant, using '$B$' coefficients to represent losses, was detailed using a differentiation and minimization method. In practice, non-linear minimization techniques with constraints are used to achieve a minimum of some function associated with the most economic operation of the system. Typical functions to be minimized are:

- total cost of system operation;
- total system losses;
- reactive requirements.

Nowadays, mathematical minimization techniques which guarantee a solution for many thousands of equations are available commercially. The emphasis is on speed, minimum data storage, accuracy, and easy data handling and interfacing. Normally, it is necessary to set up a scalar function, $f(x)$, to be minimized subject to constraints expressed as:

$$\min f(x)$$

subject to

$$g_i(x) \geqslant 0 \qquad \text{where } i = 1, 2 \ldots m$$

and

$$h_j(x) = 0 \qquad \text{where } j = 1, 2 \ldots p$$

In the case of a power system, $f(x)$ is often expressed as a cost function comprising the sum of generator costs plus the system losses; thus.

$$f(x) = \sum_{i=1}^{G} C_i P_i + \sum_{b=1}^{B} C_L P_L$$

where $C_i$ is cost per unit of output of generation $P_i$ and $C_L P_L$ is cost of losses in each circuit of $B$ branches, subject to

$$\left.\begin{array}{c} \sum P_K = 0 \\ \text{and} \\ \sum Q_K = 0 \end{array}\right\} \text{at each node } k \text{ (Kirchoff's laws)}$$

and the following limits or constraints apply:

$$\left.\begin{array}{l} {}_{\min}P_i < P_i < P_{i\max} \\ {}_{\min}Q_i < Q_i < Q_{\max} \end{array}\right\} \quad \text{each generator } i$$

$$_{\min}V_k < V_k < V_{k\max} \qquad \text{each node } k$$

$$S_{mn} < S_{mn\max} \qquad \text{for each circuit } m \text{ to } n$$

Additional constraints can be as follows:

$$_{\min}T_t < T_t < T_{t\max} \qquad \text{for each tap-change transformer t}$$
$$_{\min}Q_c < Q_c < Q_{c\max} \qquad \text{for each compensator c}$$

Limits could also be added to ensure transient stability in the form:

$$_{\min}\theta_{mn} < \theta < \theta_{mn\max} \qquad \text{where } \theta \text{ is an angle between node } m \text{ and node } n$$

More generally, it may be necessary to ensure that power $P$ and var $Q$ flows across defined boundaries in a network do not exceed pre-determined limits for security reasons. For example:

$$\sum_{mn\,\mathrm{Set}S} P_{mn}^s \leqslant P_{\mathrm{SpeC}} \qquad \text{where } P_{mn}^S \text{ is a set of flows whose sum must not exceed } P_{\mathrm{SpeC}}.$$

It can readily be seen that obtaining an optimum (minimum) solution is a complicated task, even with modern computers. The mathematics appropriate to this form of optimization is continually evolving and improving, including the use of novel techniques such as fuzzy logic, artificial neural networks, and genetic algorithms. Consequently, the employment of a computer library or commercial packages best suited to individual requirements is recommended.

# Problems

**6.1.** A single-phase distributor has the following loads at the stated distances from the supply end: 10kW at 10 m, 10 kW at 0.9 p.f. lagging at 16 m, 5 kW at 0.8 p.f. lagging at 91.5 m, and 20 kW at 0.95 p.f. lagging at 137 m. The loads may be assumed to be constant current at their nominal voltage values (240 V). If the supply voltage is 250 V and the maximum voltage drop is 5 per cent of the nominal value, determine the nearest commercially available conductor size.
(Answer: $r + 0.376x = 0.7 \times 10^{-3}$)

**6.2** In the d.c. network shown in Figure 6.12, calculate the voltage at node B by inverting the admittance matrix. Check the answer by Thevenin's theorem.
(Answer: 247.7 V)

**6.3** In the interconnected network shown in Figure 6.13, calculate the current in feeder BC.
(Answer: 26.75 A)

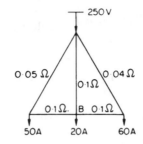

**Figure 6.12** Network for Problem 6.2

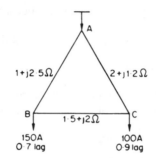

**Figure 6.13** Network for Problem 6.3

**6.4** In the network shown in Figure 6.14, the loads are represented by constant impedances $1 + j1$ p.u. Determine the current distributions in the network (a) when the transformer has its nominal ratio; (b) when the transformer is tapped up to 10 per cent.
(Note: determine the distribution with the off-nominal voltage alone and use superposition.)
(Answer: Transformer branch (a) 0; (b) $0.0735 - j0.075$ (p.u.))

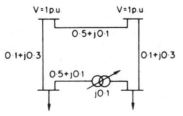

Each load represented by $Z = 1 + j1$ p.u.

**Figure 6.14** Network for Problem 6.4

**6.5** Enumerate the information which may be obtained from a load-flow study. Part of a power system is shown in Figure 6.15. The line-to-neutral reactances and values of real

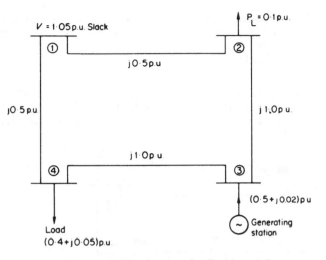

**Figure 6.15** System for Problem 6.5

and reactive power (in the form $P \pm jQ$) at the various stations are expressed as per unit values on a common MVA base. Resistance may be neglected. By the use of an iterative method suitable for a digital computer, calculate the voltages at the stations after the first iteration without the use of an accelerating factor.

(Answer: $V_2 1.03333 - j0.03333$ p.u.; $V_3 1.11666 + j0.23333$ p.u.; $V_4 1.05556 + j0.00277$ p.u.)

**6.6** A 400 kV interconnected system is supplied from bus A, which may be considered to be an infinite busbar. The loads and line reactances are as indicated in Figure 6.16.

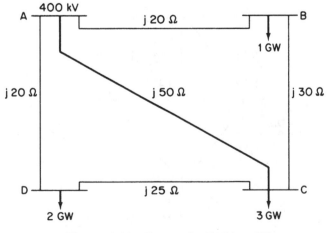

**Figure 6.16** System for Problem 6.6

Determine the flow of power in each line. In line AC a phase shift transformer is installed which gives a phase advance of 10° at C with respect to A. Calculate the new power flow in line AC.
(Answer: $P_{AC}$1.35 GW; $P_{AC}$1 GW)

**6.7** Determine the voltage at bus (2) and the reactive power at bus (3) as shown in Figure 6.17, after the first iteration of a Gauss–Seidel low-flow method. Assume the initial voltage to be $1\angle0°$ p.u. All the quantities are in per unit on a common base.
(Answer: $0.99 - j0.0133$)

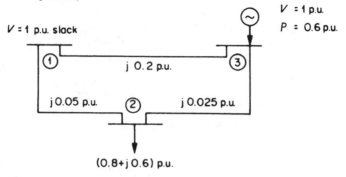

**Figure 6.17** System for Problem 6.7

**6.8** In Figure 6.18 the branch reactances and busbar loads are given in per unit on a common base. Branch resistance is neglected.

Explain briefly why an iterative method is required to determine the busbar voltages of this network.

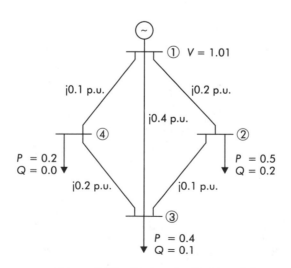

**Figure 6.18** System for Problem 6.8

Form the bus admittance matrix for this network.

Using busbar 1 as the slack (reference) busbar, carry out the first iteration of a Gauss–Seidel load-flow algorithm to determine the voltage at all busbars. Assume the initial voltages of all busbars to be 1.01 p.u.

(Answer: $V_2 = 0.997 - j0.03$, $V_3 = 1.01 - j0.01$, $V_4 = 1.01 - j0.026$)

   (*From E.C. Examination, 1995*)

# 7
# *Fault Analysis*

## 7.1 Introduction

An essential part of the design of a power supply network is the calculation of the currents which flow in the components when faults of various types occur. In a fault survey, faults are applied at various points in the network and the resulting currents are obtained by digital computation. The magnitude of the fault currents give the engineer the current settings for the protection to be used and the ratings of the circuit breakers. In some circumstances, the effect of open circuits may need investigation.

The types of fault commonly occurring in practice are illustrated in Figure 7.1, and the most common of these is the short circuit of a single conductor to ground or earth. Often, the path to earth contains resistance $R_F$ in the form of an arc, as shown in Figure 7.1(f). Although the single-line-to-earth fault is the most common, calculations are frequently performed with the three-line, balanced short circuit (Figure 7.1(d) and (e)). This is the most severe fault and also the most amenable to calculation. The causes of faults are summarized in Table 7.1, which gives the distribution of faults, due to various causes, on the U.K. National Grid System in a typical year. Table 7.2 shows the components affected. In tropical countries the incidence of lightning is much greater than in the U.K., resulting in larger numbers of faults.

As well as fault current $I_f$, fault MVA is frequently used as a rating; this is obtained from the expression $\sqrt{(3)}V_L I_F \times 10^{-6}$, where $V_L$ is the nominal line voltage of the faulted part. The fault MVA is often referred to as the *fault level*. The calculation of fault currents can be divided into the following two main types:

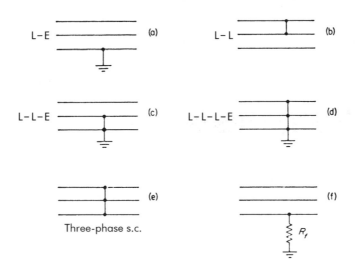

**Figure 7.1**   Common types of fault or short circuit (s.c.) L = line, E = earth

**Table 7.1**   Causes of overhead-line faults, British system 66 kV and above

|  | Faults/160 km of line/year |
|---|---|
| Lightning | 1.59 |
| Dew, fog, frost | 0.15 |
| Snow, ice | 0.01 |
| Gales | 0.24 |
| Salt spray | 0.01 |
| Total | 2 faults per 160 km, giving a total of 232 faults on system/year |

**Table 7.2**   Distribution of faults, NGC system

| Type | Number of faults/year |
|---|---|
| Overhead lines | 289 |
| Cables | 67 |
| Switchgear | 56 |
| Transformers | 59 |
| Total | 471 |

1. Faults short-circuiting all three phases when the network remains balanced electrically. For these calculations, normal single-phase equivalent circuits may be used as in ordinary load-flow calculations.

2. Faults other than three-phase short circuits when the network is electrically unbalanced. To facilitate these calculations a special method for dealing with unbalanced networks is used, known as the method of *symmetrical components*.

The main objects of fault analysis may be stated as follows:

1. to determine maximum and minimum three-phase short-circuit currents;

2. to determine the unsymmetrical fault current for single and double line-to-earth faults, line-to-line faults, and sometimes for open-circuit faults;

3. investigation of the operation of protective relays;

4. determination of rated rupturing capacity of breakers;

5. to determine fault-current distribution and busbar-voltage levels during fault conditions.


## 7.2  Calculation of Three-Phase Balanced Fault Currents

The action of synchronous generators on three-phase short circuits has been described in Chapter 3. There it was seen that, dependent on the time elapsing from the incidence of the fault, either the transient or the subtransient reactance should be used to represent the generator. For specifying switchgear, the value of the current flowing at the instant at which the circuit contacts open is required. It has been seen, however, that the initially high fault current, associated with the subtransient reactance, decays with the passage of time. Modern air-blast circuit breakers usually operate in 2.5 cycles and $SF_6$ breakers within 1.5 cycles of 60 Hz or 50 Hz alternating current, and are associated with extremely fast protection. Older circuit breakers and those on lower voltage networks usually associated with relatively cruder protection can take in the order of 8 cycles or more to operate. In calculations it is usual to use the subtransient reactance of generators and to ignore the effects of induction motors. The calculation of fault currents ignores the direct-current component, the magnitude of which depends on the instant in the cycle that the short circuit occurs. If the circuit breaker opens a reasonable time after the incidence of the fault, the direct-current component will have decayed considerably. With fast-acting circuit breakers the actual current to be interrupted is increased by the direct-current component and it must be taken into account. To allow for the direct-current component of the fault current the symmetrical r.m.s. value is modified by the use of multiplying factors such as the following:

- 8-cycle circuit breaker opening time, multiply by 1;
- 3-cycle circuit breaker opening time, multiply by 1.2;
- 2-cycle circuit breaker opening time, multiply by 1.4.

Consider an initially unloaded generator with a short circuit across the three terminals, as shown in Figure 7.2. The generated voltage per phase is $E$ and therefore the short-circuit current is $[E/Z(\Omega)]A$, where $Z$ is either the transient or subtransient impedance. If $Z$ is expressed in per unit notation,

$$Z \text{ (p.u.)} = \frac{I_{FL}Z(\Omega)}{E}$$

$$\therefore Z \text{ } (\Omega) = \frac{EZ \text{ (p.u.)}}{I_{FL}}$$

Taking $I_{FL}$ (full-load or rated current) and $E$ as base values of voltage and current, the short-circuit current

$$I_{s.c.} = \frac{E}{Z \text{ } (\Omega)}$$

$$= \frac{EI_{FL}}{EZ \text{ (p.u.)}} = \frac{I_{FL}}{Z \text{ (p.u.)}} \tag{7.1}$$

Also, the three-phase short-circuit volt-amperes

$$= \sqrt{(3)}V_L I_{s.c.} = \frac{\sqrt{(3)}V_L I_{FL}}{Z \text{ (p.u.)}}$$

$$= \frac{\text{Base volt-amperes}}{Z \text{ (p.u.)}} \tag{7.2}$$

Hence the short-circuit level is immediately obtained if the impedance $Z$ (p.u.) from the source of the voltage to the point of the fault is known.

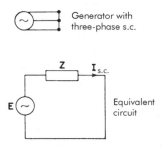

**Figure 7.2** Voltage source with short circuit and equivalent circuit

## Example 7.1

An 11.8 kV busbar is fed from three synchronous generators having the following ratings and reactances,

$$20\,\text{MVA}, X'0.08\,\text{p.u.};\qquad 60\,\text{MVA}, X'0.1\,\text{p.u.};\qquad 20\,\text{MVA}, X'0.09\,\text{p.u.}$$

Calculate the fault current and MVA if a three-phase symmetrical fault occurs on the busbars. Resistance may be neglected. The voltage base will be taken as 11.8 kV and the VA base as 60 MVA.

## Solution

The transient reactance of the 20 MVA machine on the above bases is $(60/20) \times 0.08$, i.e. 0.24 p.u., and of the 20 MVA machine $(60/20) \times 0.09$, i.e. 0.27 p.u. These values are shown in the equivalent circuit in Figure 7.3. As the generator e.m.f.s are assumed to be equal, one source may be used (Figure 7.3(c)). The equivalent reactance is

$$X_{eq} = \frac{1}{1/0.24 + 1/0.27 + 1/0.1} = 0.056\,\text{p.u.}$$

Therefore fault MVA

$$= \frac{60}{0.056} = 1071\,\text{MVA}$$

and fault current

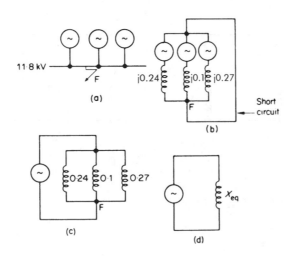

(a)

(b)

(c)

(d)

**Figure 7.3** Line diagram and equivalent circuits for Example 7.1

$$= \frac{1071 \times 10^6}{\sqrt{(3)} \times 11\,800}$$

$$= 52\,402\,\text{A}$$

## Example 7.2

In the network shown in Figure 7.4, a three-phase fault occurs at point F. Calculate the fault MVA at F. The per unit values of reactance all refer to a base of 100 MVA. Resistance may be neglected.

## Solution

The equivalent single-phase network of generator and line reactances is shown in Figure 7.4(b). This is replaced by the network shown in Figure 7.4(c) by the use of the delta–

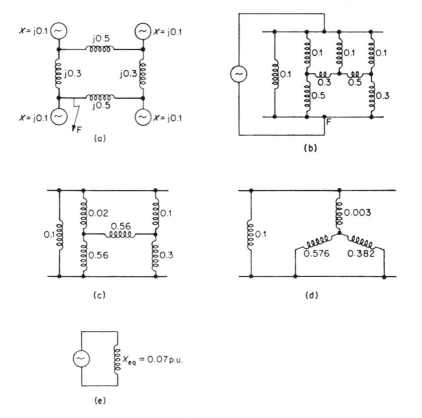

**Figure 7.4** Line diagram and equivalent circuits for Example 7.2

star transformation. A further transformation is carried out on the network in Figure 7.4(d) to give the final single equivalent reactance,

$$X_{eq} = 0.07\,\text{p.u.}$$

The fault level at point F

$$= \frac{100}{0.07} = 1430\,\text{MVA}$$

This value is based on the assumption of symmetrical fault current. Allowing for a multiplying factor of 1.4, the fault level is 2000 MVA.

### 7.2.1　Current limiting reactors

The impedances presented to fault currents by transformers and machines when faults occur on substation or generating station busbars are low. To reduce the high fault current which would do considerable damage mechanically and thermally, artificial reactances are sometimes connected between bus sections. These current-limiting reactors usually consist of insulated copper strip embedded in concrete formers; this is necessary to withstand the high mechanical forces produced by the current in neighbouring conductors. The position in the circuit occupied by the reactor is a matter peculiar to individual designs and installations (see Figure 7.5(a) and (b)).

### *Example 7.3*

A 400 kV power system contains three substations A, B, and C having fault levels (GVA) of 20, 20, and 30, respectively.

The system is to be reinforced by three lines each of reactance j5 Ω connecting together the three substations as shown in Figure 7.6(a). Calculate the new fault level at C (three-phase symmetrical fault). Neglect resistance.

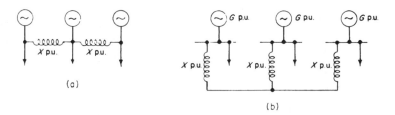

(a)　　　(b)

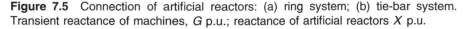

**Figure 7.5** Connection of artificial reactors: (a) ring system; (b) tie-bar system. Transient reactance of machines, *G* p.u.; reactance of artificial reactors *X* p.u.

*Solution*

The equivalent circuits for the new system are shown in Figure 7.6(a) and (b). The substations are represented by a voltage source (400 kV) in series with the value of reactance to give the specified initial fault level, e.g. for A, the fault current (before reinforcement)

$$= \frac{20 \times 10^9}{\sqrt{3} \times 400\,000} \, \text{A}$$

Hence the effective reactance 'behind' A when subject to a three-phase short circuit

$$= \frac{400\,000}{\sqrt{3}} \times \frac{\sqrt{(3)} \times 400\,000}{20 \times 10^9} = j\,8\,\Omega$$

Similarly, the effective reactance at $B = j8\,\Omega$ and at $C = j\,5.33\,\Omega$.

The 5 $\Omega$ mesh is transformed into a star with arms of value 1.66 $\Omega$ and the equivalent circuit with the fault on C after reinforcement is shown in Figure 7.6(b). Fault current at C with equivalent reactance j2.9 $\Omega$ gives:

$$= \frac{400\,000}{\sqrt{3}} \times \frac{1}{j2.9} = -j79.635\,\text{kA}$$

Fault level at C

$$= \sqrt{3} \times 400 \times 79.6\,\text{MVA} = 55\,148\,\text{MVA}$$

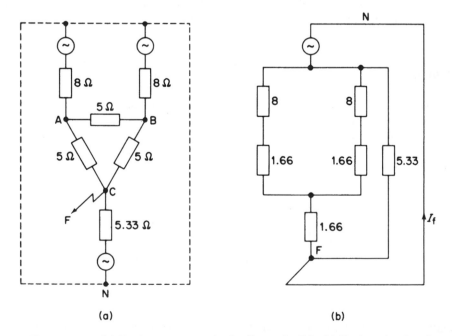

(a)                                              (b)

**Figure 7.6**   (a) Equivalent networks for Example 7.3. (b) Reduced network

Note that this value is very high and that the result of reinforcement is always to increase the fault levels. Maximum rating of a 400 kV circuit breaker is about 50 GVA, hence either a busbar must be run split between two incoming feeders or current limiters must be installed.

## 7.3 Method of Symmetrical Components

This method formulates a system of three separate phasor systems which, when superposed, give the unbalanced conditions in the circuit. It should be stressed that the systems to be discussed are essentially artificial and used merely as an aid to calculation. The various sequence-component voltages and currents do not exist as physical entities in the network, although they could be monitored by special filters.

The method postulates that a three-phase unbalanced system of voltages and currents may be presented by the following three separate systems of phasors:

1. a balanced three-phase system in the normal a–b–c (red–yellow–blue) sequence, called the *positive phase-sequence* system;

2. a balanced three-phase system of reversed sequence, i.e. a–c–b (red–blue–yellow), called the *negative phase-sequence* system;

3. three phasors equal in magnitude and phase revolving in the positive phase rotation, called the *zero phase-sequence* system.

In Figure 7.7 an unbalanced system of currents is shown with the corresponding system of symmetrical components. If each of the red-phase phasors are added, i.e. $\mathbf{I}_{1R} + \mathbf{I}_{2R} + \mathbf{I}_{0R}$, the resultant phasor will be $I_R$ in magnitude and direction; similar reasoning holds for the other two phases.

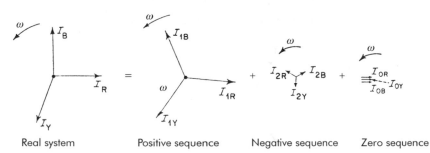

| Real system | Positive sequence | Negative sequence | Zero sequence |

**Figure 7.7** Real system and corresponding symmetrical components

To express the phasors algebraically, use is made of the complex operator 'a' (sometimes denoted by $\lambda$ or **h**) which denotes a phase-shift operation of $+120°$ and a multiplication of unit magnitude; i.e.

$$V\angle\phi \times \mathbf{a} = V\angle\phi \times 1\angle 120° = V\angle\phi + 120°, \mathbf{a} = e^{j2\pi/3} \text{ and } \mathbf{a}^3 = e^{j3\times 2\pi/3} = 1$$

Also,

$$\mathbf{a}^2 + \mathbf{a} = (-0.5 - j0.866) + (-0.5 + j0.866) = -1$$
$$\therefore \ \mathbf{a}^3 + \mathbf{a}^2 + \mathbf{a} = 0$$

and

$$1 + \mathbf{a} + \mathbf{a}^2 = 0$$

For positive-sequence phasors, taking the red phasor as reference,

$$\mathbf{I}_{1R} = \mathbf{I}_{1R}e^{j0} = \text{ reference phasor} \qquad \text{(Figure 7.8)}$$
$$\mathbf{I}_{1Y} = \mathbf{I}_{1R}(-0.5 - j0.866) = \mathbf{a}^2\mathbf{I}_{1R}$$

and

$$\mathbf{I}_{1B} = \mathbf{I}_{1R}(-0.5 + j0.866) = \mathbf{a}\mathbf{I}_{1R}$$

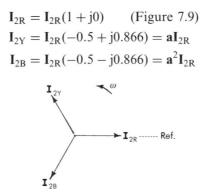

**Figure 7.8**   Positive-sequence phasors

For negative-sequence quantities,

$$\mathbf{I}_{2R} = \mathbf{I}_{2R}(1 + j0) \qquad \text{(Figure 7.9)}$$
$$\mathbf{I}_{2Y} = \mathbf{I}_{2R}(-0.5 + j0.866) = \mathbf{a}\mathbf{I}_{2R}$$
$$\mathbf{I}_{2B} = \mathbf{I}_{2R}(-0.5 - j0.866) = \mathbf{a}^2\mathbf{I}_{2R}$$

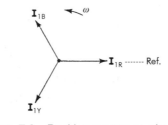

**Figure 7.9**   Negative-sequence phasors

Returning to the original unbalanced system of currents, $\mathbf{I_R}$, $\mathbf{I_Y}$, and $\mathbf{I_B}$,

$$\mathbf{I_R} = \mathbf{I_{1R}} + \mathbf{I_{2R}} + \mathbf{I_{0R}}$$

$$\mathbf{I_Y} = \mathbf{I_{1Y}} + \mathbf{I_{2Y}} + \mathbf{I_{0Y}} = \mathbf{a}^2\mathbf{I_{1R}} + \mathbf{a}\mathbf{I_{2R}} + \mathbf{I_{0R}}$$

$$\mathbf{I_B} = \mathbf{I_{1B}} + \mathbf{I_{2B}} + \mathbf{I_{0B}} = \mathbf{a}\mathbf{I_{1R}} + \mathbf{a}^2\mathbf{I_{2R}} + \mathbf{I_{0R}}$$

Hence in matrix form,

$$
\begin{bmatrix} \mathbf{I_R} \\ \mathbf{I_Y} \\ \mathbf{I_B} \end{bmatrix} =
\begin{bmatrix} 1 & 1 & 1 \\ 1 & \mathbf{a}^2 & \mathbf{a} \\ 1 & \mathbf{a} & \mathbf{a}^2 \end{bmatrix}
\begin{bmatrix} \mathbf{I_{0R}} \\ \mathbf{I_{1R}} \\ \mathbf{I_{2R}} \end{bmatrix}
\qquad (7.3)
$$

Inverting the matrix,

$$
\begin{bmatrix} \mathbf{I_{0R}} \\ \mathbf{I_{1R}} \\ \mathbf{I_{2R}} \end{bmatrix} = \tfrac{1}{3}
\begin{bmatrix} 1 & 1 & 1 \\ 1 & \mathbf{a} & \mathbf{a}^2 \\ 1 & \mathbf{a}^2 & \mathbf{a} \end{bmatrix}
\begin{bmatrix} \mathbf{I_R} \\ \mathbf{I_Y} \\ \mathbf{I_B} \end{bmatrix}
\qquad (7.4)
$$

The above also holds for voltages, i.e.

$$[\mathbf{E_{actual}}] = [\mathbf{T_s}][\mathbf{E_{1,2,0}}]$$

where $[\mathbf{T}_s]$ is the symmetrical component transformation matrix,

$$
\begin{bmatrix} 1 & 1 & 1 \\ 1 & \mathbf{a}^2 & \mathbf{a} \\ 1 & \mathbf{a} & \mathbf{a}^2 \end{bmatrix}
$$

In a three-wire system the instantaneous voltages and currents add to zero and therefore no single-phase component is required. A fourth wire or connection to earth must be provided for single-phase currents to flow. In a three-wire system the zero phase-sequence components are replaced by zero in equations (7.3) and (7.4). Also,

$$\mathbf{I_{0R}} = \frac{\mathbf{I_R} + \mathbf{I_Y} + \mathbf{I_B}}{3} = \frac{\mathbf{I_N}}{3}$$

where $\mathbf{I_N}$ is the neutral current.

$$\therefore \ \mathbf{I_N} = 3\mathbf{I_{0R}} = 3\mathbf{I_{0Y}} = 3\mathbf{I_{0B}}$$

In the application of this method it is necessary to calculate the symmetrical components of the current in each line of the network and then to combine them to obtain the actual values. The various phase-sequence values are obtained by considering a network derived from the actual network in which only a particular sequence current flows; for example, in a zero-sequence network, only zero-sequence currents and voltages exist. The positive-sequence network is identical with the real, balanced equivalent network, i.e. it is the same as used for three-phase symmetrical short-circuit studies. The negative-sequence network is almost the same as the real one except that the values of impedance used for rotating machines are different and there are no generated voltages from an ideal machine. The zero-sequence network is considerably different from the real one.

The above treatment assumes that the respective sequence impedances in each of the phases are equal, i.e. $\mathbf{Z}_{1R} = \mathbf{Z}_{1Y} = \mathbf{Z}_{1B}$, etc. Although this covers most cases met in practice, unequal values may occur in certain circumstances, e.g. an open circuit on one phase. The following equation applies for the voltage drops across the phase impedances:

$$\begin{bmatrix} \mathbf{V}_R \\ \mathbf{V}_Y \\ \mathbf{V}_B \end{bmatrix} = [\mathbf{T}_S] \begin{bmatrix} \mathbf{V}_{0R} \\ \mathbf{V}_{1R} \\ \mathbf{V}_{2R} \end{bmatrix} = \begin{bmatrix} \mathbf{Z}_R & & \\ & \mathbf{Z}_Y & \\ & & \mathbf{Z}_B \end{bmatrix} [\mathbf{T}_S] \begin{bmatrix} \mathbf{I}_{0R} \\ \mathbf{I}_{1R} \\ \mathbf{I}_{2R} \end{bmatrix}$$

From which,

$$\mathbf{V}_{0R} = \tfrac{1}{3}\mathbf{I}_{1R}(\mathbf{Z}_R + \mathbf{a}^2\mathbf{Z}_Y + \mathbf{a}\mathbf{Z}_B) + \tfrac{1}{3}\mathbf{I}_{2R}(\mathbf{Z}_R + \mathbf{a}\mathbf{Z}_Y + \mathbf{a}^2\mathbf{Z}_B) + \tfrac{1}{3}\mathbf{I}_{0R}(\mathbf{Z}_R + \mathbf{Z}_Y + \mathbf{Z}_B)$$

and, similarly, expressions $\mathbf{V}_{1R}$ and $\mathbf{V}_{2R}$ may be obtained.

It is seen that the voltage drop in each sequence is influenced by the impedances in all three phases. If, as previously assumed, $\mathbf{Z}_R = \mathbf{Z}_Y = \mathbf{Z}_B = \mathbf{Z}$, then the voltage drops become $\mathbf{V}_{1R} = \mathbf{I}_{1R}\mathbf{Z}$, etc., as before.

## 7.4    Representation of Plant in the Phase-Sequence Networks

### 7.4.1    The synchronous machine (see Table 3.1)

The positive-sequence impedance $Z_1$ is the normal transient or subtransient value. Negative-sequence currents set up a rotating magnetic field in the opposite direction to that of the positive-sequence currents and which rotates round the rotor surface at twice the synchronous speed; hence the effective impedance $(Z_2)$ is different from $Z_1$. The zero-sequence impedance $Z_0$ depends upon the nature of the connection between the star point of the windings and the earth and the single-phase impedance of the stator windings in parallel. Resistors or reactors are frequently connected between the star point and earth for reasons usually connected with protective gear and the limitation of overvoltages. Normally, the only voltage sources appearing in the networks are in the positive-sequence one, as the generators only generate positive-sequence e.m.f.s.

### 7.4.2    Lines and cables

The positive- and negative-sequence impedances are the normal balanced values. The zero-sequence impedance depends upon the nature of the return path through the earth if no fourth wire is provided. It is also modified by the presence of an earth wire on the towers that protects overhead lines against lightning strikes. In the absence of detailed information the following rough guide to the value of $Z_0$ may be used. For a single-circuit line $(Z_0/Z_1) = 3.5$ with no earth wire and $Z_0/Z_1 = 2$ with one earth wire. For a double-circuit line, $(Z_0/Z_1) = 5.5$. For underground cables, $(Z_0/Z_1) = 1 - 1.25$ for single-core and 3–5 for three-core cables.

## 7.4.3 Transformers

The positive- and negative-sequence impedances are the normal balanced ones. The zero-sequence connection of transformers is, however, complicated, and depends on the nature of the connection of the windings. Table 7.3 lists the zero-sequence representation of transformers for various winding arrangements. Zero-sequence currents in the windings on one side of a transformer must produce the corresponding ampere-turns in the other. But three in-phase

**Table 7.3**

| Connections of windings | | Representation per phase | Comments |
|---|---|---|---|
| Primary | Secondary | | |
| | | $Z_0$ | Zero-sequence currents free to flow in both primary and secondary circuits |
| | | $Z_0$ | No path for zero-sequence currents in primary circuits |
| | | $Z_0$ | Single-phase currents can circulate in the delta but not outside it |
| | | $Z_0$ | No flow of zero-sequence currents possible |
| | | $Z_0$ | No flow of zero-sequence currents possible |
| | | $Z_0$ | Tertiary winding provides path for zero-sequence currents |

currents cannot flow in a star connection without a connection to earth. They can circulate round a delta winding, but not in the lines outside it. Owing to the mutual impedance between the phases, $Z_0 \neq Z_1$. For a three-limb transformer, $Z_0 < Z_1$; for a five-limb transformer, $Z_0 > Z_1$.

An example showing the nature of the three sequence networks for a small transmission link is shown in Figure 7.10.

## 7.5   Types of Fault

In the following, a single voltage source in series with an impedance is used to represent the power network as seen from the point of the fault. This is an extension of Thevenin's theorem to three-phase systems. It represents the general method used for manual calculation, i.e. the successive reduction of

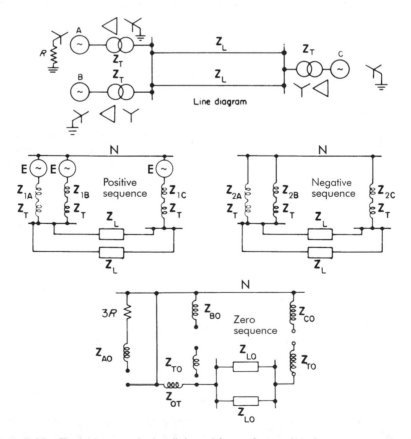

**Figure 7.10**   Typical transmission link and form of associated sequence networks

the network to a single impedance and voltage or current source. The network is assumed to be initially on no-load before the occurrence of the fault, and linear, so that superposition applies.

### 7.5.1  Single-line-to-earth fault

The three-phase circuit diagram is shown in Figure 7.11, where the three phases are on open-circuit at their ends.

Let $I_1$, $I_2$, and $I_0$ be the symmetrical components of $I_R$ and let $V_1$, $V_2$, and $V_0$ be the components of $V_R$. For this condition, $V_R = 0$, $I_B = 0$, and $I_Y = 0$. Also, $Z_R$ includes components $Z_1$, $Z_2$, and $Z_0$.

From equation (7.4)

$$I_0 = \tfrac{1}{3}(I_R + I_B + I_Y)$$
$$I_1 = \tfrac{1}{3}(I_R + aI_Y + a^2I_B)$$
$$I_2 = \tfrac{1}{3}(I_R + a^2I_Y + aI_B)$$

Hence,

$$I_0 = \frac{I_R}{3} = I_1 = I_2 \qquad (\text{as } I_B = I_Y = 0)$$

Also,

$$V_R = E - I_1Z_1 - I_2Z_2 - I_0Z_0 = 0$$

Eliminating $I_0$ and $I_2$, we obtain

$$E - I_1(Z_1 + Z_2 + Z_0) = 0$$

hence

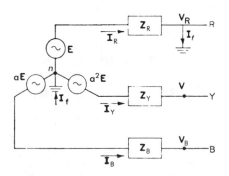

**Figure 7.11**  Single line-to-earth fault—Thevenin equivalent of system at point of fault

$$I_1 = \frac{E}{Z_1 + Z_2 + Z_0} \tag{7.5}$$

The fault current,

$$I_f = I_R = 3I_1$$

so

$$I_f = \frac{3E}{Z_1 + Z_2 + Z_0} \tag{7.6}$$

The e.m.f. of the Y phase $= a^2E$, and (from equation (7.4))

$$I_Y = I_0 + a^2I_1 + aI_2$$
$$\therefore V_Y = a^2E - I_0Z_0 - a^2I_1Z_1 - aI_2Z_2$$

The pre-fault and post-fault phasor diagrams are shown in Figure 7.12, where it should be noted that only $V_{YB}$ remains at its pre-fault value.

It is usual to form an equivalent circuit to represent equation (7.5) and this can be obtained from an inspection of the equations. The circuit is shown in Figure 7.13 and it will be seen that $I_1 = I_2 = I_0$, and

$$I_1 = \frac{E}{Z_1 + Z_2 + Z_0}$$

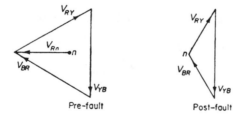

Pre-fault                    Post-fault

**Figure 7.12**   Pre- and post-fault phasor diagrams—single-line-to-earth fault

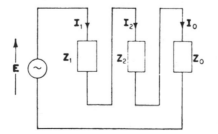

**Figure 7.13**   Interconnection of positive-, negative-, and zero-sequence networks for single-line-to-earth faults

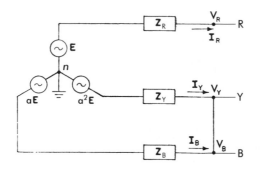

**Figure 7.14**   Line-to-line fault on phases Y and B

### 7.5.2  Line-to-line fault

In Figure 7.14, $E$ = e.m.f. per phase and the R phase is again taken as the reference phasor. In this case, $I_R = 0$, $I_Y = -I_B$, and $V_Y = V_B$.

From equation (7.4),

$$I_0 = 0$$
$$I_1 = \tfrac{1}{3}I_Y(a - a^2)$$

and

$$I_2 = \tfrac{1}{3}I_Y(a^2 - a)$$
$$\therefore I_1 = -I_2$$

As $V_Y = V_B$ (but not equal to zero),

$$a^2E - a^2I_1Z_1 - aI_2Z_2 = aE - aI_1Z_1 - a^2I_2Z_2$$
$$\therefore E(a^2 - a) = I_1[Z_1(a^2 - a) + Z_2(a^2 - a)]$$
$$\text{because } I_1 = -I_2$$

hence

$$I_1 = \frac{E}{Z_1 + Z_2} \tag{7.7}$$

This can be represented by the equivalent circuit in Figure 7.15, in which, of course, there is no zero-sequence network. If the connection between the two lines has an impedance $Z_f$ (the fault impedance), this is connected in series in the equivalent circuit.

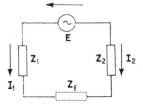

**Figure 7.15** Interconnection of sequence networks for a line-to-line fault (including fault impedance $Z_f$, if present)

### 7.5.3 Line-to-line-to-earth fault (Figure 7.16)

$$I_R = 0 \qquad V_Y = V_B = 0$$

and

$$I_R = I_1 + I_2 + I_0 = 0$$
$$\therefore \ a^2E - a^2I_1Z_1 - aI_2Z_2 - I_0Z_0 = V_Y = 0$$

and

$$aE - aI_1Z_1 - a^2I_2Z_2 - I_0Z_0 = V_B = 0$$

Hence,

$$I_1 = \frac{E}{Z_1 + [Z_2Z_0/(Z_2 + Z_0)]} \tag{7.8}$$

$$I_2 = -I_1 \cdot \frac{Z_0}{Z_2 + Z_0} \tag{7.9}$$

and

$$I_0 = -I_1 \cdot \frac{Z_2}{Z_2 + Z_0} \tag{7.10}$$

These can be represented by the equivalent circuit as shown in Figure 7.17.

The inclusion of impedances in the earth path, such as the star-point-to-earth connection in a generator or transformer, modifies the sequence diagrams. For a line-to-earth fault an impedance $Z_g$ in the earth path is represented by an impedance of $3Z_g$ in the zero-sequence network. $Z_g$ can include the impedance of the fault itself, usually the resistance of the arc. As $I_1 = I_2 = I_0$ and $3I_1$ flows through $Z_g$ in the physical system, it is necessary to use $3Z_g$ to obtain the required effect. Hence,

$$I_f = \frac{3E}{Z_1 + 3Z_g + Z_2 + Z_0}$$

Again, for a double-line-to-earth fault an impedance $3Z_g$ is connected as shown in Figure 7.18. $Z_g$ includes both machine neutral impedances and fault impedances.

The phase shift introduced by star–delta transformers has no effect on the magnitude of the fault currents, although it will affect the voltages at various

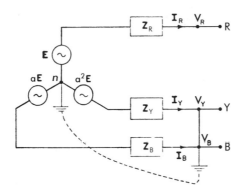

**Figure 7.16**   Line-to-line-to-earth fault

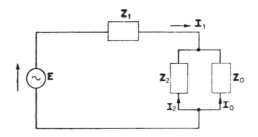

**Figure 7.17**   Interconnection of sequence networks—double line-to-earth fault

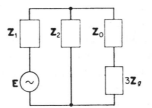

**Figure 7.18**   Modification of network in Figure 7.17 to account for neutral imped-ance $Z_g$

points. It is shown in Chapter 3 that the positive-sequence voltages and cur-rents are advanced by a certain angle and the negative-sequence quantities are retarded by the same angle for a given connection.

# 7.6   Fault Levels in a Typical System

In Figure 7.19, a section of a typical system is shown. At each voltage level the fault level can be ascertained from the reactances given. It should be noted that the short-circuit level will change with network conditions, and there will normally be two extreme values: that with all plant connected and that with

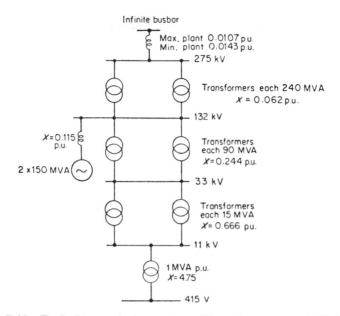

Infinite busbar

Max. plant 0.0107 p.u.
Min. plant 0.0143 p.u.

275 kV

Transformers each 240 MVA
$X = 0.062$ p.u.

132 kV

$X = 0.115$ p.u.

2 x 150 MVA

Transformers
each 90 MVA
$X = 0.244$ p.u.

33 kV

Transformers
each 15 MVA
$X = 0.666$ p.u.

11 kV

1 MVA p.u.
$X = 4.75$

415 V

**Figure 7.19**   Typical transmission system. All reactances on a 100 MVA base

the minimum plant normally connected. The short-circuit MVA at 275 kV busbars in Britain is normally 10 000 MVA, but drops to 7000 MVA with minimum plant connected. Maximum short-circuit (three-phase) levels experienced in the British system are as follows: 275 kV, 15 000 MVA; 132 kV, 3500 MVA; 33 kV, 750/1000 MVA; 11 kV, 150/250 MVA; 415 V, 30 MVA.

As the transmission voltages increase, the short-circuit currents also increase, and for the 400 kV system, circuit breakers of 35 000 MVA breaking capacity are required. In order to reduce the fault level the number of parallel paths is reduced by sectionalizing. This is usually achieved by opening the circuit breaker connecting two sections of a substation or generating station busbar. One great advantage of direct-current transmission links in parallel with the alternating-current system is that no increase in the short-circuit currents results.

### 7.6.1  Circuit parameters with faults

*Fault resistance*

The resistance of the fault is normally that of the arc and may be approximated by

$$R_a\,(\Omega) = 44\,(V \text{ in kV})/I_f \qquad \text{for} \qquad V < 110\,\text{kV}$$

and

$$R_a\,(\Omega) = 22\,(V \text{ in kV})/I_f \qquad \text{for} \qquad V > 110\,\text{kV}$$

For example, a 735 kV line would have an arc resistance of 4 Ω with a fault current of 4 kA, assuming no resistance in ground return path.

The overall grounding resistance depends on the footing resistance (resistance of tower metalwork to ground) of the towers ($R_T$) and also on the resistance per section of the ground wires ($R_s$), where present. The situation is summarized in Figure 7.20, where usually the effective grounding resistance is smaller than the individual tower footing resistance. There is normally a spread in the values of $R_T$, but normally $R_T$ should not exceed 10 Ω (ground wires present).

Typical values of fault resistance with fault location are as follows: at source, 0 Ω; on line with ground wires, 15 Ω; on line without ground wires, 50 Ω.

## X/R ratio

The range of $X/R$ values for typical voltage class (Canadian) are as follows: 735 kV, 18.9–20.4; 500 kV, 13.6–16.5; 220 kV, 2–25; 110 kV, 3–26. The $X/R$ value decreases with separation of the fault point from the source and can be substantially decreased by the fault resistance.

For distribution circuits, $X/R$ is lower, and, although data is limited, typical values are 10 (at source point) and 2–4 on an overhead line.

## *Example 7.4*

A synchronous machine A generating 1 p.u. voltage is connected through a star–star transformer, reactance 0.12 p.u., to two lines in parallel. The other ends of the lines are connected through a star–star transformer of reactance 0.1 p.u. to a second machine B, also generating 1 p.u. voltage. For both transformers, $X_1 = X_2 = X_0$.

Calculate the current fed into a double-line-to-earth fault on the line-side terminals of the transformer fed from A.

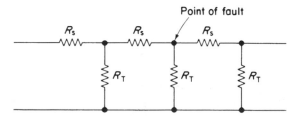

**Figure 7.20** Equivalent network of towers with footing resistance ($R_T$) and ground wire sections (resistance $R_s$)

The relevant per unit reactances of the plant, all referred to the same base, are as follows:

For each line $X_1 = X_2 = 0.30$, $X_0 = 0.70$.

|  | $X_1$ | $X_2$ | $X_0$ |
| --- | --- | --- | --- |
| Machine A | 0.30 | 0.20 | 0.05 |
| Machine B | 0.25 | 0.15 | 0.03 |

The star points of machine A and of the two transformers are solidly earthed.

## Solution

The positive-, negative-, and zero-sequence networks are shown in Figure 7.21. All per unit reactances are on the same base. From these diagrams the following equivalent reactances up to the point of the fault are obtained: $\mathbf{Z}_1 = j0.23\,\text{p.u.}$, $\mathbf{Z}_2 = j0.18\,\text{p.u.}$, and $\mathbf{Z}_0 = j0.17\,\text{p.u.}$

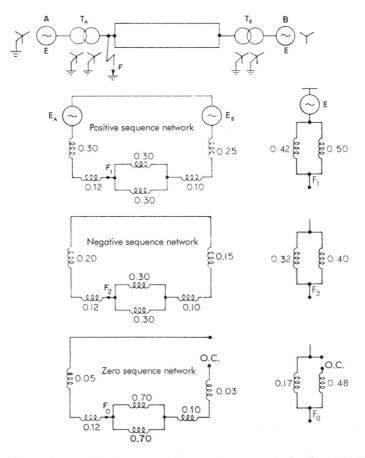

**Figure 7.21**   Line diagram and sequence networks for Example 7.4

The red phase is taken as reference phasor and the blue and yellow phases are assumed to be shorted at the fault point. From the equivalent circuit for a line-to-line fault,

$$\mathbf{I}_1 = \frac{1}{\mathbf{Z}_1 + [\mathbf{Z}_0\mathbf{Z}_2/(\mathbf{Z}_0 + \mathbf{Z}_2)]} \text{ p.u.}$$

$$= \frac{1}{j\{0.23 + [0.17 \times 0.18/(0.17 + 0.18)]\}} \text{ p.u.}$$

$$= -j3.15 \text{ p.u.}$$

$$\mathbf{I}_2 = -\mathbf{I}_1 \frac{\mathbf{Z}_0}{\mathbf{Z}_0 + \mathbf{Z}_2}$$

$$= j3.15 \times \frac{0.17}{0.17 + 0.18}$$

$$= j1.53 \text{ p.u.}$$

$$\mathbf{I}_0 = -\mathbf{I}_1 \frac{\mathbf{Z}_2}{\mathbf{Z}_0 + \mathbf{Z}_2}$$

$$= j3.15 \times \frac{0.18}{0.18 + 0.17}$$

$$= j1.62 \text{ p.u.}$$

$$\mathbf{I}_Y = \mathbf{I}_0 + a^2\mathbf{I}_{R1} + a\mathbf{I}_{R2}$$

$$= j1.62 + (-0.5 - 0.866j)(-j3.15) + (-0.5 + j0.866)(j1.53)$$

$$= -4.05 + j2.43 \text{ p.u.}$$

$$\mathbf{I}_B = j1.62 + (-0.5 + j0.866)(-j3.15) + (-0.5 - j0.866)(j1.53)$$

$$= 4.05 + j2.43.$$

$$\mathbf{I}_Y = \mathbf{I}_B = 4.72 \text{ p.u.}$$

The correctness of the first part of the solution can be checked as

$$\mathbf{I}_R = \mathbf{I}_1 + \mathbf{I}_2 + \mathbf{I}_0 = -j3.15 + j1.53 + j1.62 = 0$$

## Example 7.5

An 11 kV synchronous generator is connected to a 11/66 kV transformer which feeds a 66/11/3.3 kV three-winding transformer through a short feeder of negligible impedance. Calculate the fault current when a single-phase-to-earth fault occurs on a terminal of the 11 kV winding of the three-winding transformer. The relevant data for the system are as follows:

*Generator* $X_1 = j0.15$ p.u., $X_2 = j0.1$ p.u., $X_0 = j0.03$ p.u., all on a 10 MVA base; star point of winding earthed through a 3 $\Omega$ resistor.

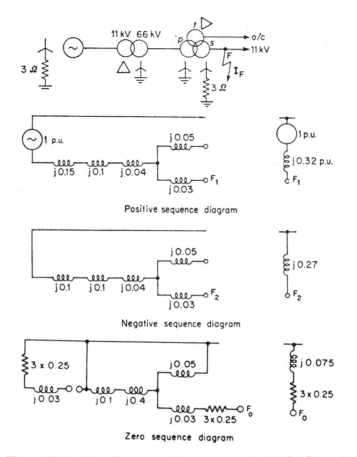

**Figure 7.22** Line diagram and sequence networks for Example 7.5

*11/66 kV Transformer* $X_1 = X_2 = X_0 = j0.1$ p.u. on a 10 MVA base; 11 kV winding delta connected and the 66 kV winding star connected with the star point solidly earthed.

*Three-winding transformer* A 66 kV winding, star connected, star point solidly earthed; 11 kV winding, star connected, star-point earthed through a 3Ω resistor; 3.3 kV winding, delta connected; the three windings of an equivalent star connection to represent the transformer have sequence impedances, 66 kV winding $X_1 = X_2 = X_0 = j0.04$ p.u., 11 kV winding $X_1 = X_2 = X_0 = j0.03$ p.u., 3.3 kV winding $X_1 = X_2 = X_0 = j0.05$ p.u., all on a 10 MVA base.

Resistance may be neglected throughout.

*Solution:*

The line diagram and the corresponding positive-, negative-, and zero-sequence networks are shown in Figure 7.23. A 10 MVA base will be used. The 3 $\Omega$ earthing resistor has the following p.u. value:

$$\frac{3 \times 10 \times 10^6}{(11)^2 \times 10^6} \quad \text{or} \quad 0.25 \, \text{p.u.}$$

Much care is needed with the zero-sequence network owing to the transformer connections. For a line-to-earth fault, the equivalent circuit shown in Figure 7.13 is used, from which

$$\mathbf{I}_1 = \mathbf{I}_2 = \mathbf{I}_0 \quad \text{and} \quad \mathbf{I}_f = \mathbf{I}_1 + \mathbf{I}_2 + \mathbf{I}_0$$

Hence

$$\mathbf{I}_f = \frac{3 \times 1}{\mathbf{Z}_1 + \mathbf{Z}_2 + \mathbf{Z}_0 + 3\mathbf{Z}_g}$$

$$= \frac{3}{j0.32 + j0.27 + j0.075 + 0.75}$$

$$= \frac{3}{0.75 + j0.66} = \frac{3}{1\underline{/41^\circ}} \, \text{p.u.}$$

$$\therefore \mathbf{I}_f = \frac{3 \times 10 \times 10^6}{\sqrt{(3)} \times 11\,000}$$

$$= 1575 \, \text{A}$$

## 7.7 Power in Symmetrical Components

The total power in a three-phase network

$$= \mathbf{V}_a \mathbf{I}_a^* + \mathbf{V}_b \mathbf{I}_b^* + \mathbf{V}_c \mathbf{I}_c^*$$

where $\mathbf{V}_a$, $\mathbf{V}_b$ and $\mathbf{V}_c$ are phase voltages and $\mathbf{I}_a$, $\mathbf{I}_b$, and $\mathbf{I}_c$ are line currents. In phase (a),

$$P_a + jQ_a = (\mathbf{V}_{a0} + \mathbf{V}_{a1} + \mathbf{V}_{a2})(\mathbf{I}_{a0}^* + \mathbf{I}_{a1}^* + \mathbf{I}_{a2}^*)$$

$$= (\mathbf{V}_{a0}\mathbf{I}_{a0}^* + \mathbf{V}_{a1}\mathbf{I}_{a1}^* + \mathbf{V}_{a2}\mathbf{I}_{a2}^*)$$

$$+ (\mathbf{V}_{a0}\mathbf{I}_{a1}^* + \mathbf{V}_{a1}\mathbf{I}_{a2}^* + \mathbf{V}_{a2}\mathbf{I}_{a0}^*)$$

$$+ (\mathbf{V}_{a0}\mathbf{I}_{a2}^* + \mathbf{V}_{a1}\mathbf{I}_{a0}^* + \mathbf{V}_{a2}\mathbf{I}_{a0}^*)$$

with similar expressions for phases (b) and (c).

In extending this to cover the total three-phase power it should be noted that

$$\mathbf{I}_{b1}^* = (\mathbf{a}^2\mathbf{I}_{a1})^* = (\mathbf{a}^2)^*\mathbf{I}_{a1}^* = \mathbf{a}\mathbf{I}_{a1}^*$$

Similarly, $\mathbf{I}_{b2}^*$, $\mathbf{I}_{c1}^*$, and $\mathbf{I}_{c2}^*$ may be replaced.

$$\text{The total power} = 3(\mathbf{V}_{a0}\mathbf{I}_{a0}^* + \mathbf{V}_{a1}\mathbf{I}_{a1}^* + \mathbf{V}_{a2}\mathbf{I}_{a2}^*) \tag{7.11}$$

i.e. $3\times$ (the sum of the individual sequence powers in any phase).

# 7.8 Systematic Methods for Fault Analysis in Large Networks

The methods described so far become unwieldy when applied to large networks and a systematic approach utilizing digital computers is used. The digital methods used for load flows are also suitable for three-phase symmetrical fault calculations, which after all are merely load flows under certain network conditions. The input information must be modified so that machines are represented by the appropriate reactances. The generators are represented by their no-load voltages in series with the subtransient reactances. If loads are to be taken into account they are represented by the equivalent shunt admittance to neutral. This will involve some modification to the original load-flow admittance matrix, but the general form will be as in the normal load flow.

A current of 1 p.u. is injected at the fault point and removed at the neutral. If the voltage at the fault becomes $\mathbf{V}_f$ then $1/\mathbf{V}_f$ is the fault level (i.e. short-circuit admittance) at node $f$. Also, the flow along branch $kf$ into the fault is given by

$$\frac{(\mathbf{V}_f - \mathbf{V}_k)}{\mathbf{V}_f}\mathbf{Y}_{fk}$$

The voltages due to the injected currents can be obtained by the same numerical techniques as used for load flows. As loads are represented by admittance the problem is linear and iterative methods are not required.

Three-phase fault studies are performed in conjunction with load flows. For example, if the fault level on a solid busbar is too high the busbar will be sectioned, i.e. split into two or more sections by opening switches, and new load are flows required.

The nodal or admittance method may be applied to large networks. It is preferable, on grounds of storage and time, not to invert the matrix but to use Gauss elimination methods. The computation efficiency may also be improved by utilizing the sparsity of the $\mathbf{Y}$ matrix. The mesh or loop (impedance matrix) method may be used, although the matrix is not so easily formed.

The following example illustrates a method suitable for determination of balanced three-phase fault currents in a large system by means of a digital computer.

## Example 7.6

Determine the fault current in the system shown in Figure 7.4 for the balanced fault shown.

## Solution

As already stated, generators are represented by their voltages behind the transient reactance, and normally the system is assumed to be on no-load before the occurrence of the three-phase balanced fault. The voltage sources and transient reactances are converted into current sources and the admittance matrix is formed (including the transient reactance admittances). The basic equation $[Y][V] = [I]$ is formed and solved with the constraint that the voltage at the fault node is zero. A current may be injected into the faulted node, as explained in the previous section, or the voltage made zero and the remaining nodal voltages calculated.

The system in Figure 7.4 is replaced by the equivalent circuit shown in Figure 7.23, in which $\mathbf{V_D} = 0.0$ p.u.

Nodal admittance matrix

$$-j \begin{bmatrix} 15.33 & -2.00 & & -3.33 \\ -2.00 & 15.33 & -3.33 & \\ & -3.33 & 15.33 & -2.00 \\ -3.33 & & -2.00 & 15.33 \end{bmatrix} \begin{bmatrix} \mathbf{V_A} \\ \mathbf{V_B} \\ \mathbf{V_C} \\ 0 \end{bmatrix} = -j \begin{bmatrix} 10 \\ 10 \\ 10 \\ \mathbf{I_F} \end{bmatrix}$$

As all the **Y** values are reactive, the j values will be omitted for simplicity. If the bottom row is eliminated as explained in Chapter 6, we get:

$$\begin{bmatrix} 23 & -3 & 0 \\ -3 & 23 & -5 \\ 0 & -5 & 23 \end{bmatrix} \cdot \begin{bmatrix} \mathbf{V_A} \\ \mathbf{V_B} \\ \mathbf{V_C} \end{bmatrix} = \begin{bmatrix} 15 \\ 15 \\ 15 \end{bmatrix}$$

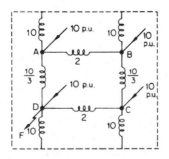

**Figure 7.23** Equivalent circuit of the system of Figure 7.4. All values are admittances (i.e. $-jY$). The generators, i.e. 1 p.u. voltage behind $-j10$ p.u. admittance, transform to $-j10$ p.u. current sources in parallel with $-j10$ p.u. admittance

from which

$$\begin{bmatrix} 520 & -115 \\ -5 & 23 \end{bmatrix} \cdot \begin{bmatrix} V_B \\ V_C \end{bmatrix} = \begin{bmatrix} 385 \\ 15 \end{bmatrix}$$

Thus,

$$V_B = 0.92950 \qquad V_C = 0.85510 \qquad \text{and} \qquad V_A = 0.77495 \,\text{p.u.}$$

and

$$I_F = -j\left(10 + \frac{0.77495}{0.3} + \frac{0.88510}{0.5}\right)\text{p.u.} = -j14.2985\,\text{p.u.}$$

Fault MVA $= 14.2985 \times 100 = 1429.85\,\text{MVA}$

## 7.9   Bus Impedance (Short-Circuit Matrix) Method

This method is used for computer fault analysis and has the following advantages:

1. Matrix inversion is avoided, resulting in savings in computer storage and time.
2. The matrices for the sequences quantities are determined only once and retained for later use; they are readily modified for system changes.
3. Mutual impedances between lines are readily handled.
4. Subdivisions of the main system may be incorporated.

The system is represented by the usual symmetrical component sequence networks and, frequently, the positive and negative impedances are assumed to be identical. Balanced phase impedances for all items of plant are assumed as are equal voltages for all generators. For simplicity, in this treatment, mutual coupling between lines will be neglected and load currents assumed negligible. The method will be explained using the positive-sequence network as this will be most familiar to the reader.

In the system shown in Figure 7.24 the voltage source supplies a common bus, and four busbars of interest in the passive network are identified. The network loop matrix, i.e. $[e] = [Z][i]$, is set up in terms of the various loop currents which must include those flowing through busbars (1)–(4) to the neutral bus, i.e. these busbars are short-circuited to N. In the matrix equation so formed only nodes 1–4 are of interest, and by the process of partitioning, described in the section on load flows, a new equation is obtained involving only currents from these busbars. $E = 1\,\text{p.u.}$

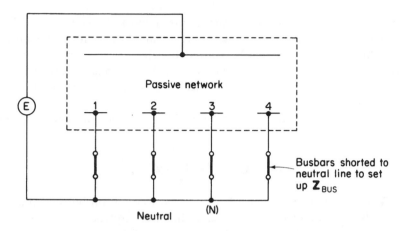

**Figure 7.24**  Identification of busbars of Interest In a large network—single-phase equivalent network

$$
\begin{array}{ccc}
\mathbf{(e)} & (\mathbf{Z_A}) & (\mathbf{Z_B}) \\
\begin{bmatrix} 1.0 \\ 1.0 \\ 1.0 \\ 1.0 \\[6pt] 0 \\ \cdot \\ \cdot \\ 0 \end{bmatrix}
=
\left[\begin{array}{cccc|c}
Z_{11} & Z_{12} & Z_{13} & Z_{14} & \cdots Z_{1n} \\
Z_{21} & Z_{22} & Z_{23} & Z_{24} & \cdots Z_{2n} \\
Z_{31} & Z_{32} & Z_{33} & Z_{34} & \cdots Z_{3n} \\
Z_{41} & Z_{42} & Z_{43} & Z_{44} & \cdots Z_{4n} \\ \hline
\cdot & \cdot & \cdot & \cdot & \cdots \\
\cdot & \cdot & \cdot & \cdot & \cdots \\
\cdot & \cdot & \cdot & \cdot & \cdots \\
Z_{n1} & Z_{n2} & Z_{n3} & Z_{n4} & \cdots Z_{n}
\end{array}\right]
\cdot
\begin{bmatrix} i_1 \\ i_2 \\ i_3 \\ i_4 \\[6pt] \cdot \\ \cdot \\ \cdot \\ i_n \end{bmatrix} \\
& (\mathbf{Z_C}) \qquad (\mathbf{Z_D}) \cdot
\end{array}
$$

i.e.

$$[\mathbf{e}] = [\mathbf{Z}_{sc}]\,[\mathbf{i}]$$

where

$$\mathbf{Z}_{sc} = \mathbf{Z}_A - \mathbf{Z}_B \mathbf{Z}_D^{-1} \mathbf{Z}_C$$

The new matrix is called the *bus impedance* or short-circuit matrix and the equation is given by

$$
\begin{bmatrix} 1.0 \\ 1.0 \\ 1.0 \\ 1.0 \end{bmatrix}
=
\begin{bmatrix}
z_{11} & z_{12} & z_{13} & z_{14} \\
z_{21} & z_{21} & z_{23} & z_{24} \\
z_{31} & z_{32} & z_{33} & z_{34} \\
z_{41} & z_{42} & z_{43} & z_{44}
\end{bmatrix}
\cdot
\begin{bmatrix} I_1 \\ I_2 \\ I_3 \\ I_4 \end{bmatrix}
\qquad (7.12)
$$

This matrix may be represented by the simple network shown in Figure 7.25, in which the buses of interest are shown short-circuited to the neutral. Consider a fault on bus (1) only, currents $I_2$, $I_3$, and $I_4$ will be zero and, from equation $e = zi$, $1.0 = z_{11} I_1$, where $I_1$ = fault current with three-phase symmetrical fault on 1. Similarly, the currents with balanced faults on each of the other buses may be easily determined.

The voltages at the remaining busbars with a balanced fault on bus (1) may also be readily determined, e.g.

$$1.0 - V_{n2} = Z_{21} I_1 = Z_{21} \frac{1.0}{Z_{11}}$$

where $V_{n2}$ is the voltage on bus (2) with a short circuit on bus (1), and so on.

For unbalanced faults on the busbars, the three sequence networks are used in the normal way. In Figure 7.26 the connection of the three reduced-sequence networks for a line-to-ground fault on bus (1) is illustrated, where superscripts $+$, $-$ and 0 denote the sequence impedances. As usual, the sequence networks refer to phase (a) of the real network.

For a L–E fault on bus (1),

$$I_1 = 3/(2z_{11}^+ + z_{11}^0)$$

and

$$I_1^+ = I_1^- = I_1^0 = 1/(2z_{11}^+ + z_{11}^0)$$

The assumption is made that the positive- and negative-sequence impedances are equal. This aids computation and does not produce major errors.

Also, the voltage on bus (3),

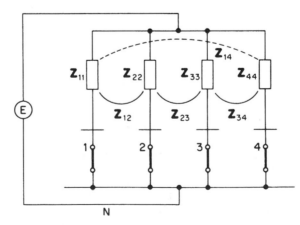

**Figure 7.25**   Equivalent circuit of the $Z_{bus}$ matrix equation—single-phase equivalent, i.e. balanced-fault conditions

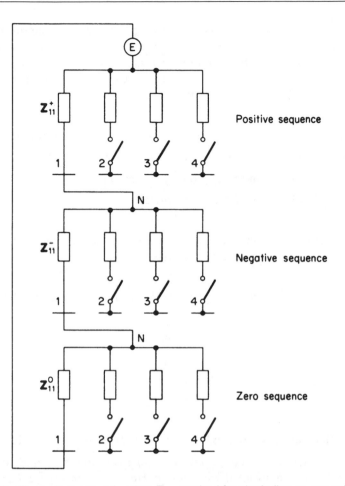

**Figure 7.26** Equivalent network using $Z_{bus}$ method for single-line-to ground fault on busbar (1) of the network in Figure 7.24

$$V_3^+ = 1 - \frac{z_{31}^+}{(2z_{11}^+ + z_{11}^0)}$$

$$V_3^- = -\frac{z_{31}^-}{2z_{11}^+ + z_{11}^0}$$

$$V_3^0 = -\frac{z_{31}^0}{(2z_{11}^+ + z_{11}^0)}$$

and

$$V_3 = V_3^+ + V_3^- + V_3^0 = 1 - \left(\frac{2z_{12}^+ + z_{13}^0}{2z_{11}^+ + z_{11}^0}\right)$$

The bus impedance matrix has so far been derived from the full-loop impedance matrix obtained either by inspection or by systematic methods. An alternative approach, which is more economical in computer usage, is to form the matrix one step at a time as each component of the network is added. These additions fall into four categories as follows:

1. A new generator bus or a new bus with a direct connection to the neutral.
2. A generator (or connection to neutral) connected to an existing bus.
3. Connection from an existing bus to a new bus.
4. Connection between two existing buses.

To illustrate the addition of buses, a $2 \times 2$ matrix describing buses (1) and (2) will be assumed to have already been assembled.

1. New bus with connection to neutral through $z_L$ (generator) (see Figure 7.27).

$$
\text{New } z_B =
\begin{bmatrix}
z_{11} & z_{12} & 0 \\
z_{21} & z_{22} & 0 \\
0 & 0 & z_L
\end{bmatrix}
\quad \text{Old}
$$

Note that currents through buses (1) and (2) have no interaction in bus (3).

2. Connection to neutral of generator from an existing bus. $z_L$ is connected to bus (2) and is associated with loop current $i_3$, as in Figure 7.28. The new matrix is

$$
\begin{matrix}
& (A) & (B) \\
& \textcircled{1} \quad \textcircled{2} & \textcircled{3} \\
\end{matrix}
$$

$$
\begin{bmatrix}
z_{11} & z_{12} & z_{12} \\
z_{21} & z_{22} & z_{22} \\
z_{21} & z_{22} & z_{22} + z_L
\end{bmatrix}
\begin{matrix}
(i_1) \\
(i_2) \\
(i_3)
\end{matrix}
$$

$$
\begin{matrix}
(C) & (D)
\end{matrix}
$$

Note that the $i_1$ influences loop $i_3$ via $z_{12}$. The internal loop is eliminated by applying $z = z_A - z_B z_D^{-1} z_C$ to the partitioned matrix, giving a $2 \times 2$ matrix.

3. Connection of impedance between an existing bus and a new bus. New line $z_L$ connected to bus (2). Consider loop current $i_3$ as shown in Figure 7.29.

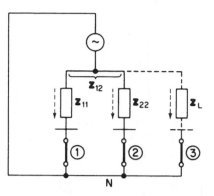

**Figure 7.27** New busbar with connection to neutral through $z_L$, e.g. generator

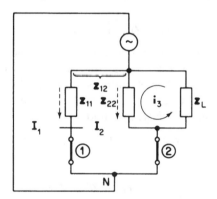

**Figure 7.28** Connection to neutral of generator from existing bus (2)

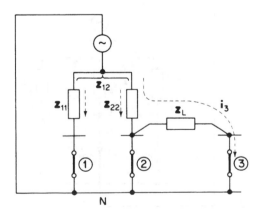

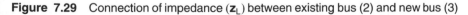

**Figure 7.29** Connection of impedance ($z_L$) between existing bus (2) and new bus (3)

$$
\mathbf{z}_{\text{bus}} = \begin{array}{ccc} \text{①} & \text{②} & \text{③} \end{array}
\left[
\begin{array}{cc|c}
\mathbf{z}_{11} & \mathbf{z}_{12} & \mathbf{z}_{12} \\
\mathbf{z}_{21} & \mathbf{z}_{22} & \mathbf{z}_{22} \\
\hline
\mathbf{z}_{21} & \mathbf{z}_{22} & \mathbf{z}_{22} + \mathbf{z}_{\text{L}}
\end{array}
\right]
$$

4. Connection of impedance (line, transformer, etc.) between two existing buses. Consider the new link $\mathbf{z}_{\text{L}}$ between existing buses (2) and (3) in Figure 7.30 and loop current $\mathbf{i}_{\text{L}}$.

$$
\begin{array}{cccc} \text{①} & \text{②} & \text{③} & \quad\quad \text{Ⓛ} \end{array}
\left[
\begin{array}{ccc|c}
\mathbf{z}_{11} & \mathbf{z}_{12} & \mathbf{z}_{13} & (\mathbf{z}_{1\text{L}} = \mathbf{z}_{12} - \mathbf{z}_{13}) \\
\mathbf{z}_{21} & \mathbf{z}_{22} & \mathbf{z}_{23} & (\mathbf{z}_{2\text{L}} = \mathbf{z}_{22} - \mathbf{z}_{23}) \\
\mathbf{z}_{31} & \mathbf{z}_{32} & \mathbf{z}_{33} & (\mathbf{z}_{3\text{L}} = \mathbf{z}_{23} - \mathbf{z}_{33}) \\
\hline
\mathbf{z}_{\text{L}1} & \mathbf{z}_{\text{L}2} & \mathbf{z}_{\text{L}3} & \mathbf{z}_{\text{L}} + \mathbf{z}_{33} + \mathbf{z}_{22} - 2\mathbf{z}_{23}
\end{array}
\right]
$$

Note that $\mathbf{i}_{\text{L}}$ in $\mathbf{z}_{22}$ induces a voltage in the $\mathbf{z}_{33}$ branch due to $\mathbf{z}_{23}$. Similarly, it induces in $\mathbf{z}_{22}$ when flowing in $\mathbf{z}_{33}$, giving an opposing double voltage $2\mathbf{i}_{\text{L}} \times \mathbf{z}_{23}$.

As $\mathbf{i}_{\text{L}}$ traverses the loop it induces into $\mathbf{z}_{11}$ due to $\mathbf{z}_{12}$ and $\mathbf{z}_{13}$. Hence $\mathbf{z}_{\text{L}1} = \mathbf{z}_{12} - \mathbf{z}_{13}$ or, generally, the mutual between the loop current $\mathbf{i}_{\text{L}}$ and node $i$ with a line added between $k$ and $m$ is

$$
\mathbf{z}_{\text{L}i} = \mathbf{z}_{i\text{L}} = \mathbf{z}_{ki} - \mathbf{z}_{mi} \qquad (i \neq L)
$$

e.g. for node 3,

$$
\mathbf{z}_{\text{L}3} = \mathbf{z}_{23} - \mathbf{z}_{33}
$$

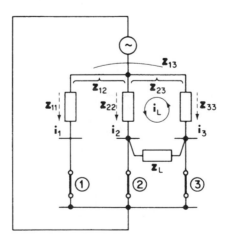

**Figure 7.30** Connection of impedance ($\mathbf{z}_{\text{L}}$) between two existing buses, (2) and (3)

Again, the internal loop is eliminated by partitioning, leaving a $3 \times 3$ matrix.

It can be shown that a *single-loop* elimination may be accomplished by operation on the individual elements as follows. Let $z'_{mn}$ be an element in the new bus impedance (reduced) matrix (i.e. column $n$, row $m$) and $z_{mn}$ an element in the original loop matrix. Then,

$$z'_{mn} = z_{mn} - z_{mL} z_{LL}^{-1} z_{Ln} \tag{7.13}$$

For example, in the above $4 \times 4$ matrix the new term, $z'_{23} = z_{23} - z_{2L} z_{LL}^{-1} z_{L3}$

$$= z_{23} - (z_{22} - z_{23})(z_L + z_{33} + z_{22} - 2z_{23})^{-1}(z_{23} - z_{33})$$

Similarly, the other eight terms are obtained.

If, in existing networks, lines or generators are to be removed, this can be achieved as outlined above, but with the use of negative impedances. The same general approach is made to systems with coupling between parallel lines and also the splitting of large networks into several smaller networks.

## Example 7.7

For the system shown in Figure 7.31, assemble the $Z_{bus}$ matrix, branch by branch.

## Solution

First consider the j0.2 branch (Figure 7.31(b)) and then add the j0.1 branch (Figure 7.31(c)), giving the matrix

$$
\begin{array}{cc}
① & ② \\
\end{array}
$$

$$
\begin{bmatrix}
j0.2 & 0 \\
0 & j0.1
\end{bmatrix}
$$

Finally, the j0.4 branch connectes nodes 1 and 2 and the matrix is

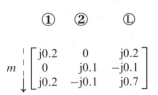

$$
n
$$

$$
\begin{array}{ccc}
① & ② & Ⓛ \\
\end{array}
$$

$$
m
\begin{bmatrix}
j0.2 & 0 & j0.2 \\
0 & j0.1 & -j0.1 \\
j0.2 & -j0.1 & j0.7
\end{bmatrix}
$$

Eliminate the internal loop, using equation (7.13).

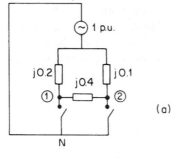

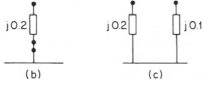

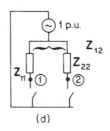

**Figure 7.31** Network for Example 7.7

$$\mathbf{Z}_{11}^1 = j0.2 - (j0.2)(j0.7)^{-1}(j0.2) = j0.143$$
$$\mathbf{Z}_{12}^1 = j0.1 - (-j0.1)(j0.7)^{-1}(-j0.1) = j0.086$$
$$\mathbf{Z}_{12}^1 = 0 - (j0.2)(j0.7)^{-1}(-j0.1) = j0.029$$
$$\mathbf{Z}_{21}^1 = j0.029$$

Therefore

$$
\begin{array}{cc}
① & ② \\
\end{array}
$$
$$
\mathbf{Z}_{\text{bus}} = j \begin{bmatrix} 0.143 & 0.029 \\ 0.029 & 0.086 \end{bmatrix}
$$

and

$$|\mathbf{E}| = |\mathbf{Z}_{\text{bus}}||\mathbf{I}|$$

(see Figure 7.31(d)).

With a three-phase short circuit on bus (1)

$$\mathbf{I}_1 = \frac{1.0}{\mathbf{z}_{11}} = \frac{1.0}{j0.143} = 6.99 \text{ p.u.}$$

$$(\mathbf{I}_2 = 0)$$

Similarly, for a fault on bus (2),

$$I_2 = \frac{1.0}{z_{22}}$$

$$I_2 = \frac{1}{j0.086} = 11.63 \, \text{p.u.}$$

# 7.10 Neutral Grounding

## 7.10.1 Introduction

From the analysis of unbalanced fault conditions it has been seen that the connection of the transformer and generator neutrals greatly influences the fault currents and voltages. In most high-voltage systems the neutrals are solidly grounded, i.e. connected directly to the ground, with the exception of generators which are grounded through a resistance to limit stator fault currents. The advantages of solid grounding are as follows:

1. Voltages to ground are limited to the phase voltage.
2. Intermittent ground faults and high voltages due to arcing faults are eliminated.
3. Sensitive protective relays operated by earth fault currents clear these faults at an early stage.

The main advantage in operating with neutrals isolated is the possibility of maintaining a supply with a ground fault on one line which places the remaining conductors at line voltage above ground. Also, interference with telephone circuits is reduced because of the absence of zero-sequence currents. With normal balanced operation the neutrals of an unground or isolated system are held at ground potential because of the presence of the system capacitance to earth. For the general case shown in Figure 7.32, the following analysis applies:

$$\frac{V_{ag}}{Z_{ag}} + \frac{V_{bg}}{Z_{bg}} + \frac{V_{cg}}{Z_{cg}} = 0 \tag{7.14}$$

Also,

$$V_{ag} = V_{an} + V_{ng}$$

where

$$V_{an} = \text{voltage of line (a) to neutral}$$
$$V_{ng} = \text{voltage of neutral to ground}$$

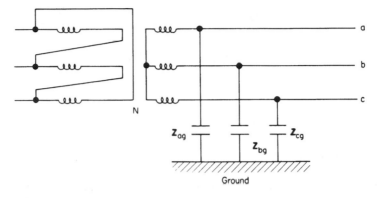

**Figure 7.32**   Line-to-ground capacitances in an unground system

Similarly,

$$V_{bg} = V_{bn} + V_{ng}$$

and

$$V_{cg} = V_{cn} + V_{ng}$$

Substituting in equation (7.14) and separating terms,

$$\frac{V_{an}}{Z_{ag}} + \frac{V_{bn}}{Z_{bg}} + \frac{V_{cn}}{Z_{cg}} + V_{ng}\left(\frac{1}{Z_{ag}} + \frac{1}{Z_{bg}} + \frac{1}{Z_{cg}}\right) = 0 \tag{7.15}$$

The equation

$$\left(\frac{1}{Z_{ag}} + \frac{1}{Z_{bg}} + \frac{1}{Z_{cg}}\right) = Y_g$$

gives the ground capacitance admittance of the system.

## 7.10.2  Arcing faults

Consider the single-phase system in Figure 7.33 at the instant when the instantaneous voltages are $v$ on line (a) and $-v$ on line (b), where $v$ is the maximum instantaneous voltage. The sudden occurrence of a fault to ground causes line (b) to assume a potential of $-2v$ and line (a) to become zero. Because of the presence of both $L$ and $C$ in the circuit, the sudden change in voltages by $v$ produces a high-frequency oscillation of peak magnitude $2v$ superimposed on the power frequency voltages (see Chapter 10) and line (a) reaches $-v$ and line (b) $-3v$, as shown in Figure 7.34. These oscillatory voltages attenuate quickly due to the resistance present. The current in the arc to earth on line (a) is approximately $90°$ ahead of the fundamental voltage, and when it is zero the

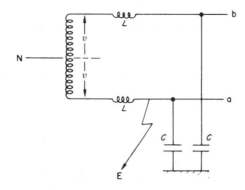

**Figure 7.33** Single-phase system with arcing fault to ground

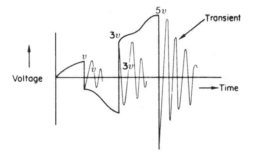

**Figure 7.34** Voltage on line (a) of Figure 7.33

voltage will be at a maximum. Hence, if the arc extinguishes at the first current zero, the lines remain charged at $-v$ for (a) and $-3v$ for (b). The line potentials now change at power frequency until (a) reaches $-3v$, when the arc could restrike causing a voltage change of $-3v$ to 0, resulting in a transient over-voltage of $+3v$ in line (a) and $+5v$ in line (b). This process could continue and the voltage build up further, but the resistance present usually limits the peak voltage to under $4v$. A similar analysis may be made for a three-phase circuit, again showing that serious overvoltages may occur with arcing faults because of the inductance and shunt capacitance of the system.

This condition may be overcome in an isolated neutral system by means of an *arc suppression* or Petersen coil. The reactance of this coil, which is connected between the neutral and ground, is made in the range 90–110 per cent of the value required to neutralize the capacitance current. The phasor diagram for the network of Figure 7.35(a) is shown in Figure 7.35(b) if the voltage drop across the arc is neglected.

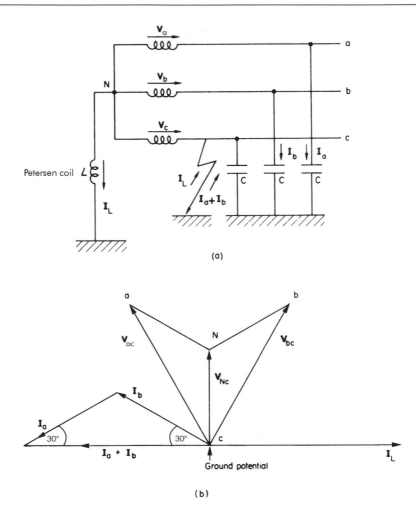

Figure 7.35   (a) System with arc suppression coil. (b) Phasor diagram of voltages and currents in part (a)

$$\mathbf{I}_a = \mathbf{I}_b = \sqrt{(3)}V\omega C$$

and

$$\mathbf{I}_a + \mathbf{I}_b = \sqrt{(3)} \times \sqrt{(3)}V\omega C = 3V\omega C$$

also,

$$\mathbf{I}_L = \frac{V}{\omega L}$$

For compensation of the arc current,

$$\frac{V}{\omega L} = 3V\omega C$$

$$\therefore \; L = \frac{1}{3\omega^2 C}$$

and

$$X_L = \frac{1}{3\omega C} \tag{7.16}$$

This result may be obtained by analysis of the ground fault by means of symmetrical components.

Generally, isolated neutral systems give rise to serious arcing-fault voltages if the arc current exceeds the region of 5–10 A, which covers most systems operating above 33 kV. If such systems are to be operated with isolated neutrals, arc-suppression coils should be used. Most systems at normal transmission voltages have grounded neutrals.

## 7.11 Interference with Communication Circuits— Electromagnetic Compatibility (EMC)

When power and telephone lines run in parallel, under certain conditions voltages sufficient to cause high noise levels may be induced into the communication circuits. This may be caused by electromagnetic and electrostatic unbalance in the power lines, especially if harmonics are present. The major problem, however, is due to faults to ground producing large zero-sequence currents in the power line, which inductively induce voltage into the neighbouring circuits. The value of induced voltage depends on the spacing, resistivity of the earth immediately below, and the frequency. If the telephone wires are a twisted pair or are situated close together and transposed, no voltage is induced between the communication conductors. However, a voltage can exist between the pair of wires and ground. These induced longitudinal voltages can be controlled by connecting the communication circuits to ground through an inductance which produces little attenuation at communication frequencies of 400–3500 Hz.

Capacitive coupling can occur if open communication circuits are run along the same route as power lines. Interference from underground cable circuits is much less (10 per cent) than that from overhead lines.

Because of right-of-way constraints, telephone and power distribution lines run parallel along the same street in many urban areas. However, the interference in rural areas is often greater because communication lines and plant may be unshielded or have higher shield resistances, and unlike urban areas there is no extensive network of water and gas pipes to share the ground return currents.

Resistive coupling between power and communication circuits can exist:

- because of physical contact between them;
- via paths through the soil between telephone and power grounds.

Various formulae exist to calculate the value of mutual inductance—in H/m between circuits with earth return. These assume that $\varepsilon_r$ (soil) is unity, displacement currents are much less than conduction currents, and the length of conductor is infinite. A typical formula is that due to Carson (1926).

During line-to-ground faults, induced voltages into communication circuits may be sufficient to be a shock hazard to personnel. Although in transmission circuits, equal current loading may be assumed in the phases, this is not the case in the lower voltage distribution circuits where significant residual currents may flow. Most telephone circuit standards now require up to 15 kV isolation if communication circuits are to be connected into substations for monitoring, control, and communication purposes. Particular care must be taken with bonding the sheath of communication circuits brought into buildings where the power distribution system is also bonded to earth and to the building structure.

# Problems

**7.1**  Four 11 kV generators designated A, B, C, and D each have a subtransient reactance of 0.1 p.u. and a rating of 50 MVA. They are connected in parallel by means of three 100 MVA reactors which join A to B, B to C, and C to D; these reactors have per unit reactances of 0.2, 0.4, and 0.2, respectively. Calculate the volt-amperes and the current flowing into a three-phase symmetrical fault on the terminals of machine B. Use a 50 MVA base.
(Answer: 940 MVA; 49 300 A)

**7.2**  Two 100 MVA, 20 kV turbogenerators (each of transient reactance 0.2 p.u.) are connected, each through its own 100 MVA, 0.1 p.u. reactance transformer, to a common 132 kV busbar. From this busbar, a 132 kV feeder, 40 km in length, supplies an 11 kV load through a 132/11 kV transformer of 200 MVA rating and reactance 0.1 p.u. If a balanced three-phase short circuit occurs on the low-voltage terminals of the load transformer, determine, using a 100 MVA base, the fault current in the feeder and the rating of a suitable circuit breaker at the load end of the feeder. The feeder impedance per phase is $(0.035 + j0.14)\,\Omega/\text{km}$.
(Answer: 482 MVA)

**7.3**  Two 60 MVA generators of transient reactance 0.15 p.u. are connected to a busbar designated A. Two identical machines are connected to another busbar B. A feeder is supplied from A through a step-up transformer rated at 30 MVA with 10 per cent reactance.

Calculate the reactance of a reactor to connect A and B if the fault level due to a three-phase fault on the feeder side of the transformer is to be limited to 240 MVA. Calculate also the voltage on A under this condition if the generator voltage is 13 kV (line).
(Answer: $X = 0.075\,\text{p.u.}$; $V_A = 10.4\,\text{kV}$)

**7.4**  A single line-to-earth fault occurs on the red phase at the load end of a 66 kV transmission line. The line is fed via a transformer by 11 kV generators connected to a common busbar. The line side of the transformer is connected in star with the star point earthed and the generator side is in delta. The positive-sequence reactances of the transformer and line are j10.9 Ω and j44 Ω, respectively, and the equivalent positive- and negative-sequence reactances of the generators, referred to the line voltage, are j18 Ω and j14.5 Ω, respectively. Measured up to the fault the total effective zero-sequence reactance is j150 Ω. Calculate the fault current in the lines if resistance may be neglected. If a two-line-to-earth fault occurs between the blue and yellow lines, calculate the current in the yellow phase.
(Answer: 392 A; 488 A)

**7.5**  A single-line-to-earth fault occurs in a radial transmission system. The following sequence impedances exist between the source of supply (an infinite busbar) of voltage 1 p.u. to the point of the fault: $Z_1 = 0.3 + j0.6$ p.u., $Z_2 = 0.3 + j0.55$ p.u., $Z_0 = 1 + j0.78$ p.u. The fault path to earth has a resistance of 0.66 p.u. Determine the fault current and the voltage at the point of the fault.
(Answer: $I_f = 0.732$ p.u.; $V_f = (0.43 - j0.23)$ p.u.)

**7.6**  Develop an expression, in terms of the generated e.m.f. and the sequence impedances, for the fault current when an earth fault occurs on phase (A) of a three-phase generator, with an earthed star point. Show also that the voltage to earth of the sound phase (B) at the point of fault is given by

$$V_B = \frac{-j\sqrt{3}E_A[Z_2 - aZ_0]}{Z_1 + Z_2 + Z_0}$$

Two 30 MVA, 6.6 kV synchronous generators are connected in parallel and supply a 6.6 kV feeder. One generator has its star point earthed through a resistor of 0.4 Ω and the other has its star point isolated. Determine: (a) the fault current and the power dissipated in the earthing resistor when an earth fault occurs at the far end of the feeder on phase (A); and (b) the voltage to earth of phase (B). The generator phase sequence is ABC and the impedances are as follows:

|  | Generator p.u./p.h. | Feeder Ω/p.h. |
|---|---|---|
| To positive-sequence currents | j0.2 | j0.6 |
| To negative-sequence currents | j0.16 | j0.6 |
| To zero-sequence currents | j0.06 | j0.4 |

Use a base of 30 MVA.
(Answer: (a) 5459∠ − 52.6° A; 11.92 MW; (b) (−1402 + j2770) V)

**7.7**  An industrial distribution system is shown schematically in Figure 7.36. Each line has a reactance of j0.4 p.u. calculated on a 100 MVA base; other system parameters are given in the diagram. Choose suitable short-circuit ratings for the oil circuit breakers, situated at substation A, from those commercially available, which are given in the table below.

| Short circuit (MVA) | 75 | 150 | 250 | 350 |
|---|---|---|---|---|
| Rated current (A) | 100 | 300 | 400 | 600 |

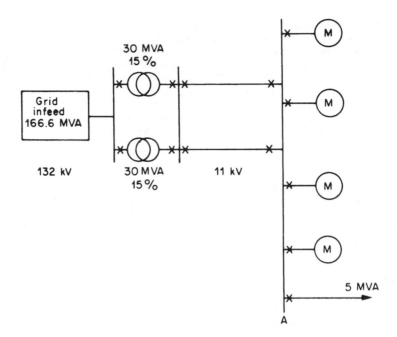

**Figure 7.36**   System for Problem 7.7

The industrial load consists of a static component of 5 MVA and four large induction motors each rated at 6 MVA. Show that only three motors can be started simultaneously given that, at starting, each motor takes five times full-load current at rated voltage, but at 0.3 p.f.

**7.8**   Explain how the Method of Symmetrical Components may be used to represent any 3 p.h. current phasors by an equivalent set of balanced phasors.

A chemical plant is fed from a 132 kV system which has a 3 p.h. symmetrical fault level of 4000 MVA. Three 15 MVA transformers, connected in parallel, are used to step down to an 11 kV busbar from which six 5 MVA, 11 kV motors are supplied. The transformers are delta–star connected with the star point of each 11 kV winding, solidly earthed. The transformers each have a reactance of 10 per cent on rating and it may be assumed that $X_1 = X_2 = X_0$. The initial fault contribution of the motors is equal to five times rated current with 1.0 p.u. terminal voltage.

Using a base of 100 MVA,

   (a) calculate the fault current (in A) for a line-to-earth short circuit on the 11 kV busbar with no motors connected;
   (b) calculate the 3 p.h. symmetrical fault level (in MVA) at the 11 kV busbar if all the motors are operating and the 11 kV busbar voltage is 1.0 p.u.
       (Answer: (a) 21.97 kA; (b) 544.45 MVA)
*(From E.C. Examination, 1996)*

**7.9**   Describe the effect on the output current of a synchronous generator following a solid three-phase fault on its terminals.

For the system shown in Figure 7.37 calculate (using symmetrical components):

(a) the current flowing in the fault for a three-phase fault at busbar A;
(b) the current flowing in the fault for a one-phase-to-earth fault at busbar B;
(c) the current flowing in the faulted phase of the overhead line for a one-phase-to-earth fault at busbar B.

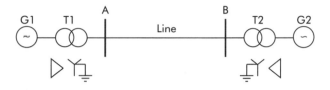

Generators G1 and G1: $X_1'' = X_2'' = j0.1$ p.u.; 11 kV
Transformers T1 and T2: $X_1 = X_2 = X_0 = j0.1$ p.u.; 11/275 kV
(Earthed star–delta)
Line: $Z_1 = Z_2 = j0.05$ p.u., $Z_0 = j0.1$ p.u.; 275 kV
(All p.u. values are quoted on a base of 100 MVA)

**Figure 7.37** Circuit for Problem 7.9

Assume the pre-fault voltage of each generator is 1 p.u. and calculate the symmetrical fault currents (in amps) immediately after each fault occurs.
(Answer: (a) 1.89 kA; (b) 2.18 kA; (c) 0.89 kA)
(*From E.C. Examination, 1997*)

**7.10** Why is it necessary to calculate short-circuit currents in large electrical systems?
A generator rated at 400 MW, 0.8 power factor, 20 kV has a star-connected stator winding which is earthed at its star point through a resistor of 1 Ω. The generator reactances, in per unit on rating, are:

$$X_1 = 0.2 \qquad X_2 = 0.16 \qquad X_0 = 0.14$$

The generator feeds a delta–star-connected generator transformer rated at 550 MVA which steps the voltage up to a 275 kV busbar. The transformer star-point is solidly earthed and the transformer reactance is 0.15 p.u. on its rating. The 275 kV busbar is connected only to the transformer. Assume that for the transformer $X_1 = X_2 = X_0$.
Using a base of 500 MVA calculate the base current and impedance of each voltage level.
Calculate the fault current in amperes for:

(a) a 275kV busbar three-line fault;
(b) a 275 kV single-line-to earth fault on the busbar;
(c) a 20 kV three-line fault on the generator terminals;
(d) a 20 kV single-line-to-earth fault on the generator terminals.

(Answer: (a) 3.125 kA; (b) 4.1 kA; (c) 72.15 kA; (d) 11.44 kA)
(*From E.C. Examination, 1995*)

# 8
# *System Stability*

## 8.1   Introduction

The stability of a system of interconnected dynamic components is its ability to return to normal or stable operation after having been subjected to some form of disturbance. The study of stability is one of the main concerns of the control engineer whose methods are applied to electric power systems.

When the rotor of a synchronous generator advances beyond a certain critical angle, the magnetic coupling between the rotor and the stator fails. The rotor, no longer held in synchronism with the rotating field of the stator currents, rotates relative to the field and pole slipping occurs. Each time the poles traverse the angular region where stability obtains, synchronizing forces attempt to pull the rotor into synchronism. It is general practice to disconnect the machine from the system if it commences to slip poles. However, a generator having lost synchronism may operate successfully as an induction generator for some time and then be resynchronized; the possibility of allowing controlled pole-slipping for limited periods is thus possible. It should be remembered, however, that an induction generator takes its excitation requirements from the network, which must be capable of supplying the requisite reactive power, and the rotor must not overheat and fail.

There are two forms of instability in power systems: the loss of synchronism between synchronous machines, and the stalling of asynchronous loads. Synchronous stability may be divided into two regimes: steady-state and transient. Reference to the former has already been made when discussing sychronous-machine characteristics. It is basically the ability of the power system, when operating under given load conditions, to retain synchronism when subject to *small* disturbances, such as the continual changes in load or generation and the switching out of lines. It is most likely to result from the changes in source-to-load impedance resulting from changes in the network configuration or system state. Often, this is referred to as *dynamic stability*.

Transient stability is concerned with sudden and large changes in the network condition, such as those brought about by faults. The maximum power transmittable, the stability limit under a fault condition, is less than that for the corresponding steady-state condition.

The stability of an asynchronous load is controlled by the voltage across it; if this becomes lower than a critical value, induction motors may become unstable and stall. This is, in effect, the voltage instability problem already mentioned. In a power system it is possible for either synchronous or voltage instability to occur. The former is more onerous and hence has been given much more attention in the past. Recently, with the increasing use of static compensators, the study of voltage collapse has become important (see Section 8.8)

## 8.2  Equation of Motion of a Rotating Machine

Before the equation is considered, a revision of the definitions of certain quantities is given. The kinetic energy absorbed by a rotating mass $= \frac{1}{2}I\omega^2$ joules, and the angular momentum is $M = I\omega$ joule-seconds per radian where $\omega$ is the synchronous speed of the rotor (radians/second) and $I$ is the moment of inertia (kilogram-metre$^2$). The inertia constant $(H)$ is defined as the stored energy at synchronous speed per volt-ampere of the rating of the machine. In power systems the unit of energy is taken as the kilojoule or megajoule. If $G$ megavolt-amperes is the rating of the machine, then

$GH$ =stored energy (megajoules) or kinetic energy

$\quad = \frac{1}{2}I\omega^2 = \frac{1}{2}M\omega$ and as $\omega = 180/\pi$ (pole pairs) or $360f$ electrical

$\quad$ degrees per second, where $f$ is the system frequency (Hz)

then

$$GH = \tfrac{1}{2}M\ (360f)$$

and

$$M = GH/180f \text{ megajoule-seconds/electrical degree}$$

The inertia constant $H$ for steam or gas turbogenerators decreases from 10 kW-s/kVA for machines up to 30 MVA to values in the order of 4 kW-s/kVA for large machines, the value decreasing as the capacity increases. For salient-pole water-wheel machines $H$ depends on the number of poles; for machines in the range 200–400 r.p.m. the value increases from about 2 kW-s/kVA at 10 MVA rating to 3.5 kW-s/kVA at 60 MVA. A mean value for synchronous motors is 2 kW-s/kVA. The net accelerating torque on the rotor of a machine, $\Delta T =$ mechanical torque input $-$ electrical torque output $= I\,\mathrm{d}^2\delta/\mathrm{d}t^2$.

$$\therefore \quad \frac{\mathrm{d}^2\delta}{\mathrm{d}t^2} = \frac{\Delta T}{I} = \frac{(\Delta T \omega)\omega}{2 \times I\omega^2/2}$$

$$= \frac{\Delta P \cdot \omega}{2 \times \text{kinetic energy}}$$

where $\Delta P =$ net power corresponding to $\Delta T$, i.e. $P_{\text{mech}} - P_{\text{elect}}$.

$$\therefore \quad \frac{\mathrm{d}^2\delta}{\mathrm{d}t^2} = \frac{\Delta P}{M} \tag{8.1}$$

In equation (8.1) a negative change in power output results in an increase in $\delta$. Sometimes, $\Delta P$ is considered as the change in *electrical power output* and increase in $\Delta P_{\text{elect}}$ results in increase in angle $\delta$. As the power input is assumed to be constant, the equation of motion now becomes

$$\frac{\mathrm{d}^2\delta}{\mathrm{d}t^2} = -\frac{\Delta P}{M} \quad \text{or} \quad M\frac{\mathrm{d}^2\delta}{\mathrm{d}t^2} + \Delta P = 0$$

## 8.3 Steady-State Stability—Theoretical Considerations

The power system forms a group of interconnected electromechanical elements, the motion of which may be represented by the appropriate differential equations. With large disturbances in the system the equations are non-linear, but with small changes the equations may be linearized with little loss of accuracy. The differential equations having been determined, the characteristic equation of the system is then formed, from which information regarding stability is obtained. The solution of the differential equation of the motion is of the form

$$\delta = k_1 e^{a_1 t} + k_2 e^{a_2 t} + \cdots + k_n e^{a_n t}$$

where $k_1, k_2 \ldots k_n$ are constants of integration and $a_1, a_2 \ldots a_n$ are the roots of the characteristic equation as obtained through the well-known eigenvalue methods. If any of the roots have positive real terms then the quantity $\delta$ increases continuously with time and the original steady condition is not re-established. The criterion for stability is therefore that all the real parts of the roots of the characteristic equation, i.e. eigenvalues, be negative; imaginary parts indicate the presence of oscillation. Figure 8.1 shows the various types of motion. The determination of the roots is readily obtained through an eigenvalue analysis package, but indirect methods for predicting stability have been established, e.g. the Hurwitz–Routh criterion in which stability is predicted without the actual solution of the characteristic equation. No

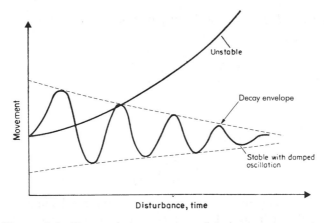

**Figure 8.1**   Types of response to a disturbance on a system

information regarding the degree of stability or instability is obtained, only that the system is, or is not, stable. One advantage of using eigenvalues is that the characteristics of the control loops associated with governors and automatic voltage regulators may be incorporated in the general treatment.

For a generator connected to an infinite busbar through a network of zero resistance it has been shown in equation (3.1) that $P = (VE/X)\sin \delta$. With operation at $P_0$ and $\delta_0$ (Figure 8.2), we can write

$$M\frac{\mathrm{d}^2 \Delta \delta}{\mathrm{d}t^2} = -\Delta P = -\Delta\delta\left(\frac{\partial P}{\partial \delta}\right)_0$$

where change in $P$ causing increase in $\delta$ is positive and refers to small changes in the load angle $\delta$ such that linearity may be assumed.

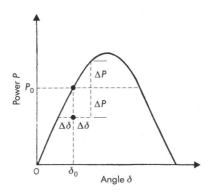

**Figure 8.2**   Small disturbance—initial operation on power-angle curve at $P_0, \delta_0$. Linear movement assumed about $P_0, \delta_0$

$$\therefore Ms^2\Delta\delta + \left(\frac{\partial P}{\partial\delta}\right)_0 \Delta\delta = 0 \tag{8.2}$$

where

$$s \equiv \mathrm{d}/\mathrm{d}t$$

Here, $Ms^2 + (\partial P/\partial\delta)_0 = 0$ is the characteristic equation which has two roots

$$\pm\sqrt{\frac{-(\partial P/\partial\delta)_0}{M}}$$

When $(\partial P/\partial\delta)_0$ is positive, both roots are imaginary and the motion is oscillatory and undamped; when $(\partial P/\partial\delta)_0$ is negative, both roots are real, and positive and negative, respectively, and stability is lost. At $\delta = 90°$, $(\partial P/\partial\delta)_{90} = 0$, and the system is at the limit. If damping is accounted for, the equation becomes

$$Ms^2\Delta\delta + K_d s\Delta\delta + \left(\frac{\partial P}{\partial\delta}\right)\Delta\delta = 0 \tag{8.3}$$

and the characteristic equation is

$$Ms^2 + K_d s + \left(\frac{\partial P}{\partial\delta}\right) = 0 \tag{8.4}$$

where $K_d$ is the damping coefficient, assumed to be constant, independent of $\delta$.

Again, if $(\partial P/\partial\delta)$ is negative, stability is lost. The frequency of the oscillation is given by the roots of the characteristic equation.

If the excitation of the generator is controlled by a fast-acting automatic voltage regulator without appreciable dead zone, the excitation voltage $E$ is increased as increments of load are added. Hence the actual power-angle curve pertaining is no longer that for constant $E$ (refer to Chapter 3) and the change of power may be obtained by linearizing the $P$–$V$ characteristic at the new operating point (1), when

$$\Delta P = \left(\frac{\partial P}{\delta E}\right)_1 \Delta E$$

The complete equation of motion is now

$$Ms^2\Delta\delta + K_d s\Delta\delta + \left(\frac{\delta P}{\partial\delta}\right)_1 \Delta\delta + \left(\frac{\partial P}{\partial E}\right)_1 \Delta E = 0 \tag{8.5}$$

Without automatic voltage control the stability limit is reached when $\delta = 90°$; with control the criterion is obtained from the characteristic equation of (8.5).

## Example 8.1

A synchronous generator of reactance 1.5 p.u. is connected to an infinite busbar system ($V = 1$ p.u.) through a line and transformers of total reactance 0.5 p.u. The no-load voltage of the generator is 1.1 p.u. and the inertia constant $H = 5$ MW-s per MVA. All per unit values are expressed on the same base; resistance and damping may be neglected. Calculate the frequency of the oscillations set up when the generator operates at a load angle of $60°$ and is subjected to a small disturbance. The system frequency is 50 Hz.

## Solution

The nature of the movement is governed by the sign of the quantity under the root sign in the equation for $s_1$ and $s_2$ (equation (8.4)). This changes when $K_d^2 = 4M(\partial P/\partial \delta)_1$; in this case $K_d = 0$. The roots of the characteristic equations give the frequency of oscillation; when $\delta_0 = 60°$,

$$\left(\frac{\partial P}{\partial \delta}\right)_{60°} = \frac{1.1 \times 1}{2}\cos 60$$

$$= 0.275 \,\text{p.u.}$$

$$s_1 \text{ and } s_2 = \pm j\sqrt{\left(\frac{\partial P}{\partial \delta}\right) \cdot \frac{1}{M}}$$

$$= \pm j\sqrt{\frac{0.275}{5 \times \left(\frac{1}{\pi \times 50}\right)}}$$

$$= \pm j\sqrt{8.64}$$

Therefore frequency of oscillation

$$= 2.94 \text{rad/s} = \frac{2.94}{2\pi} \text{ Hz}$$

$$= 0.468 \text{ Hz}$$

and the periodic time

$$= \frac{1}{0.468} = 2.14 \,\text{s}$$

## The stability of a two-machine system

The terms $M_1$ and $M_2$ refer to machines 1 and 2 which are connected in parallel through an impedance.

The equations of motion are (for small changes, damping neglected)

$$M_1 s^2 \Delta\delta_1 + \left(\frac{\partial P_1}{\partial \delta_{12}}\right) \Delta\delta_{12} = 0$$

and

$$M_2 s^2 \Delta\delta_2 + \left(\frac{\partial P_2}{\partial \delta_{12}}\right) \Delta\delta_{12} = 0$$

As

$$\Delta\delta_1 - \Delta\delta_2 = \Delta\delta_{12}$$

then

$$s^2 \Delta\delta_{12} + \left[\frac{(\partial P_1/\partial \delta_{12})}{M_1} - \frac{(\partial P_2/\partial \delta_{12})}{M_2}\right] \Delta\delta_{12} = 0$$

The characteristic equation has two roots,

$$s_1, s_2 = \pm j \sqrt{\frac{(\partial P_1/\partial \delta_{12})}{M_1} - \frac{(\partial P_2/\partial \delta_{12})}{M_2}}$$

Stability is assured if the quantity under the square root is positive. Hence the stability limit for small disturbances is not the same as the maximum power limit discussed in Chapter 3 and is, in fact, always larger. The difference, however, is never large, and when one machine is effectively an infinite busbar, the two limits coincide.

## Effects of governor action

In the above analysis the oscillations set up with small changes in load on a system have been considered and the effects of governor operation ignored. After a certain time has elapsed the governor control characteristics commence to influence the powers and oscillations, as explained in Section 4.3. It is now the practice to represent both the excitation system and the governor system with the dynamic equations of the generator in the control-state space form, from which the eigenvalues of the complete system with feedback can be determined. Using well-established control design techniques, appropriate feedback paths and time constants can be established for a range of generating conditions and disturbances, thereby assuring adequate dynamic stability margins. Further information on these design processes can be found in advanced-control textbooks. Here we will consider the simpler but important aspects of so-called steady-state stability.

## 8.4   Steady-State Stability—Practical Considerations

The steady-state stability limit is the maximum power that can be transmitted in a network between sources and loads when the system is subject to small disturbances. The power system is, of course, constantly subjected to small changes as load variations occur. To obtain the limiting value of power, small increments of load are added to the system; after each increment the generator excitations are adjusted to maintain constant terminal voltages and a load flow is carried out. Eventually, a condition of instability is reached.

The stability limits of synchronous machines have been discussed in Chapter 3. It was seen that provided the generator operates within the 'safe area' of the performance chart (Figure 3.15), stability is assured; usually, a 20 per cent margin of safety on the var absorb side is allowed and the limit is extended by the use of automatic voltage regulators. Often, the performance charts are not used directly and the generator-equivalent circuit employing the synchronous impedance is used. The normal operating load angle for modern machines is in the order of 60 electrical degrees, and for the limiting value of 90° this leaves 30° to cover the transmission network. In a complex system a reference point must be taken from which the load angles are measured; this is usually a point where the direction of power flow reverses.

The simplest criterion for steady-state synchronous stability is $(\partial P/\partial \delta) > 0$, i.e. the synchronizing coefficient must be positive. The use of this criterion involves the following assumptions:

1.  generators are represented by constant impedances in series with the no-load voltages;

2.  the input torques from the turbines are constant;

3.  changes in speed are ignored;

4.  electromagnetic damping in the generators is ignored;

5.  the changes in load angle $\delta$ are small.

The degree of complexity to which the analysis is taken has to be decided, e.g. the effects of machine inertia, governor action, and automatic voltage regulators can be included; these items, however, greatly increase the complexity of the calculations. The use of the criterion $(\partial P/\partial \delta) = 0$ alone gives a pessimistic or low result and hence an inbuilt factor of safety.

In a system with several generators and loads, the question as to where the increment of load is to be applied is important. A conservative method is to assume that the increment applies to one machine only, determine the stability, and then repeat for each of the other machines in turn. Alternatively, the power outputs from all but the two generators having the largest load angles are kept constant.

For calculations made without the aid of computers it is usual to reduce the network to the simplest form that will keep intact the generator nodes. The values of load angle, power, and voltage are then calculated for the given conditions, $\partial P/\partial \delta$ determined for each machine, and, if positive, the loading is increased and the process repeated.

In a system consisting of a generator supplying a load through a network of lines and transformers of effective reactance, $X_T$ the value of $(dP/d\delta) = (EV/X_T)\cos \delta$, where $E$ and $V$ are the supply- and receiving-end voltages and $\delta$ the total angle between the generator rotor and the phasor of $V$. The power transmitted is obviously increased with higher system voltages and lower reactances, and it may be readily shown that line series capacitance increases the stability limit. The determination of $dP/d\delta$ is not very difficult if the voltages at the loads can be assumed to be constant or if the loads can be represented by impedances. Use can be made of the $P-V$, $Q-V$ characteristics of the load if the voltages change appreciably with the redistribution of the power in the network; this process, however, is extremely tedious.

## *Example 8.2*

For the system shown in Figure 8.3, investigate the steady-state stability.

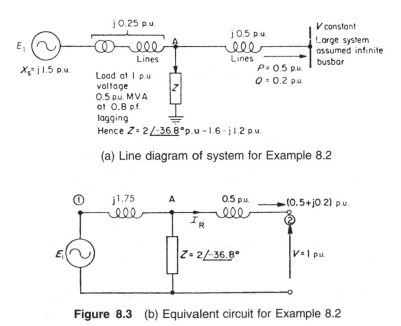

(a) Line diagram of system for Example 8.2

**Figure 8.3** (b) Equivalent circuit for Example 8.2

*Solution*

Voltage at the load is given by

$$V_A = \sqrt{\left[\left(V + \frac{QX}{V}\right)^2 + \left(\frac{PX}{V}\right)^2\right]}$$

$$= \sqrt{\left[\left(1 + \frac{0.2 \times 0.5}{1.0}\right)^2 + \frac{(0.5 \times 0.2)^2}{1.0}\right]}$$

$$= 1.105 \text{ p.u. at } 5.19° \text{ to } V$$

Reactive power absorbed by line A to (2)

$$= I_R^2 X = \left(\frac{P^2 + Q^2}{V^2}\right)X = \left(\frac{0.5^2 + 0.2^2}{1.0^2}\right)0.5$$

$$= 0.145 \text{ p.u.}$$

The actual load taken by A (if represented by an impedance) is given by

$$\frac{V_A^2}{Z} = \frac{1.105^2}{2\underline{/-36.8°}} = 0.61\underline{/36.8°} = (0.49 + j0.37) \text{ p.u.}$$

Total load supplied by link from generator to A

$$= (0.5 + 0.49) + j(0.2 + 0.145 + 0.37)$$
$$= (0.99 + j0.715) \text{ p.u.}$$

Internal voltage of generator $E_1$

$$= \sqrt{\left[\left(1.105 + \frac{0.715 \times 1.75}{1.105}\right)^2 + \left(\frac{0.99 \times 1.75}{1.105}\right)^2\right]}$$

$$= \sqrt{[5.006 + 2.458]} = 2.73 \text{ at } 35.02° \text{ to } V_A$$

Hence, angle between $E_1$ and $V$ is

$$35.02 + 5.19 = 40.21°$$

Since this angle is much less than 90°, the system is stable.
(Note that the formula $P = (E_1 V/X)\sin\delta$ is not valid here as the load at A provides a resistance in the equivalent network. If the angle between $E_1$ and $V$ by the approximate calculation shown above is nearly equal to 90°, then the power will need to be calculated from the exact equations as in Chapter 3.)

# 8.5 Transient Stability—Consideration of Rotor Angle

Transient stability is concerned with the effect of large disturbances. These are usually due to faults, the most severe of which is the three-phase short circuit which governs the transient stability limits in the U.K. Elsewhere, limits are based on other types of fault, notably the single-line-to-earth fault which is by far the most frequent in practice.

When a fault occurs at the terminals of a synchronous generator the power output of the machine is greatly reduced as it is supplying a mainly inductive circuit. However, the input power to the generator from the turbine has not time to change during the short period of the fault and the rotor endeavours to gain speed to store the excess energy. If the fault persists long enough the rotor angle will increase continuously and synchronism will be lost. Hence the time of operation of the protection and circuit breakers is all-important.

An aspect of importance is the use of *autoreclosing* circuit breakers. These open when the fault is detected and automatically reclose after a prescribed period (usually less than 1 s). If the fault persists the circuit breaker reopens and then recloses as before. This is repeated once more, when, if the fault still persists, the breaker remains open. Owing to the transitory nature of most faults, often the circuit breaker successfully recloses and the rather lengthy process of investigating the fault and switching in the line is avoided. The length of the autoreclose operation must be considered when assessing transient stability limits; in particular, analysis must include the movement of the rotor over this period and not just the first swing, as is often the case. If, in equation (8.1), both sides are multiplied by $2(d\delta/dt)$, then

$$\left(2\frac{d\delta}{dt}\right)\left(\frac{d^2\delta}{dt^2}\right) = \frac{d}{dt}\left(\frac{d\delta}{dt}\right)^2 = \frac{2\Delta P}{M}\left(\frac{d\delta}{dt}\right)$$

$$\therefore \left(\frac{d\delta}{dt}\right)^2 = \frac{2}{M}\int_{\delta_0}^{\delta}\Delta P\,d\delta \tag{8.6}$$

The rotor will swing until its angular velocity is zero, in which case the machine remains stable; if $d\delta/dt$ does not become zero the rotor will continue to move and synchronism is lost. The integral of $\Delta P\,d\delta$ in equation (8.6) represents an area on the $P$–$\delta$ diagram. Hence the criterion for stability is that the area between the $P$–$\delta$ curve and the line representing the power input $P_0$ in Figure 8.2 must be zero. This is known as the *equal-area criterion*. It should be noted that this is based on the assumption that synchronism is retained or lost on the first swing or oscillation of the rotor, which in certain cases is subject to doubt. Physically, the criterion means that the rotor must be able to return to the system all the energy gained from the turbine during the acceleration period.

A simple example of the equal-area criterion may be seen by an examination of the switching out of one of two parallel lines which connect a generator to an infinite busbar (Figure 8.4). For stability to be retained the two

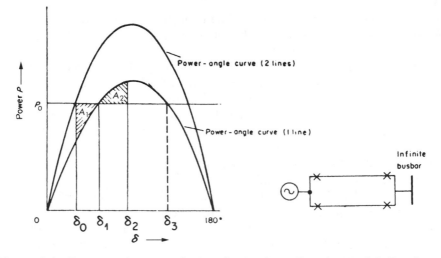

**Figure 8.4**  Power-angle curves for one line and two lines in parallel. Equal-area criterion. Resistance neglected

shaded areas ($A_1$ and $A_2$) are equal and the rotor comes initially to rest at angle $\delta_2$, after which it oscillates until completely damped. In this particular case the initial operating power and angle could be increased to such values that the shaded area between $\delta_0$ and $\delta_1$ ($A_1$) could be equal to the area between $\delta_1$ and $\delta_3$, where $\delta_3 = 180 - \delta_1$; this would be the condition for maximum input power. Beyond $\delta_3$, energy would again be absorbed by the rotor from the turbine.

The power-angle curves pertaining to a fault on one of two parallel lines are shown in Figure 8.5. The fault is cleared in a time corresponding to $\delta_1$ and the shaded area $\delta_0$ to $\delta_1$ ($A_1$) indicates the energy stored. The rotor swings until it reaches $\delta_2$ when the two areas $A_1$ and $A_2$ are equal. In this particular case $P_0$ is the maximum operating power for a fault clearance time corresponding to $\delta_1$, and, conversely, $\delta_1$ is the *critical clearing angle* for $P_0$. If the angle $\delta_1$ is decreased it is possible to increase the value of $P_0$ without loss of synchronism. The general case where the clearing angle $\delta_1$ is not critical is shown in Figure 8.6. Here, the rotor swings to $\delta_2$, where the shaded area from $\delta_0$ to $\delta_1$ ($A_1$) is equal to the area $\delta_1$ to $\delta_2$ ($A_2$). Critical conditions are reached when $\delta_2 = 180 - \sin^{-1}(P_0/P_2)$. The time corresponding to the critical clearing angle is called the *critical clearing time* for the particular (normally full-load) value of power input. The time is of great importance to protection and switchgear engineers as it is the maximum time allowable for their equipment to operate without instability occurring. The critical clearing angle for a fault on one of two parallel lines may be determined as follows:

Applying the equal-area criterion to Figure 8.5:

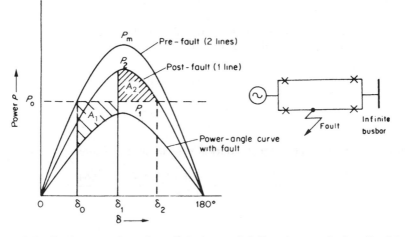

**Figure 8.5** Fault on one line of two lines in parallel. Equal-area criterion. Resistance neglected. $\delta_1$ is critical clearing angle for input power $P_0$

$$\int_{\delta_0}^{\delta_1} (P_0 - P_1 \sin \delta)\, d\delta + \int_{\delta_1}^{\delta_2} (P_0 - P_2 \sin \delta)\, d\delta = 0$$

$$\therefore\ [P_0\delta + P_1 \cos \delta]_{\delta_0}^{\delta_1} + [P_0\delta + P_2 \cos \delta]_{\delta_1}^{\delta_2} = 0$$

from which,

$$\delta_2 = 180 - \sin^{-1}\left(\frac{P_0}{P_2}\right)$$

and

$$\cos \delta_1 = \frac{P_0(\delta_0 - \delta_2) + P_1 \cos \delta_0 - P_2 \cos \delta_2}{P_1 - P_2} \tag{8.7}$$

Hence the critical clearing angle $\delta_1$ is determined.

## *Example 8.3*

A generator operating at 50 Hz delivers 1 p.u. power to an infinite busbar through a network in which resistance may be neglected. A fault occurs which reduces the maximum power transferable to 0.4 p.u., whereas before the fault this power was 1.8 p.u. and after the clearance of the fault it is 1.3 p.u. By the use of the equal-area criterion, determine the critical clearing angle.

## *Solution*

The appropriate load-angle curves are shown in Figure 8.6. $P_0 = 1$ p.u., $P_1 = 0.4$ p.u., $P_2 = 1.3$ p.u., and $P_m = 1.8$ p.u.

$$\delta_0 = \sin^{-1}\left(\frac{1}{1.8}\right) = 33.8 \text{ electrical degrees}$$

$$\delta_2 = 180 - \sin^{-1}\left(\frac{1}{1.3}\right) = 180 - 50.28° = 129.72°$$

Applying equation (8.7) (note that electrical degrees must be expressed in radians),

$$\cos\delta_1 = \frac{1(0.59 - 2.26) + 0.4 \times 0.831 - 1.3(-0.64)}{0.4 - 1.3}$$

$$= 0.562$$

$$\therefore \delta_1 = 55.8°$$

In a large system it is usual to divide the generators and loads into a single equivalent generator connected to an equivalent motor. The main criterion is that the machines should be electrically close when forming an equivalent generator or motor. If stability with faults in various places is investigated, the position of the fault will decide the division of machines between the equivalent generator and motor. A power system (including generation) at the receiving end of a long line would constitute an equivalent motor if not large enough to be an infinite busbar. It should be noted that a three-phase short circuit on the generator terminals or on a closely connected busbar absorbs zero power and prevents the generator outputting any power to the system. Consequently, $P_1 = 0$ in Figure 8.6.

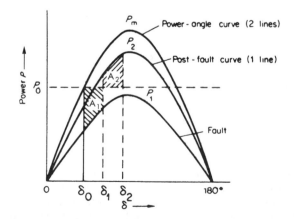

**Figure 8.6**  Situation as in Figure 8.5, but $\delta_1$ not critical

*Reduction to simple system*    With a number of generators connected to the same busbar the inertia constant ($H$) of the equivalent machine,

$$H_e = H_1 \frac{S_1}{S_b} + H_2 \frac{S_2}{S_b} \cdots H_n \frac{S_n}{S_b}$$

where $S_1 \ldots S_n$ = MVA of the machines and $S_b$ = base MVA. This is obtained by equating the stored energy of the equivalent machine to the total of the individual machines. For example, consider six identical machines connected to the same busbar, each having an $H$ of 5 MW-s/MVA and rated at 60 MVA. Making the base MVA equal to the combined rating of the machines, the effective value of

$$H = 5 \times \frac{60}{6 \times 60} \times 6 = 5 \frac{\text{MW-s}}{\text{MVA}} \qquad .$$

It is important to remember that the inertia of the loads must be included; normally, this will be the sum of the inertias of the induction motors and their mechanical loads connected.

Two synchronous machines connected by a reactance may be reduced to one equivalent machine feeding through the reactance to an infinite busbar system. The properties of the equivalent machine are found as follows.

The equation of motion for the two-machine system is

$$\frac{d^2\delta}{dt^2} = \frac{\Delta P_1}{M_1} - \frac{\Delta P_2}{M_2} = \left(\frac{1}{M_1} + \frac{1}{M_2}\right)(P_0 - P_m \sin \delta)$$

where $\delta$ is the relative angle between the machines. Note that

$$\Delta P_1 = -\Delta P_2 = (P_0 - P_m \sin \delta)$$

where $P_0$ is the input power and $P_m$ the maximum transmittable power.

For a single generator of $M_e$ and the same input power connected to the infinite busbar system,

$$M_e \frac{d^2\delta}{dt^2} = P_0 - P_m \sin \delta$$

therefore,

$$M_e = \frac{M_1 M_2}{M_1 + M_2} \tag{8.8}$$

This equivalent generator has the same mechanical input as the actual machines and the load angle $\delta$ it has with respect to the busbar is the angle between the rotors of the two machines.

Often, the maximum powers transferable before, during, and after a fault need to be calculated from the system configuration reduced to a network between the relevant generators. The use of network reduction by nodal elimination is most valuable in this context; it only remains then for the transfer reactances to be calculated, as any shunt impedance at the reduced nodes does not influence the power transferred. With unbalanced faults more power is transmitted during the fault period than with three-phase short circuits and the stability limits are higher.

### Effect of automatic voltage regulators and governors

These may be represented in the equation of motion as follows,

$$M\frac{\mathrm{d}^2\delta}{\mathrm{d}t^2} + K_\mathrm{d}\frac{\mathrm{d}\delta}{\mathrm{d}t} = (P_0 - \Delta P_0) - P_\mathrm{e} \qquad (8.9)$$

where $K_\mathrm{d}$ = damping coefficient;

$P_0$ = power input;

$\Delta P_0$ = change in input power due to governor action;

$P_\mathrm{e}$ = electrical power output modified by the voltage regulator.

Equation (8.9) is best solved by digital computer.

## 8.6  Transient Stability—Consideration of Time

### The swing curve

In the previous section, attention has been mainly directed towards the determination of rotor angular position; in practice, the corresponding times are more important. The protection engineer requires allowable times rather than angles when specifying relay settings. The solution of equation (8.1) with respect to time is performed by means of numerical methods and the resulting time–angle curve is known as the swing curve. A simple step-by-step method will be given in detail and references will be made to more sophisticated methods suitable for digital computation.

   In this method the change in the angular position of the rotor over a short time interval is determined. In performing the calculations the following assumptions are made:

1.  The accelerating power $\Delta P$ at the commencement of a time interval is considered to be constant from the middle of the previous interval to the middle of the interval considered.

2.  The angular velocity is constant over a complete interval and is computed for the middle of this interval.

These assumptions are probably better understood by reference to Figure 8.7. From Figure 8.7,

$$\omega_{n-\frac{1}{2}} - \omega_{n-\frac{3}{2}} = \frac{\mathrm{d}^2\delta}{\mathrm{d}t^2}\Delta t = \frac{\Delta P_{n-1}}{M}\Delta t$$

The change in $\delta$ over the $(n-1)$th interval, i.e. from times $(n-1)$ to $(n-2)$

$$= \delta_{n-1} - \delta_{n-2} = \Delta\delta_{n-1} = \omega_{n-\frac{3}{2}}\Delta t$$

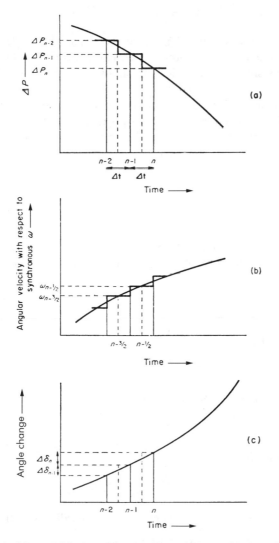

**Figure 8.7** (a), (b), and (c) Variation of $\Delta P$, $\omega$, and $\Delta\delta$ with time. Illustration of step-by-step method to obtain $\delta$ time curve

as $\omega_{n-\frac{3}{2}}$ is assumed to be constant. Over the $n$th interval,

$$\Delta\delta_n = \delta_n - \delta_{n-1} = \omega_{n-\frac{1}{2}}\Delta t$$

From the above,

$$\Delta\delta_n - \Delta\delta_{n-1} = \Delta t\left(\omega_{n-\frac{1}{2}} - \omega_{n-\frac{3}{2}}\right)$$

$$= \Delta t \cdot \Delta t \cdot \frac{\Delta P_{n-1}}{M}$$

$$\therefore \Delta\delta_n = \Delta\delta_{n-1} + \frac{\Delta P_{n-1}}{M}(\Delta t)^2 \tag{8.10}$$

It should be noted that $\Delta\delta_n$ and $\Delta\delta_{n-1}$ are the *changes* in angle.

Equation (8.10) is the basis of the numerical method. The time interval $\Delta t$ used should be as small as possible (the smaller $\Delta t$, however, the larger the amount of labour involved), and a value of $0.05\,$s is frequently used. Any change in the operational condition causes an abrupt change in the value of $\Delta P$. For example, at the commencement of a fault ($t = 0$), the value of $\Delta P$ is initially zero and then immediately after the occurrence it takes a definite value. When two values of $\Delta P$ apply, the mean is used. The procedure is best illustrated by an example.

## Example 8.4

In the system described in Example 8.3 the inertia constant of the generator plus turbine is 2.7 p.u. Obtain the swing curve for a fault clearance time of 125 ms.

## Solution

$H = 2.7\,$p.u., $f = 50\,$Hz, $G = 1\,$p.u.

$$\therefore M = \frac{HG}{180f} = 3 \times 10^{-4}\,\text{p.u.}$$

A time interval $\Delta t = 0.05\,$s will be used. Hence

$$\frac{(\Delta t)^2}{M} = 8.33$$

The initial operating angle

$$\delta_0 = \sin^{-1}\left(\frac{1}{1.8}\right) = 33.8°$$

Just before the fault the accelerating power $\Delta P = 0$. Immediately after the fault,

$$\Delta P = 1 - 0.4\sin\delta_0$$
$$= 0.78\,\text{p.u.}$$

The first value is that for the middle of the preceding period and the second is for the middle of the period under consideration. The value to be taken for $\Delta P$ at the commencement of this period is $(0.780/2)$, i.e. $0.39$ p.u. At $t = 0, \delta = 33.8°$.

$$\Delta t_1 = 0.05 \,\text{s} \qquad \Delta P = 0.39 \,\text{p.u.}$$

$$\therefore \ \Delta \delta_n = \Delta \delta_{n-1} + \frac{(\Delta t)^2}{M} \Delta P_{n-1}$$

$$\therefore \ \Delta \delta_n = 8.33 \times 0.39 = 3.25°$$

$$\therefore \ \delta_{0.05} = 33.8 + 3.25 = 37.05°$$

$\Delta t_2 :$
$$\Delta P = 1 - 0.4 \sin 37.05° = 0.76 \,\text{p.u.}$$

$$\therefore \ \Delta \delta_n = 3.25 + (8.33 \times 0.76) = 9.58°$$

$$\therefore \ \delta_{0.1} = 37.05 + 9.58 = 46.63°$$

$\Delta t_3 :$
$$\Delta P = 1 - 0.4 \sin 46.63° = 0.71 \,\text{p.u.}$$

$$\Delta \delta_n = 9.58 + (8.33 \times 0.71) = 15.49°$$

$$\delta_{0.15} = 46.63 + 15.49 = 62.12°$$

The fault is cleared after a period of $0.125\,\text{s}$. As this discontinuity occurs in the middle of a period $(0.1–0.15\,\text{s})$, no special averaging is required (Figure 8.8).
If, on the other hand, the fault is cleared in $0.15\,\text{s}$, an averaging of two values would be required.
From $t = 0.15\,\text{s}$ onwards,

$$P = 1 - 1.3 \sin \delta \qquad \text{(note change to } P_2 \text{ curve of Figure 8.6)}$$

$\Delta t_4 :$
$$\Delta P = 1 - 1.3 \sin 62.12° = -0.149 \,\text{p.u.}$$

$$\Delta \delta_n = 15.49 + 8.33(-0.149) = 14.25°$$

$$\therefore \ \delta_{0.2} = 14.25 + 62.12 = 76.37°$$

$\Delta t_5 :$
$$\Delta P = 1 - 1.3 \sin 76.4° = -0.26 \,\text{p.u.}$$

$$\Delta \delta_n = 14.25 - (8.33 \times 0.26) = 12.08°$$

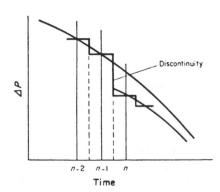

**Figure 8.8** Discontinuity of $\Delta P$ in middle of a period of time

and                                         $\delta_{0.25} = 88.39°$

$\Delta t_6$                                $\Delta P = 1 - 1.3 \sin 88.39 = -0.3 \, \text{p.u.}$

                                            $\Delta \delta_n = 12.09 - 2.5 = 9.59°$

and                                         $\delta_{0.3} = 97.98°$

If this process is continued, $\delta$ commences to decrease and the generator remains stable. If computed by hand, a tabular calculation is recommended, as shown in Figure 8.9.

| t(s) | $P = P_1 \sin \delta$ | $\Delta P$ | $\frac{(\Delta t)^2}{M} \cdot \Delta P$ | $\Delta \delta_n °$ | $\delta_n °$ | |
|------|-----------------------|------------|------------------------------------------|---------------------|--------------|--|
| 0−   | 1.0                   | 0.00       | —                                        | —                   | 33.80        | |
| 0+   |                       | 0.78       | —                                        | —                   | 33.80        | |
| 0.05 | 0.23                  | 0.39       | 3.25                                     | 3.25                | 37.05        | |
| 0.10 | 0.24                  | 0.76       | 6.33                                     | 9.58                | 46.63        | |
| 0.15 | 0.29                  | 0.71       | 5.91                                     | 15.49               | 62.12        | |
| 0.20 | 1.149                 | −0.149     | −1.24                                    | 14.25               | 76.37        | |
| 0.25 | 1.26                  | −0.26      | −2.17                                    | 12.08               | 88.39        | |
| ⋮    | ⋮                     | ⋮          | ⋮                                        | ⋮                   | ⋮            | |
| ⋮    | ⋮                     | ⋮          | ⋮                                        | ⋮                   | ⋮            | |
| ⋮    | ⋮                     | ⋮          | ⋮                                        | ⋮                   | ⋮            | |

**Figure 8.9**   Tabular calculation of $\delta_n$

This calculation should be continued for at least the peak of the first swing, but if switching or autoreclosing is likely to occur somewhere in the system, the calculation of $\delta°_n$ needs to be continued until oscillations are seen to be dying away. A typical swing curve shown in Figure 8.10 illustrates this situation.

Different curves will be obtained for other values of clearing time. It is evident from the way the calculation proceeds that for a sustained fault, $\delta$ will continuously increase and stability will be lost. The critical clearing time should be calculated for conditions which allow the least transfer of power from the generator. Circuit breakers and the associated protection operate in times dependent upon their design; these times can be in the order of a few cycles of alternating voltage. For a given fault position a faster clearing time implies a greater permissible value of input power $P_0$. A typical relationship between the critical clearing time and input power is shown in Figure 8.11 — this is often referred to as the stability boundary. The critical clearing time increases with increase in the inertia constant ($H$) of turbine generators. Often, the first swing of the machine is sufficient to indicate stability.

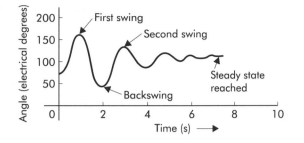

**Figure 8.10** Typical swing curve for generator

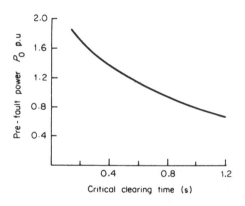

**Figure 8.11** Typical stability boundary

## 8.7   Transient Stability Calculations by Computer

It is obvious that a digital computer program can be readily written to carry out the simple studies of Section 8.6. If a load-flow program is readily available, then improved accuracy will be obtained if, for each value of $\delta_n$, the actual power output of the generators is calculated. At the same time, the effect of the excitation system and the governor movement can be included. Such calculations make use of numerical integration packages based on mathematical concepts. Techniques such as trapezoidal integration provide fast and sufficiently accurate results for many stability studies; more accurate techniques, such as Runge–Kutta (fourth order), predictor–corrector routines, etc., can be employed if the improved accuracy and longer run times can be economically justified. Most commercial stability programs offer various options for inclusion of generator controls, system switching and reclosing, compensator modelling, and transformer tap-change operation, according to some input criteria. Packages dealing with 1000 generators, 2000 lines, and 1500 nodes are available.

# 8.8 Stability of Loads Leading to Voltage Collapse

In Chapter 5 the power–voltage characteristics of a line supplying a load were considered. It was seen that for a given load power-factor, a value of transmitted power was reached, beyond which further decreases of the load impedance produce greatly reduced voltages, i.e. voltage instability. If the load is purely static, e.g. represented by an impedance, the system will operate stably at these lower voltages. Sometimes, in load-flow studies, this lower voltage condition is unknowingly obtained and unexpected load flows result. If the load contains non-static elements, such as induction motors, the nature of the load characteristics is such that beyond the critical point the motors will run down to a standstill or stall. It is therefore of importance to consider the stability of composite loads that will normally include a large proportion of induction motors.

The process of voltage collapse may be seen from a study of the $V$–$Q$ characteristics of an induction motor (see Figure 3.48), from which it is seen that below a certain voltage the reactive power consumed increases with decrease in voltage until $(dQ/dV) = \infty$, when the voltage collapses. In the power system the problem arises owing to the impedance of the connection between the load and infinite busbar and is obviously aggravated when this impedance is high, i.e. connection is electrically weak. The usual cause of an abnormally high impedance is the loss of one line of two or more forming the connection. It is profitable, therefore, to study the process in its basic form— that of a load supplied through a reactance from a constant voltage source (Figure 8.12)

Already, two criteria for load instability have been given, i.e. $(dP/d\delta) = 0$ and $(dQ/dV) = \infty$; from the system viewpoint, voltage collapse takes place when $(dE/dV) = 0$ or $(dV/dE) = \infty$. Each value of $E$ yields a corresponding value for $V$, and the plot of $E$–$V$ is shown in Figure 8.13; also the plot of $V$–$X$ for various values of $E$ is shown in Figure 8.14. In these graphs the critical operating condition is clearly shown and the improvement produced by higher values of $E$ is apparent, indicating the importance of the system-operating voltage from the load viewpoint.

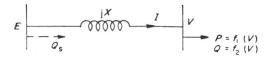

**Figure 8.12** System with a load dependent on voltage as follows: $P = f_1(V)$ and $Q = f_2(V)$; $Q_s = $ supply vars $= Q + I^2 X$; $E = $ supply voltage

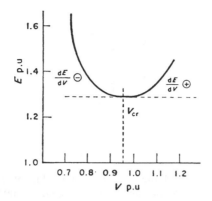

**Figure 8.13** The *E–V* relationship per system in Figure 8.12. $V_{cr}$ = critical voltage after which instability occurs

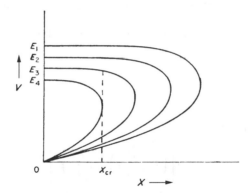

**Figure 8.14** The *V–X* relationship. Effect of change in supply voltage *E*. $X_{cr}$ = critical reactance of transmission link

In the circuit shown in Figure 8.12, and from equation 2.11,

$$E = V + \frac{QX}{V} \qquad \text{if} \qquad \frac{PX}{V} \ll \frac{V^2 + QX}{V}$$

$$\therefore \frac{\mathrm{d}E}{\mathrm{d}V} = 1 + \left(\frac{\mathrm{d}Q}{\mathrm{d}V} \cdot XV - QX\right)\frac{1}{V^2}$$

which is zero at the stability limit and negative in the unstable region.
*At the limit,*

$$\frac{\mathrm{d}E}{\mathrm{d}V} = 0 \qquad \text{and} \qquad \frac{\mathrm{d}Q}{\mathrm{d}V} = \left(\frac{QX}{V^2} - 1\right)\frac{V}{X} = \frac{Q}{V} - \frac{V}{X}$$

From equation (2.11),

$$\frac{Q}{V} = \frac{E}{X} - \frac{V}{X}$$

$$\therefore \quad \frac{dQ}{dV} = \frac{E}{X} - \frac{2V}{X} \tag{8.11}$$

## *Example 8.5*

A load is supplied from a 275 kV infinite busbar through a line of reactance 70 Ω phase-to-neutral. The load consists of a constant power demand of 200 MW and a reactive power demand $Q$ which is related to the load voltage $V$ by the equation:

$$(V - 0.8)^2 = 0.2(Q - 0.8)$$

This is shown in Figure 8.15, where the base quantities for $V$ and $Q$ are 275 kV and 200 MVAr, respectively.

   Examine the voltage stability of this system, indicating clearly any assumptions made in the analysis.

## *Solution*

It has been shown that

$$E = \sqrt{\left[\left(V + \frac{QX}{V}\right)^2 + \left(\frac{PX}{V}\right)^2\right]}$$

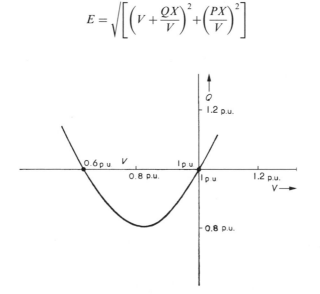

**Figure 8.15**   Example 8.5. Reactive power–voltage characteristic

If

$$\frac{PX}{V} < \frac{V^2 + QX}{V} \qquad \text{then} \qquad E = \frac{V^2 + QX}{V}$$

and

$$\frac{\mathrm{d}E}{\mathrm{d}V} = 1 + \left(\frac{\mathrm{d}Q}{\mathrm{d}V}XV - QX\right)\frac{1}{V_2} = 0$$

In this problem

and

$$\left.\begin{array}{c} P = 200\,\text{MW} \\ 1\,\text{p.u.} \quad Q = 200\,\text{MVAr} \end{array}\right\} \text{ three-phase values}$$

$$X = \frac{70 \times 200}{275^2}\text{p.u.}$$

$$= 0.185\,\text{p.u.}$$

$$E = V + \frac{QX}{V}$$

$$\therefore\ 1 = V + \frac{0.185}{V}\left(\frac{(V-0.8)^2}{0.2} + 0.8\right)$$

$$\therefore\ V^2 = V - 0.925V^2 - 0.59 + 1.48V - 0.148$$

giving

$$1.925V^2 - 2.48V + 0.74 = 0$$

$$\therefore\ V = \frac{+2.48 \pm \sqrt{(2.48^2 - 4 \times 1.925 \times 0.74)}}{2 \times 1.925}$$

$$= \frac{2.48 \pm 0.67}{3.85}$$

Taking the upper value, $V = 0.818\,\text{p.u.}$,

$$Q \text{ at this } V = \frac{(0.818 - 0.8)^2}{0.2} + 0.8$$

$$= 0.8016\,\text{p.u.}$$

$$\frac{\mathrm{d}E}{\mathrm{d}V} = 1 + [10(0.818 - 0.8) \times 0.185 \times 0.818 - 0.8016 \times 0.185/V^2]$$

$$= 1 + \left(\frac{0.027 - 0.148}{0.818^2}\right)$$

which is positive, i.e. the system is stable.

(Note: $PX/V \approx 0.16$ and $(V^2 + QX)/V \approx 1$; therefore approximation is reasonable.)

When the reactance between the source and load is very high, the use of tap-changing transformers is of no assistance. Large voltage drops exist in the supply lines and the 'tapping-up' of transformers increases these because of the increased supply currents. Hence the peculiar effects from tap-changing noticed when conditions close to a voltage collapse have occurred in practice, i.e. tapping-up reduces the secondary voltage, and vice versa. One symptom of the approach of critical conditions is sluggishness in the response of tap-changing transformers.

The above form of instability is essentially a form of steady-state instability, and a full assessment of the latter should not only take into account the synchronous criterion $(\mathrm{d}P/\mathrm{d}\delta) \geqslant 0$, although this is the more likely to occur, but also the criterion for load instability $(\mathrm{d}V/\mathrm{d}E) = \infty$.

Recent studies into the voltage collapse phenomenon in interconnected power systems have shown how difficult it is to predict its occurrence or, more usefully, estimate the margin available to collapse at any given operating condition. One approach is to increase all loads at constant power factor until the load flow diverges. Unfortunately, this does not provide a realistic case since (1) any increase in real-power load will result in reserve generation being brought on line at various nodes of the system not previously used for generation input; and (2) as voltage falls, loads behave non-linearly (see Figure 8.15) but our knowledge of load behaviour below about 0.9 p.u. voltage is extremely sparse. Consequently, only worst-case situations can be assumed if voltage margins are to be estimated. Further data from actual situations and improved study methods, e.g. using eigenvalues, are being sought.

## 8.9 Further Aspects

### 8.9.1 Faults on the feeders to induction motors

A more common cause of the stalling of induction motors (or the low-voltage releases operating and removing them from the supply) occurs when the supply voltage is either zero or very low for a brief period because of a fault on the supply system, commonly known as 'voltage sag'. When the supply voltage is restored the induction motors accelerate and endeavour to attain their previous operating condition. In accelerating, however, a large current is taken, and this, plus the fact that the system impedance has increased due to the loss of a line, results in a depressed voltage at the motor terminals. If this voltage is too low the machines will stall or cut out of circuit. An uninterruptible power supply (UPS) or voltage injection from a compensator could help to alleviate this problem.

### 8.9.2 Steady-state instability due to voltage regulators

Consider a generator supplying an infinite busbar through two lines, one of which is suddenly removed. The load angle of the generator is instantaneously unchanged and therefore the power output decreases due to the increased system reactance, thus causing the generator voltage to rise. The automatic voltage regulator of the generator then weakens its field to maintain constant voltage, i.e. decreases the internal e.m.f., and synchronous instability may result.

### 8.9.3 Dynamic stability

The control circuits associated with generator AVRs, although improving steady-state stability, can introduce problems of poorly damped response and even instability. For this reason, dynamic stability studies are performed, i.e. steady-state stability analysis, including the automatic control features of the machines (see Section 8.5). The stability is assessed by determining the response to small step changes of rotor angle, and hence the machine and control-system equations are often linearized around the operating point, i.e. constant machine parameters and linear AVR characteristic. The study usually extends over several seconds of real-system time.

## 8.10   Multimachine Systems and Energy-Type Functions

In networks interconnecting many generators and dynamic loads, it is a complicated task, even with a fast digital computer, to solve a set of dynamic equations to establish if a given state of the system is stable or unstable following a credible disturbance. Obviously, many hundreds of studies may need to be run with different faults and the output data assessed in some way. It must be remembered that if each dynamic machine (generator or load) is to be represented by its swing trajectory, starting from an initial steady-state angle (as in Figure 8.10), then a multitude of trajectories up to 5 s will be presented by the computer output. Some machines will probably not be affected by the disturbance and their angles remain within a few degress of their initial angle. Others will show oscillations but their mean angle could gradually diverge from the more or less unaffected machines—these diverging machines would indicate that the system is unstable and, if continued, would split up, by action of the interconnecting circuit-protection systems, into two or more 'islands'. A common feature of such studies often shows that machine angles oscillate at different frequencies but that they *all* gradually change their angles in unison from the initial angles, implying that the system is stable but that, subsequent to the initiating disturbance, the frequency of the whole system is either increasing or decreasing.

To enable a digital study to assess its own results, criteria need to be built into the software. One of these is whether or not the system angles are within a norm. Another is, in the case of instability, which parts of the system are splitting away from other parts. To achieve this assessment, a concept known as 'Centre of Inertia' (COI) is employed. This is similar to determining the centre of gravity of a mechanical system by writing

$$M_{\text{tot}} \cdot \delta_{\text{COI}} = \sum_{i=1}^{i=n} M_i \delta_i \tag{8.12}$$

where $M_{\text{tot}}$ is the total area angular momentum $= \sum_{i=1}^{i-n} M_i$ and $\delta_{\text{COI}}$ is the average angle of the machines in the system.

Consequently, we now have, for an area of the system, that

$$M_{\text{tot}} \cdot \frac{\mathrm{d}^2 \delta_{\text{COI}}}{\mathrm{d}t^2} + \Delta P_{\text{area}} = 0 \qquad \text{(see equation (8.1))}$$

where $\Delta P_{\text{area}}$ is the combined power being input (negative if output) into that area. By use of conditional statements in the software used for post-analysis of a stability study, areas of the system which swing together, i.e. whose angles are within specified limits around the $\delta_{\text{COI}}$ (known as *coherency*), can be identified. Note also that the power transfer across the area boundary $\Delta P_{\text{area}}$ before the disturbance can also be established for each area, once the vulnerable areas are known. To ensure stability, further studies need to be undertaken with revised generator scheduling such that critical $\Delta P_{\text{area}}$ flows are reduced to a sustainable level. System operators should then observe these area flow limits to ensure stability under credible system contingencies. In most systems, by proper design of transient machine controllers or reinforcement of the transfer capability of the vulnerable circuits, only a few critical areas remain where power transfer is limited by stability considerations. For example, in the U.K., the Scotland–England transfer over two double-circuit lines is often limited in this way; similarly under some conditions, circuits from North Wales to the rest of the National Grid system are flow-restricted.

### 8.10.1  Transient Energy Functions (TEF)

A fast stability study in large systems is essential to establish viable operating conditions. The use of Lyapunov-type functions for this purpose is left for advanced study. However, the equal-area criterion (EAC) is a form of energy function which can be used as a screening tool to enable a fast assessment of multimachine stability to be made.

Figure 8.16 shows three curves similar to those of Figure 8.5. From equation (8.8) we have that

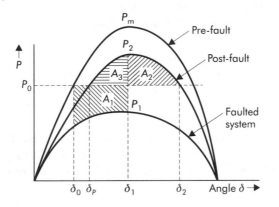

**Figure 8.16** Equal-area criterion applied to an equivalent machine connected to infinite busbar through a system

$$M_e \frac{d^2\delta}{dt^2} = P_0 - P_2 \sin \delta$$

We have also that

$$\frac{d\delta}{dt} = \omega \quad \text{and} \quad \frac{d\omega}{dt} = \frac{(P_0 - P_2 \sin \delta)}{M_e}$$

noting that $\omega$ is the incremental speed from the steady-state speed $\omega_s$. At any moment, the *energy* in this system consists of two components, namely:

$M_e\omega$        the incremental kinetic energy stored in the rotating masses of the machines at any instant;

$(P_0 - P_2 \sin \delta)$      the potential energy due to the excess electrical power at any instant.

The energy balance in the system during the disturbance period can be obtained by integrating the two energy components over the period. However, we must remember that prior to the disturbance under steady-state conditions the energy put into the system equalled the energy being taken out. Our development of the EAC only took into account the energy increments or decrements from steady-state conditions; therefore we need only concern ourselves with the *incremental* kinetic and potential energy functions. We can compute a value for the total transient energy in the system, TEF, by using

$$\text{TEF} = \int_0^{\omega_1} M_e\omega \, d\omega - \int_{\delta_p}^{\delta_1} (P_0 - P^2 \sin \delta) \, d\delta \quad \text{(Joules)} \qquad (8.13)$$

where $\omega_1$ is the incremental angular speed at angle $\delta_1$ and $\delta_P$ is the angle at which the system settles down following the disturbance; both angles are shown in Figure 8.16.

In fact, for any intermediate value of $\omega$ or $\delta$ the value of TEF is an indication that there is a surplus or deficiency of incremental energy and that the system is still in a dynamic or oscillation state. Only when $\delta = \delta_P$ and $\omega = 0$, i.e. no increment on the steady-state synchronous speed, will the oscillations have died away and the system be stable.

If the system is to regain a steady-state condition, the area $A_2$ must equal $A_1$, and the maximum angle attained at the limiting conditions is $\delta_2$, where the incremental speed $\omega$ would again be zero. Under these conditions we see that

$$\text{TEF}_{\text{limit}} = - \int_{\delta_P}^{\delta_1} (P_0 - P_2 \sin \delta) \, d\delta \qquad \text{since } \omega = 0$$

*TEF*$_{\text{limit}}$ represents the maximum energy that the system can gain due to a disturbance and subsequently dissipate by transfer over the system, if it is to remain stable. Knowing the value of TEF$_{\text{limit}}$ from the above equation enables a quick determination of stability by calculating TEF as the disturbance proceeds. This can be done readily as the step-by-step integration proceeds because values of $\omega$ and $\delta$ will be available. Provided that the value of TEF remains less than TEF$_{\text{limit}}$ the system is stable and will settle down to a new angle $\delta_P$. It is usual to compute TEF at fault clearance and assume afterwards that no more energy is added (rather, it is dissipated by system damping and losses), thereby checking that TEF $<$ TEF$_{\text{limit}}$.

It is worth noting that in Figure 8.16:

1. the kinetic energy-like term $\int_0^{\omega_1} M_e \omega \, d\omega$ is the area $A_1$;
2. the potential energy-like term $\int_{\delta_P}^{d_1} (P_0 - P_2 \sin \delta) \, d\delta$ is the area $A_3$;
3. TEF is equal to areas $A_1 + A_3$;
4. TEF$_{\text{limit}}$ is the area $A_2 + A_3$;
5. Equating TEF and TEF$_{\text{limit}}$ produces

$$A_1 + A_3 = A_2 + A_3 \qquad \text{i.e. } A_1 = A_2$$

which is the EAC for the system.

This shows that for stability the total energy gained during a disturbance must equal the capability of the system to transfer that energy away from the system afterwards. Considerable effort has been put into the determination of dynamic system boundaries through improved calculations of the TEF.

## 8.10.2 *Improvement of system stability*

Apart from the use of fast-acting AVRs the following techniques are in use:

1. Reduction of fault clearance times, 80 ms is now the norm with $SF_6$ circuit breakers and high-speed protection (see Chapter 11).

2. Turbine fast-valving by bypass valving—this controls the accelerating power by closing steam valves. Valves which can close or open in 0.2 s are available. CCGTs can reduce power by fuel control within 0.2 to 0.5 s.

3. Dynamic braking by the use of shunt resistors across the generator terminals; this limits rotor swings. The switching of such resistors can be achieved by thyristors.

4. High-speed reclosure (independent pole tripping) in long (point-to-point) lines. In highly interconnected systems the increase in overall clearance times on unsuccessful reclosure makes this technique of dubious value. Delayed autoreclose (DAR) schemes with delays of 12–15 s are preferable if the voltage sag can be tolerated.

5. Increased use of H.V. direct-current links using thyristors and GTOs also alleviates stability problems.

6. Semiconductor-controlled static compensators enabling oscillations following a disturbance to be damped out.

7. Energy-storage devices, e.g. batteries, superconducting magnetic energy stores (SMES) with fast control, providing the equivalent of a UPS.

# Problems

**8.1** A round-rotor generator of synchronous reactance 1 p.u. is connected to a transformer of 0.1 p.u. reactance. The transformer feeds a line of reactance 0.2 p.u. which terminates in a transformer (0.1 p.u. reactance) to the *LV* side of which a synchronous motor is connected. The motor is of the round-rotor type and of 1 p.u. reactance. On the line side of the generator transformer a three-phase static reactor of 1 p.u. reactance per phase is connected via a switch. Calculate the steady-state power limit with and without the reactor connected. All per unit reactances are expressed on a 10 MVA base and resistance may be neglected. The internal voltage of the generator is 1.2 p.u. and of the motor 1 p.u.
(Answer: 5 MW and 3.14 MW for shunt reactor)

**8.2** In the system shown in Figure 8.17, two equivalent round-rotor generators feed into a load which may be represented by the constant impedance shown. Determine $Z_{11}$, $Z_{12}$, and $Z_{22}$ for the system and calculate $\delta_{12}$ if $P_1$ is 1.2 p.u. All per unit values are expressed on the same base and the resistance of the system (apart from the load) may be neglected.
(Answer: $Z_{11} = 0.78\angle87.8°$; $Z_{12} = 1.47\angle112°$; $Z_{22} = 0.22\angle62.5°$)

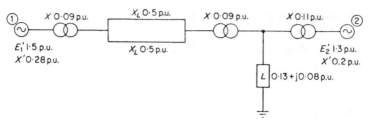

**Figure 8.17**  Line diagram of system in Problem 8.2

**8.3**  A hydroelectric generator feeds a load through two 132 kV lines in parallel, each having a total reactance of 70 Ω phase-to-neutral. The load consists of induction motors operating at three-quarters of full load and taking 30 MVA. A local generating station feeds directly into the load busbars. Determine the parameters of an equivalent circuit, consisting of a single machine connected to an infinite busbar through a reactance, which represents the above system when a three-phase symmetrical fault occurs halfway along one line.

The machine data are as follows:

| | |
|---|---|
| Hydroelectric generator: | rating 60 MW at power factor 0.9 lagging; transient reactance 0.3 p.u.; inertia constant 3 kW-s/kVA. |
| Induction motors: | composite transient reactance 3 p.u.; inertia constant 1 kW-s/kVA. |
| Local generator: | rating 30 MVA; transient reactance 0.15 p.u. inertia constant 10 kW-s/kVA. |

(Answer: $X = 9.9$ p.u., $M = 0.000121$ p.u., on a 100 MVA base)

**8.4**  For the system described in Example 8.3 determine the swing curves for fault clearance times of 250 and 500 ms.

**8.5**  The $P$–$V$, $Q$–$V$ characteristics of a substation load are as follows:

| $V$ | 1.05 | 1.025 | 1 | 0.95 | 0.9 | 0.85 | 0.8 | 075 |
|---|---|---|---|---|---|---|---|---|
| $P$ | 1.03 | 1.017 | 1 | 0.97 | 0.94 | 0.92 | 0.98 | 0.87 |
| $Q$ | 1.09 | 1.045 | 1 | 0.93 | 0.885 | 0.86 | 0.84 | 0.85 |

The substation is supplied through a link of total reactance 0.8 p.u. and negligible resistance. With nominal load voltage, $P = 1$ and $Q = 1$ p.u. By determining the supply voltage–received voltage characteristic, examine the stability of the system by the use of $dE/dV$. All quantities are per unit.

**8.6**  An induction motor load is supplied through a transformer of 0.1 p.u. reactance from a receiving-end substation which is supplied from a generator through a transmission link of total reactance 0.3 p.u. Plant data are as follows: generator $X_s = 1.1$, $X' = 0.3$. Induction motor load $X_s = 0.2$, $R_2 = 0.03$, mechanical power output is independent of speed. All per unit values are on a 60 MVA base. Examine the stability of the load when the motor takes 50 MW when the generator has (a) no AVR; (b) a continuously acting AVR.

**8.7** A large synchronous generator, of synchronous reactance 1.2 p.u., supplies a load through a link comprising a transformer of 0.1 p.u. reactance and an overhead line of initially 0.5 p.u. reactance; resistance is negligible. Initially, the voltage at the load busbar is 1 p.u. and the load $P + jQ$ is $(0.8 + j0.6)$ p.u. regardless of the voltage. Assuming the internal voltage of the generators is to remain unchanged, determine the value of line reactance at which voltage instability occurs.
(Answer: Unstable when $X = 2.0$ p.u.)

**8.8** A load is supplied from an infinite busbar of voltage 1 p.u. through a link of series reactance $X$ p.u. and of negligible resistance and shunt admittance. The load consists of a constant power component of 1 p.u. at 1 p.u. voltage and a per unit reactive power component ($Q$) which varies with the received voltage ($V$) according to the law

$$(V - 0.8)^2 = 0.2(Q - 0.8)$$

All per unit values are to common voltage and MVA bases.
  Determine the value of $X$ at which the received voltage has a unique value and the corresponding magnitude of the received voltage.
  Explain the significance of this result in the system described. Use approximate voltage-drop equations.
(Answer: $X = 0.25$ p.u.; $V = 0.67$ p.u.)

**8.9** Explain the criterion of stability based on the equal-area diagram.
  A synchronous generator is connected to an infinite busbar via a generator transformer and a double-circuit overhead line. The transformer has a reactance of 0.15 p.u. and the line an impedance of $0 + j0.4$ p.u. per circuit. The generator is supplying 0.8 p.u. power at a terminal voltage of 1 p.u. The generator has a transient reactance of 0.2 p.u. All impedance values are based on the generator rating and the voltage of the infinite busbar is 1 p.u.

(a) Calculate the internal transient voltage of the generator.
(b) Determine the critical clearing angle if a three-phase solid fault occurs on the sending (generator) end of one of the transmission line circuits and is cleared by disconnecting the faulted line.

(Answer: (a) $0.98 + j0.28$ p.u.; (b) $65.2°$
(*From E.C. Examination, 1995*)

**8.10** A 500 MVA generator with 0.2 p.u. reactance is connected to a large power system via a transformer and overhead line which have a combined reactance of 0.3 p.u. All p.u. values are on a base of 500 MVA. The amplitude of the voltage at both the generator terminals and at the large power system is 1.0 p.u. The generator delivers 450 MW to the power system.
  Calculate

(a) the reactive power in MVAr supplied by the generator at the transformer input terminals;
(b) the generator internal voltage;
(c) the critical clearing angle for a 3 p.h. short circuit at the generator terminals.

(Answer: (a) 62 MVAr; (b) 1.04 p.u.; (c) 86°)
(*From E.C. Examination, 1996*)

# 9
# Direct-Current Transmission

## 9.1  Introduction

Alternating-current transmission using three phases is the well-established method of transmitting large blocks of electrical energy over both short and long distances. However, there is a limit to the distance that a.c. can be transmitted by overhead lines because of the surge impedance limitation producing high voltages at the receiving end unless some form of compensation is employed. For long overhead lines an alternative form of transmission is use of direct current. If undersea crossings greater than 45 km are required, then, because of the charging current limitations in a.c. cables, d.c. is the only alternative.

The choice of d.c. is, however, based on economics. With the use of high-voltage, high-current semiconductor devices, converter stations and their controls are becoming cheaper and more reliable, and comparable to many a.c. components and systems. Consequently, the main technical reasons for high-voltage direct-current (h.v.d.c.) transmission are:

1. interconnection of two large a.c. systems without having to ensure synchronism and stability margins between them (the U.K.–France cross-channel link of 2000 MW is of this nature);

2. interconnection between systems of different frequency (e.g. the connection between north and south islands in Japan, which have 50 Hz and 60 Hz systems, respectively);

3. long overland transmission of high powers where a.c. transmission towers, insulators and conductors are more expensive than using h.v.d.c. (e.g. Nelson River scheme in Manitoba—total of 4000 MW over 600+ km).

The main advantages of h.v.d.c. compared with h.v.a.c. are:

1.   two conductors, positive and negative to ground, are required instead of three, thereby reducing tower or cable costs;

2.   the direct voltage can be designed equivalent to the peak of the alternating voltage for the same amount of insulation to ground (i.e. $V_{d.c.} = \sqrt{2}V_{a.c.}$)

3.   the voltage stress at the conductor surface can be reduced on d.c., thereby reducing corona loss, audible emissions, and radio interference;

4.   h.v.d.c. infeeds do not raise the short-circuit capacity required of switch-gear on the a.c. sides;

5.   fast control of converters can be used to damp out connected a.c. system oscillations.

Disadvantages of h.v.d.c. are:

1.   the higher cost of converter stations compared with an a.c. transformer substation;

2.   the need to provide filters and associated equipment to ensure acceptable waveform and power factor on a.c. side;

3.   limited ability to form multiterminal d.c. networks because of the need for coordinated controls and no acceptable d.c. circuit breaker.

## 9.2  Semiconductor Valves for High-Voltage Direct-Current Converters

The rapid growth in the use of direct current since about 1980 has been due to the development of high-voltage, high-current semiconductor devices. These have taken over from the previously used complex and expensive mercury arc valve, employing a mercury pool as cathode and a high-voltage graded column of anodes, with the whole enclosed in steel and ceramic to provide a vacuum-tight enclosure. Nowadays, the semiconductor units are stacked to form a group which is able to withstand the designed voltages and to pass the desired maximum currents—this group still being termed a 'valve' in h.v.d.c. parlance.

Since h.v.d.c. converters operate connected to active a.c. systems, line commutation is the cheapest and most reliable method to employ, requiring devices which can be triggered on and turned off by reducing the current through them to zero. The main device used for this purpose is the thyristor.

## 9.2.1  Thyristors

Thyristors are manufactured from pure silicon wafers and are four-layer versions of the simple rectifier *P–N* junction, as shown in Figure 9.1(a). The *P* layer in the middle is connected to a gate terminal biassed such that the whole unit can be prevented from passing current, even when a positive voltage exists on the anode. By applying a positive pulse to the gate, conduction can be started, after which the gate voltage control has no effect until the main forward current falls below its latching value (see Figure 9.1(c)). This current must be kept below the latching value for typically $100\,\mu s$ before the thyristor is able to regain its voltage hold-off properties. (Note that in forward conduction there is still a small voltage across the *P–N* junctions, implying that power is being dissipated—hence the semiconductor devices must be cooled and their losses accounted for.)

In practice, many devices, each of rating 5000 V and up to 2000 A, are stacked in a valve to provide a rating of, say, 200 kV, 2000 A. Valves are connected in series to withstand direct voltages up to 500 or 600 kV to earth on each 'pole'. Each semiconductor thyristor can be 8–10 cm in diameter and 2 cm depth with its anode and cathode terminals. A typical device is shown in Figure 9.2(a) and a valve in Figure 9.2(b).

A recent development has been the insulated gate bipolar thyristor (IGBT) which is a development of the mosfet, in which removal of current from the gate switches off the through current, thereby allowing power to be switched on or off throughout an a.c. cycle. This provides a means of controlling currents in relation to the voltage in an a.c. system so that vars may be generated or absorbed. Current injection into a.c. systems for power-flow control is achieved using these devices. Ratings up to 4 kV and 1000 A are possible with IGBTs, although the on-state voltage drop is generally larger than with a thyristor.

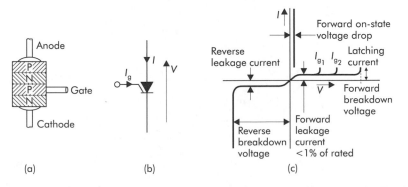

**Figure 9.1** (a) Structure of a four-layer thyristor. (b) Symbol. (c) Thyristor characteristic: $I_g$ gate current to switch thyristor on at forward voltage

(a)

(b)

**Figure 9.2** (a) High-power thyristor silicon device (*Courtesy E.P.R.I.—U.S.A.*). (b) Thyristor valves in converter station (*Couresy of I.E.E. Power Engineering Journal*)

# 9.3 Basic Converter and Direct-Current System Operation

A converter is required at each end of a d.c. line and operates as a rectifier (a.c. to d.c.) or an inverter (power transfer from d.c. to a.c.). The valves at the sending end of the link rectify the alternating current, providing direct current which is transmitted to the inverter. Here, it is converted back into alternating current which is fed into the connected a.c. system (Figure 9.3(a)). If a reversal of power flow is required, the inverter and rectifier exchange roles and the direct voltages at each end are reversed (Figure 9.3(b)). This is necessary because the direct current can flow in one direction only (anode to cathode in the valves), so to reverse the direction (or sign) of power the voltage direction must be reversed.

The alternating-current waveform injected by the inverter into the receiving-end a.c. system, and taken by the rectifier, is roughly trapezoidal in shape, and thus produces not only a fundamental sinusoidal wave but also harmonics of an order dependent on the number of valves. For a six-valve bridge the order is $6n \pm 1$, i.e. 5, 7, 11, 13, etc. Filters are incorporated to tune out harmonics up to the 25th.

In the following sections the processes of rectification and inversion will be analysed and the control of the complete link discussed.

# 9.4 Rectification

Transformer secondary connections may be arranged to give several phases for supplying the valves. Common arrangements are three, six, and twelve phases and higher numbers are possible. Six-phase is popular because of the better output-voltage characteristics. To begin with, a three-phase arrangement will be described, but most of the analysis will be for $n$ phases so that results are readily adaptable for any system. In Figure 9.4(a), a three-phase rectifier is shown and Figure 9.4(b) shows the current and voltage variation with time in the three phases of the supply transformer. With no gate control, conduction will take place between the cathode and the anode of highest potential. Hence the output-voltage wave is the thick line and the current output is continuous. In an $n$-phase system the anode changeover occurs at $(\pi/2 - \pi/n)$ degrees at a voltage $\hat{V} \sin \pi/n$ and the mean value of the direct-output voltage ($\hat{V}$ is the peak a.c. side voltage) is

$$
\begin{aligned}
V_{\mathrm{O}} &= \frac{1}{2\pi/n} \int_{\pi/2-\pi/n}^{\pi/2+\pi/n} \hat{V} \sin \omega t \, \mathrm{d}(\omega t) \\
&= \frac{\hat{V} \sin(\pi/n)}{\pi/n}
\end{aligned} \tag{9.1}
$$

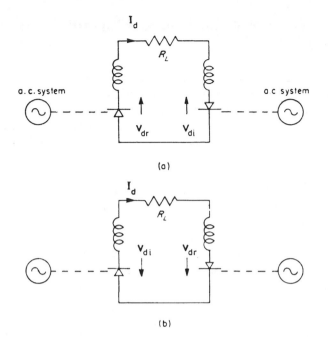

**Figure 9.3** (a) Symbolic representation of two alternating current systems connected by a direct-current link; $V_{dr}$ = direct voltage across rectifier, $V_{di}$ = direct voltage across inverter. (b) System as in part (a) but power flow reversed

For three phases,

$$V_O = \frac{\hat{V} \sin 60°}{\pi/3}$$

$$= \hat{V} \frac{3\sqrt{3}}{2\pi} = 0.83\hat{V}$$

and for six phases,

$$V_O = \frac{3\sqrt{2}}{\pi} V = \frac{3}{\pi}\hat{V} = 0.955\hat{V}$$

where $V$ = r.m.s. voltage.

Owing to the inductance present in the circuit, the current cannot change instantaneously from $+I_d$ to 0 in one anode and from 0 to $I_d$ in the next. Hence, two anodes conduct simultaneously over a period known as the commutation time or overlap angle ($\gamma$). When valve b commences to conduct, it short-circuits the (a) and (b) phases, the short-circuit current eventually becoming zero in valve a and $I_d$ in valve b. This is shown in Figure 9.5.

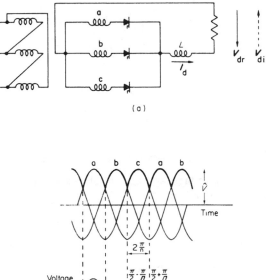

(a)

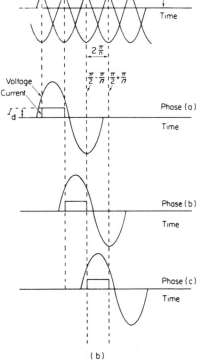

(b)

**Figure 9.4** (a) Three-phase rectifier; $V_{di}$ = voltage of inverter. (b) Waveforms of anode voltage and rectified current in each phase

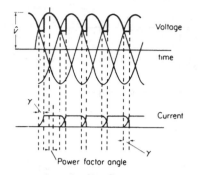

**Figure 9.5** Waveforms of voltage and current showing effect of the commutation angle $\gamma$. A lagging power factor is produced

### 9.4.1 Gate control

A positive pulse applied to a gate situated between anode and cathode controls the instant at which conduction commences, and once conduction has occurred the gate exercises no further control. In the voltage waveforms shown in Figure 9.6 the conduction in the valves has been delayed by an angle $\alpha$ by suitably delaying the application of positive voltage to the gates. Considering $n$ phases and ignoring the commutation angle $\gamma$, the new direct-output voltage with a delay angle of $\alpha$ is,

$$V'_O = \frac{1}{2\pi/n} \int_{\pi/2-\pi/n+\alpha}^{\pi/2+\pi/n+\alpha} \hat{V} \sin \omega t \, \mathrm{d}(\omega t)$$

$$= \frac{n\hat{V}}{2\pi} \int_{-\pi/n+\alpha}^{\pi/n+\alpha} \cos \omega t \, \mathrm{d}(\omega t)$$

$$= \frac{n\hat{V}}{2\pi} \cdot 2 \cdot \sin\left(\frac{\pi}{n}\right) \cos \alpha$$

$$\therefore V'_O = V_O \cos \alpha \tag{9.2}$$

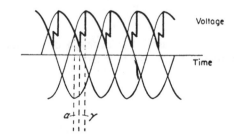

**Figure 9.6** Waveforms of rectifier with instant of firing delayed by an angle $\alpha$ by means of gate control

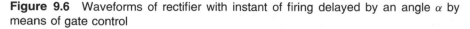

where $V_O$ is the maximum value of direct-output voltage as defined by equation (9.1).

### 9.4.2  Bridge connection

To avoid undue complexity in describing the basic operations, the converter arrangement used so far in this chapter is simple and has disadvantages in practice. Mainly because the d.c. output voltage is doubled, the bridge arrangement shown in Figure 9.7 is favoured, in which there are always two valves conducting in series. The corresponding voltage waveforms are shown in Figure 9.8 along with the currents (assuming ideal rectifier operation).

The sequence of events in the bridge connection is as follows (see Figures 9.7 and 9.8). Assume that the transformer voltage $V_A$ is most positive at the beginning of the sequence, then valve 1 conducts and the current flows through valve 1 and the load then returns through valve 6 as $V_B$ is most negative. After this period, $V_C$ becomes the most negative and current flows through valves 1 and 2. Next, valve 3 takes over from valve 1, the current still returning through valve 2. The complete sequence of valves conducting is therefore: 1 and 6; 1 and 2; 3 and 2; 3 and 4; 5 and 4; 5 and 6; 1 and 6. Control may be obtained in exactly the same manner as previously described, and the voltage waveforms with delay and commutation time accounted for are shown in Figure 9.9.

The direct-voltage output with the bridge may be calculated either by using the line voltage (phase-to-phase) in the formula for six-valve, six-phase rectification or by determining the magnitude for three-valve operation and doubling it, as both sides of the bridge contribute to the direct voltage. Hence,

$$V_0 = \frac{\hat{V} 3\sqrt{3}}{2\pi} \times 2 = \sqrt{2} V \times 3\sqrt{3}/\pi$$
$$= \sqrt{2} V_L \times 3/\pi \qquad (\text{as } V_L = \sqrt{3}\, V)$$

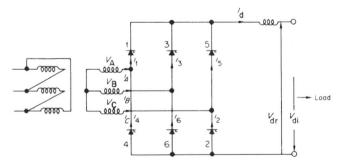

**Figure 9.7**  Bridge arrangement of valves

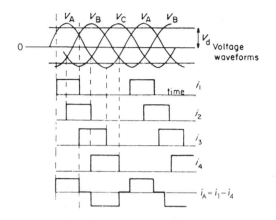

**Figure 9.8**   Idealized voltage and current waveforms for bridge arrangement

If the analysis used to obtain equation (9.2) is repeated for the bridge, it will be shown that $V_\mathrm{d} = V_0 \cos\alpha$, where $V_0$ is the maximum direct-voltage output for the bridge connection.

### 9.4.3   Current relationships in the bridge circuit

During the commutation process when two valves are conducting simultaneously, the two corresponding secondary phases of the supply transformer are short-circuited and if the voltage drop across the valves is neglected the following analysis applies.

When two phases of the transformer each of leakage inductance $L$ henries are effectively short-circuited, the short-circuit current ($i_\mathrm{s}$) is governed by the equation,

$$2L\frac{di_\mathrm{s}}{dt} = \hat{V}_\mathrm{L}\sin\omega t = \text{resultant voltage between the two phases}$$

$$\therefore i_\mathrm{s} = -\frac{\hat{V}_\mathrm{L}}{2L}\frac{\cos\omega t}{\omega} + A$$

where $A$ is a constant of integration and $\hat{V}_\mathrm{L}$ is the peak value of the line-to-line voltage. Now

$$\omega t = \alpha \qquad \text{when} \qquad i_\mathrm{s} = 0$$

$$\therefore A = \frac{\hat{V}_\mathrm{L}}{2\omega L}\cos\alpha$$

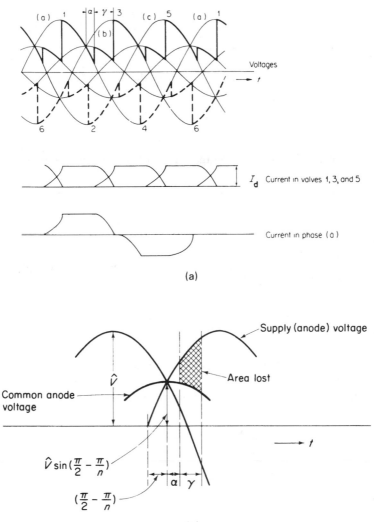

**Figure 9.9** (a) Voltage and current waveforms in the bridge connection, including commutation ($\gamma$) and delay ($\alpha$). Rectifier action. (b) Expanded waveforms showing voltage drop due to commutation

Also, when

$$\omega t = \alpha + \gamma \qquad i_s = I_d$$

$$\therefore I_d = \frac{\hat{V}_L}{2\omega L}[\cos\alpha - \cos(\alpha + \gamma)]$$

$$= \frac{V_L}{\sqrt{2}\omega L}[\cos\alpha - \cos(\alpha + \gamma)]$$

where $V_L$ = r.m.s. line-to-line voltage and, as for the bridge circuit,

$$V_0 = \frac{3\sqrt{2}}{\pi}V_L$$

and

$$I_d = \frac{\pi V_0}{3\sqrt{2}} \cdot \frac{1}{\sqrt{2}X}[\cos\alpha - \cos(\alpha + \gamma)] = \frac{\pi V_0}{6X}[\cos\alpha - \cos(\alpha + \gamma)] \qquad (9.3)$$

The mean direct-output voltage, with gate delay angle $\alpha$ only considered, has been shown to be $V_0 \cos\alpha$. With both $\alpha$ and the commutation angle $\gamma$, the voltage with $\alpha$ only will be modified by the subtraction of a voltage equal to the mean of the area under the anode voltage curve lost due to commutation (see Figure 9.5 and 9.9).

Referring to Figure 9.5 ($\alpha = 0$), the voltage drop due to commutation,

$$\Delta V_0 = \frac{\text{area lost}}{2\pi/n}$$

$$\Delta V_0 = \frac{n}{2\pi}\int_0^\gamma \hat{V}\sin\frac{\pi}{n}\sin\omega t\, d(\omega t)$$

$$= \frac{n}{2\pi}\hat{V}\sin\frac{\pi}{n}(1 - \cos\gamma) = \frac{V_0}{2}(1 - \cos\gamma)$$

When $\alpha > 0$ (see Figure 9.9(b)) the voltage drop is obtained as follows: area between input voltage wave and common anode voltage is

$$\int_{\pi/2-\pi/2+\alpha}^{\pi/2-\pi/n+\gamma+\alpha} \hat{V}\sin\omega t\, d(\omega t) - \int_\alpha^{\alpha+\gamma} \hat{V}\sin\left(\frac{\pi}{2} - \frac{\pi}{n}\right)\cos\omega t\, d(\omega t)$$

$$= \hat{V}\left[-\cos\left(\frac{\pi}{2} - \frac{\pi}{n} + \alpha + \gamma\right) + \cos\left(\frac{\pi}{2} - \frac{\pi}{n} + \alpha\right)\right]$$

$$- \sin\left(\frac{\pi}{2} - \frac{\pi}{n}\right) \sin(\alpha + \gamma) + \sin\left(\frac{\pi}{2} - \frac{\pi}{n}\right) \sin\alpha\Big]$$

$$= \hat{V}\Big[ -\sin\left(\frac{\pi}{n} - \alpha - \gamma\right) + \sin\left(\frac{\pi}{n} - \alpha\right) - \cos\frac{\pi}{n}\sin(\alpha + \gamma)$$

$$+ \cos\frac{\pi}{n}\sin\alpha\Big]$$

$$= \hat{V}\Big[ \sin\frac{\pi}{n}[\cos\alpha - \cos(\alpha + \gamma)]$$

Voltage drop (mean value of lost area)

$$= \frac{\hat{V}\sin(\pi/n)}{2(\pi/n)}[\cos\alpha - \cos(\alpha + \gamma)].$$

$$= \frac{V_0}{2}[\cos\alpha - \cos(\alpha + \gamma)]$$

The direct-voltage output,

$$V_{\mathrm{d}} = V_0\cos\alpha - \frac{V_0}{2}[\cos\alpha - \cos(\alpha + \gamma)]$$

$$= \frac{V_0}{2}[\cos\alpha + \cos(\alpha + \gamma)] \tag{9.4}$$

Adding equations (9.3) and (9.4),

$$\frac{I_{\mathrm{d}}3X}{\pi} + V_{\mathrm{d}} = V_0\cos\alpha$$

$$\therefore V_{\mathrm{d}} = V_0\cos\alpha - \frac{3XI_{\mathrm{d}}}{\pi} \tag{9.5}$$

The power factor is given approximately by

$$\cos\phi = \tfrac{1}{2}[\cos\alpha + \cos(\alpha + \gamma)] \tag{9.6}$$

Equation (9.5) may be represented by the equivalent circuit shown in Figure 9.10, the term $(3X/\pi)I_{\mathrm{d}}$ represents the voltage drop due to commutation and not a physical resistance drop. It should be remembered that $V_0$ is the

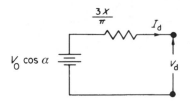

**Figure 9.10** Equivalent circuit representing operation of a bridge rectifier. Reactance per phase $X$ ($\Omega$)

theoretical maximum value of direct-output voltage and it is evident that $V_d$ can be varied by changing $V_0$ (control of transformer secondary voltage by tap changing) and by changing $\alpha$.

## 9.5 Inversion

With rectifier operation the output current $I_d$ and output voltage $V_d$ are such that power is absorbed by a load. For inverter operation it is required to transfer power from the direct-current to the alternating-current systems, and as current can flow only from anode to cathode (i.e., in the same direction as with rectification), the direction of the associated voltage must be reversed. An alternating-voltage system must exist on the primary side of the transformer, and gate control of the converters is essential.

As the bridge connection is in common use it will be used to explain the inversion process. If the bridge rectifier is given progressively greater delay the output voltage decreases, becoming zero when $\alpha$ is 90°, as shown in Figure 9.11. With further delay the average direct voltage becomes negative and the applied direct voltage (from the rectifier) forces current through the valves against this negative or back voltage. The converter thus receives power and inverts. The inverter bridge is shown in Figure 9.12(a) and the voltage and current waveforms in Figure 9.12(b). Commutation from valve 5 to valve 1 is possible only when phase (a) is positive with respect to phase (b), and the current changeover must be complete before (F) by a time ($\delta_0$) equal to the recovery time of the valves. From the current waveforms it is seen that the current supplied by the inverter to the a.c. system *leads* the voltage and hence the inverter may be considered as an absorber of vars.

The power factor $\cos\phi \simeq (\cos\delta + \cos\beta)/2$, where $\delta$ and $\beta$ are defined in Figure 9.12(a). Valve 5 is triggered at time A and as the cathode is held negative to the anode by the applied direct voltage ($V_d$), current flows, limited

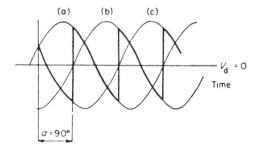

**Figure 9.11** Waveforms with operation with $\alpha = 90°$, direct voltage zero. Transition from rectifier to inverter action

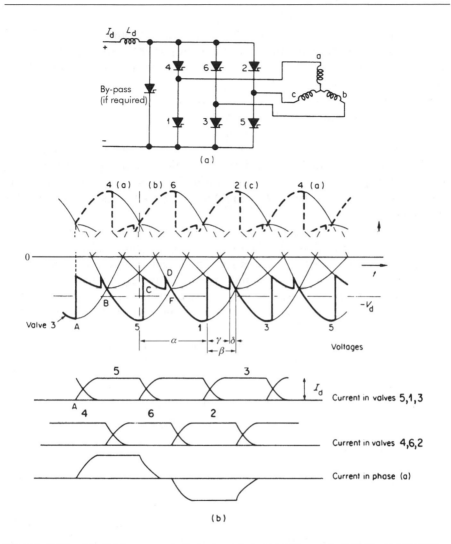

**Figure 9.12** (a) Bridge connection—inverter operation. (b) Bridge connection—inverter voltage and current waveforms

only by the circuit impedance. If the voltage drop across the valve is neglected, then the cathode and anode are at the same potential in valve 5. When time B is reached, the anode-to-cathode open-circuit voltage is zero and the valve endeavours to cease conduction. The large d.c.-side inductance $L_d$, however, which has previously stored energy, now maintains the current constant ($e = -L(di/dt)$ and if $L_d \to \infty$, $di/dt \to 0$). Conduction in valve 5 continues until time C, when valve 1 is triggered. As the anode-to-cathode voltage for valve 1 is greater than for valve 5, valve 1 will conduct, but for a time valves 5 and 1 conduct together (commutation time), the current gradually being transferred from valve 5 to valve 1 until valve 5 is non-conducting at point D. If triggering is delayed to point F, valve 5, which is still conducting, would be subject to a positively rising voltage and would continue to conduct into the positive half-cycle with breakdown of the inversion process. Hence, triggering must allow cessation of current flow before time F.

The angle ($\delta$) between the extinction of valve 1 and the point F, where the anode voltages are equal, is called the extinction angle, i.e. sufficient time must be allowed for the gate to regain control. It is usual to replace the delay angle $\alpha$ by $\beta = 180 - \alpha$, hence $\beta$ is equal also to $(\gamma + \delta)$. The minimum value of $\delta$ is $\delta_0$.

The action of the inverter is essentially that of the rectifier, but with the delay angle $\alpha$ greater than 90° the direct voltage output ($V_d$) is in a certain direction; as $\alpha$ increases, $V_d$ decreases; and when $\alpha = 90°$, $V_d = 0\,\mathrm{V}$; with further increase in $\alpha$, $V_d$ reverses and inverter action is obtained. Hence the change from rectifier to inverter action, and vice versa, is smoothly obtained by control of $\alpha$. This may be seen by consulting Figures 9.5 ($\alpha = 0$), 9.11 ($\alpha = 90°$), and 9.12(a) ($\alpha > 90°$).

Equations (9.3) and (9.4) may be used to describe inverter action. Replacing $\alpha$ by $(180 - \beta)$ and $\gamma$ by $(\beta - \delta)$ the following are obtained:

$$V_d = -[V_0 \cos \beta + I_d R_c] \tag{9.7}$$
$$V_d = -[V_0 \cos \delta - I_d R_c] \tag{9.8}$$

where

$$R_c = \frac{3X}{\pi}$$

Therefore two equivalent circuits are obtained for the bridge circuit as shown in Figure 9.13(a) for constant $\beta$ and Figure 9.13(b) for constant $\delta$.

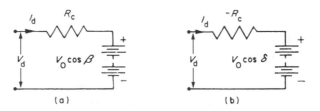

(a)     (b)

**Figure 9.13**  (a) Equivalent circuit of inverter in terms of angle $\beta$. (b) Equivalent circuit of inverter in terms of angle $\delta$

## 9.6 Complete Direct-Current Link

The complete equivalent circuit for a d.c. transmission link under steady-state operation is shown in Figure 9.14. If both inverter and rectifier operate at constant delay angles the current transmitted

$$= I_d = \frac{V_{dr} - V_{di}}{R_L}$$

or

$$I_d = \frac{V_{0r} \cos \alpha - V_{0i} \cos \beta}{R_L + R_{ci} + R_{ci}}$$

where $R_L$ is the loop resistance of the line or cable, and $R_{cr}$ and $R_{ci}$ are the effective commutation resistances of the rectifier and inverter, respectively.

The magnitude of direct current can be controlled by variation of $\alpha$, $\beta$, $V_{0r}$, and $V_{0i}$ (the last two by tap-changing of the supply transformers). Inverter control using constant delay angle has the disadvantage that if $\delta$ and hence $\beta$ are too large, excessively high reactive-power demand results (it will be seen from Figure 9.12(b) that the inverter currents are considerably out of phase with the anode voltages and hence a large requirement for reactive power is established). Also, a reduction in the direct voltage to the inverter results in an increase in the commutation angle $\gamma$, and if $\beta$ is made large to cover this the reactive-power demand will again be excessive. Therefore it is more usual to operate the inverter with a constant $\delta$, which is achieved by the use of suitable control systems (called *compounding*).

The equations governing the operation of the inverter may be summarized as follows:

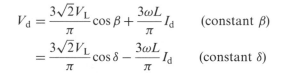

$$V_d = \frac{3\sqrt{2} V_L}{\pi} \cos \beta + \frac{3\omega L}{\pi} I_d \qquad \text{(constant } \beta)$$

$$= \frac{3\sqrt{2} V_L}{\pi} \cos \delta - \frac{3\omega L}{\pi} I_d \qquad \text{(constant } \delta)$$

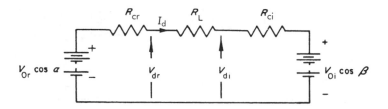

**Figure 9.14** Equivalent circuit of complete link with operation with given delay angles for rectifier and for inverter

(Note that $L$ is the leakage inductance per phase of the inverter transformer, and the power factor is $\cos \phi = (\cos \beta + \cos \delta)/2$ leading.)

The advantages of operating with $\delta$ fixed and as small as possible have been discussed; it is also advisable to incorporate a facility for constant-current operation. The rectifier and inverter change roles as the required direction of power flow dictates, and it is necessary for each device to have dual-control systems. A schematic diagram of the control systems is shown in Figure 9.15. In Figure 9.16 the full characteristics of the two converters of a link are shown, with each converter operating as rectifier and inverter in turn. In the top half of the diagram, converter (A) acts as a rectifier and the optimum characteristic with an $\alpha$ value of zero is shown. With constant-current control, $\alpha$ is increased and the output voltage–current characteristic crosses the $V_\mathrm{d} = 0$ axis at point D, below which converter (A) acts as an inverter and can be operated on constant-$\delta$ control. A similar characteristic is shown for converter (B), which commences as an inverter and with constant-current control ($\beta$ increasing) eventually changes to rectifier operation. It is seen that the current setting for converter (A) is larger than that for converter (B).

In Figure 9.17 the inverter characteristic for a power flow from A to B is shown drawn in the upper half of the graph and this will facilitate the discussion of the operational procedure.

### 9.6.1  Methods of control

To operate an inverter at a constant $\delta$, the instant of valve firing is controlled by a computer which takes into account variations in the instantaneous values of voltage and current. The computer then controls the firing times such that the extinction angle $\delta$ is slightly larger than the recovery angle of the valve. As

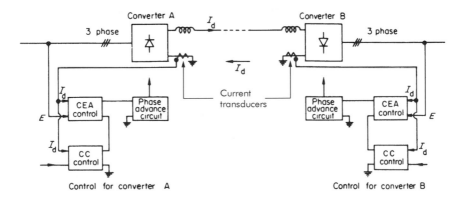

**Figure 9.15**  Schematic diagram of control of an h.v.d.c. system with ground return. CEA = constant extinction angle; CC = constant current

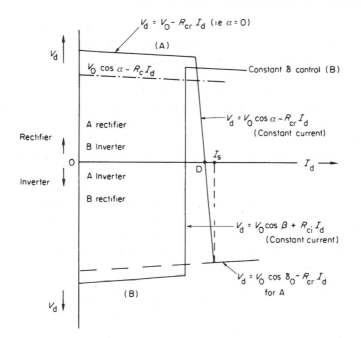

**Figure 9.16**  Voltage–current characteristics of converters with compounding

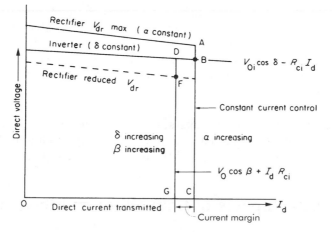

**Figure 9.17**  Inverter and rectifier operation characteristics with constant-current compounding. Point of operation is where the two characteristics intersect

the current rating of the valves should not be exceeded, some measure of current control is desirable, the ideal being constant-current operation. This may be achieved in the inverter by increasing $\beta$ beyond the constant-$\delta$ value, thus decreasing the back-voltage developed. Similarly, constant-current compounding can be incorporated in the rectifier. In Figure 9.17 the voltage–current characteristics with both inverter and rectifier current-compounding are shown, and it is evident that the transmitted current cannot increase beyond the prescribed value.

The methods of control may be summarized as follows, making reference to Figure 9.17.

1.  In the rectifier the transmitted current is regulated by varying the delay angle and hence $V_{dr}$. This is described by ABC. The rectifier tap-changing transformer is used to keep the delay angle within reasonable limits and also to give an excess in voltage over that of the inverter to cover sudden falls in rectifier output voltage which can be recovered in time by the tap-changer. It should be remembered that the greater the delay angle the greater the reactive power consumed. Control is normally performed by the rectifier, with inverter control used when necessary as follows.

2.  The inverter transformer gives the required back direct-voltage by tap-changing. Again, this is slow acting and constant-current control is required, giving a characteristic indicated by DG. The reference value of current is lower than that for the rectifier (a margin of 10–20 per cent of the rated current).

3.  Point B shows the operating condition with normal rectifier control. Should the rectifier voltage fall below the allowed margin to avoid $I_d$ becoming zero, the inverter back direct-voltage (i.e. the mean voltage of the negative anode voltages) is decreased, and operation makes place along DFG and the current is maintained at value OG. The new operating point is F and the power transmitted is smaller than before. Eventually, the rectifier tap-changer restores the original conditions. The value of the voltage margin is chosen to avoid frequent operations in the inverter control region.

To summarize, with normal operation the rectifier operates at constant current and the inverter at constant $\delta$; under emergency conditions the rectifier is at zero firing-delay (i.e. $\alpha = 0$) and the inverter is at constant current.

## 9.7 Further Aspects of Converters and Systems

The problem of obtaining a uniform voltage distribution with many thyristors in series is well known in low-voltage techniques. Because of the widely differing capacitances, the low-voltage technique of connecting *R–C* circuits in parallel with each thyristor does not suffice, and additional capacitance must be incorporated. Also, it is necessary to retain uniform transient voltage distributions with time. To achieve this, inductors with a non-linear characteristic may be used. On a sudden rise in voltage the inductor first absorbs the voltage, the capacitors are charged in the opposite direction with the inductor magnetizing current only, and the voltage rise across the thyristor is delayed. Similarly, the rate of change of current may be controlled on switching off.

Obviously, it is advantageous to have as high a voltage per thyristor as possible. This cannot be achieved by increasing the silicon wafer thickness because of the resulting increased temperature rise. The device can only block the voltage if the temperature is below 125°C. The voltage collapse across the wafer when a positive impulse is applied to the gate must not be accompanied by relatively high currents until conduction exists over the entire wafer. Only about a third of the possible reverse voltage is used for normal working because of overvoltages, etc.

The electrical circuitry associated with each thyristor is shown in Figure 9.18. The inductance limits the current rise upon firing and upon the discharge of $C_2$. The chain $R_1$, $C_1$ bypasses current from the thyristor, which is not yet or is no longer conducting, thus avoiding unacceptable voltages. At the end of the working interval the current is allowed to flow in the reverse direction until a barrier layer is established in the wafer that enables voltage to be built up and current to cease (see Figure 9.19). After this, the element can withstand negative voltage, but the current must fall to a very low value ($T_2$ in Figure 9.19) if the thyristor is to block positive voltage whilst waiting for a new gate firing impulse. Extreme care must be taken to avoid thermal overload of the thyristors because of the resulting damage. Thyristors with integrated gate

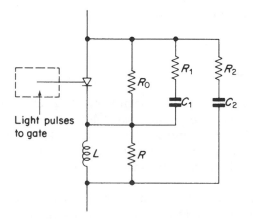

**Figure 9.18**  Circuitry associated with each thyristor

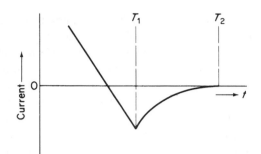

**Figure 9.19**  End of conduction period in thyristor—recovery of blocking property

circuitry pulsed through fibre-optic channels enable ready voltage isolation to be obtained.

### 9.7.1  Transmission systems

The cheapest arrangement is a single conductor with a ground return. This, however, has various disadvantages. The ground-return current results in the corrosion of buried pipes, cable sheaths, etc., due to electrolysis. With submarine cables the magnetic field set up may cause significant errors in ships' compass readings, especially when the cable runs north–south. This system is shown in Figure 9.20(a). Two variations on two-conductor schemes are shown in Figure 9.20(b) and (c). The latter has the advantage that if the ground is used in emergencies, a double-circuit system is formed to provide some security of operation. Note that in the circuits of Figure 9.20(c), transformer connections provide six-phase operation, thereby reducing harmonics generated on the a.c. sides.

### 9.7.2  Harmonics

A knowledge of the harmonic components of voltage and current in a power system is necessary because of the possibility of resonance and also the enhanced interference with communication circuits. The direct-voltage output of a converter has a waveform containing a harmonic content, which results in current and voltage harmonics along the line. These are normally reduced by a smoothing choke.

The currents produced by the converter currents on the a.c. side contain harmonics. The current waveform in the a.c. system produced by a delta–star transformer bridge converter is shown in Figures 9.9 and 9.12. The order of the

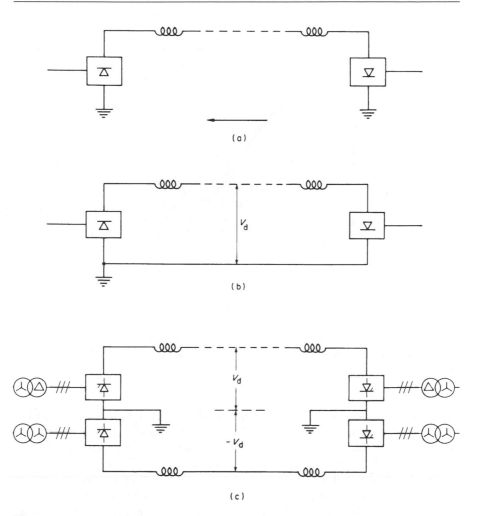

**Figure 9.20** Possible conductor arrangements for d.c. transmissions: (a) ground return; (b) two conductors, return earthed at one end; (c) double-bridge arrangement

harmonics produced is $6n \pm 1$, where $n$ is the number of valves. By the use of the Fourier series the equation for the current ($i$) is given by

$$i = \frac{2\sqrt{3}}{\pi} I_\mathrm{d} \left( \cos \omega t + \frac{1}{5} \cos 5\omega t - \frac{1}{7} \cos 7\omega t - \frac{1}{11} \cos 11\omega t + \cdots \right)$$

Figure 9.21 shows the variation of the seventh harmonic component with both commutation (overlap) angle ($\gamma$) and delay angle ($\alpha$). Generally, the harmonics decrease with decrease in $\gamma$ this being more pronounced at higher harmonics. Changes in $\alpha$ for a given $\gamma$ value do not cause large decreases in the

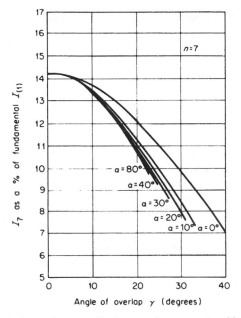

**Figure 9.21** Variation of seventh harmonic current with delay angle $\alpha$ and commutation angle (*Permission of International Journal of Electrical Engineering Education*)

harmonic components, the largest change being for $\alpha$ values between 0 and 10°. For normal operation, $\alpha$ is less than 10° and $\gamma$ is perhaps of the order of 20°; hence the harmonics are small. During faults, however, $\alpha$ may reach nearly 90°, $\gamma$ is small, and the harmonics produced are large.

The harmonic voltages and currents produced in the a.c. system by the converter current waveform may be determined by representing the system components by their reactances at the particular harmonic frequency. Most of the system components have resonance frequencies between the fifth and eleventh harmonics.

It is usual to provide filters (*L–C* shunt resonance circuits) tuned to the harmonic frequencies. A typical installation is shown in detail in Figure 9.22. At the fundamental frequency the filters are capacitive and help to meet the reactive-power requirements of the converters.

### 9.7.3 Variable compensators

Manual control of shunt units can achieve a reasonably constant voltage profile under steady-state conditions, but cannot cope with transient conditions. Similarly, synchronous compensators are too slow to help materially under

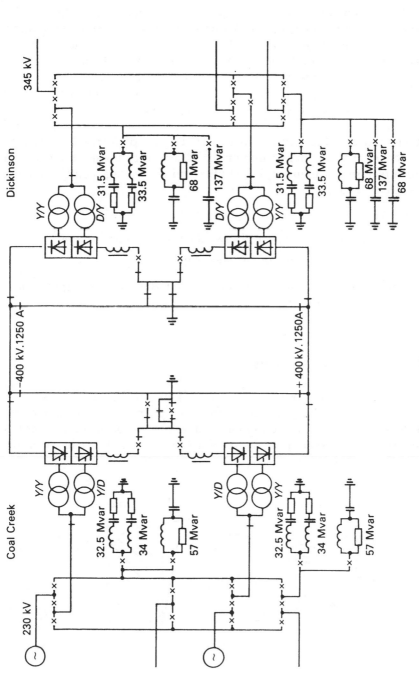

**Figure 9.22** Single-line diagram of the main circuit of an h.v.d.c. scheme showing filter banks and shunt compensation. (By kind permission of I.E.E.E.)

transient conditions and also contribute to fault currents. A method of providing fast variable-shunt compensation lies in the use of saturated reactors, which have response times of about 0.03 s. The equivalent circuit and characteristics of a saturable, iron-cored, three-phase reactor are shown in Figure 9.23. It has a high inductance and low var absorption if the voltage is below a prescribed level. Above this voltage the inductance decreases for part of the cycle and more vars are absorbed. Capacitance $C_0$ changes the slope of the $Q-V$ characteristic. The overall cost is in the same order as a synchronous compensator.

An alternative to the saturable reactor is the use of thyristor control by varying the firing angle. Many variations exist depending upon the application and speed of response required for stability and overvoltage control. Among the possible types are:

TCR:        thyristor-controlled reactor

MSR/MSC:  mechanically switched reactor or capacitor

TSC:        thyristor-switched capacitor

Typical combinations, usually supplied from the tertiary of a grid transformer at 13–15 kV, are shown in Figure 9.24.

### 9.7.4  Multiterminal systems

These are feasible, but very elaborate coordination of the control systems is necessary. Obviously, the great requirement of systems with more than two terminals is a d.c. circuit breaker. Intensive development of such a breaker has been in progress for some time, based on a vacuum interrupter in which the

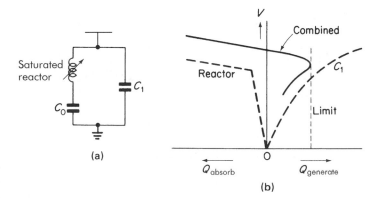

**Figure 9.23** Saturated reactor for voltage control: (a) equivalent circuit; (b) *V–Q* characteristics

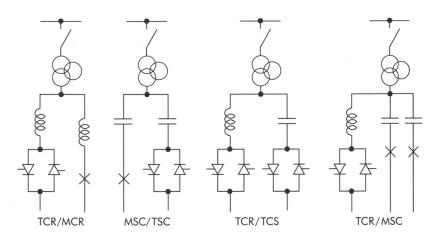

TCR/MCR        MSC/TSC            TCR/TCS            TCR/MSC

**Figure 9.24** Possible combinations of controlled reactors and capacitors for variable compensation purposes

discharge of a capacitor through a triggered vacuum gap produces an oscillation in the interrupter gap, thereby creating a current zero. However, the cost is still very high, and by coordinated control of all terminals, a brief voltage zero followed by isolator-switched and -controlled start-up of remaining circuits is considered preferable. Such systems operate in Italy Sardinia, and in California.

# Problems

**9.1**  A bridge-connected rectifier is fed from a 230 kV/120 kV transformer from the 230 kV supply. Calculate the direct voltage output when the commutation angle is 15° and the delay angle is (a) 0°; (b) 30°; (c) 60°.
(Answer: (a) 160 kV; (b) 127kV; (c) 61.5 kV)

**9.2**  It is required to obtain a direct voltage of 100 kV from a bridge-connected rectifier operating with $\alpha = 30°$ and $\gamma = 15°$. Calculate the necessary line secondary voltage of the rectifier transformer, which is nominally rated at 345 kV/150 kV; calculate the tap ratio required.
(Answer: 94 kV line; 1.6)

**9.3**  If the rectifier in Problem 9.2 delivers 800 A d.c., calculate the effective reactance $X\,(\Omega)$ per phase.
(Answer: $X = 13.0\,\Omega$)

**9.4**  A d.c. link comprises a line of loop resistance 5 Ω and is connected to transformers giving secondary voltage of 120 kV at each end. The bridge-connected converters operate as follows:

Rectifier: $\alpha = 10°$   Inverter: $\delta_0 = 10°$

$X = 15\,\Omega$   Allow 5° margin on $\delta_0$ for $\delta$

$X = 15\,\Omega$

Calculate the direct current delivered if the inverter operates on constant $\delta$ control. If all parameters remain constant, except $\alpha$, calculate the maximum direct current transmittable.
(Answer: 610; 1104 A)

**9.5** The system in Problem 9.4 is operated with $\alpha = 15°$ and on constant $\beta$ control. Calculate the direct current for $\gamma = 15°$.
(Answer: 550 A)

**9.6** For a bridge arrangement, sketch the current waveforms in the valves and in the transformer windings and relate them, in time, to the anode voltages. Neglect delay and commutation times. Comment on the waveforms from the viewpoint of harmonics.

**9.7** Draw a schematic diagram of an existing or proposed direct-current transmission scheme. Give vital parameters, e.g. voltage, rating of valves, etc.

**9.8** A direct current transmission link connects two a.c. systems via converters, the line voltages at the transformer–converter junctions being 100 kV and 90 kV. At the 100 kV end the converter operates with a delay angle of 10°, and at the 90 kV end the converter operates with a $\delta$ of 15°. The effective reactance per phase of each converter is 15 Ω and the loop resistance of the link is 10 Ω. Determine the magnitude and direction of the power delivered if the inverter operates on constant-$\delta$ control. Both converters consist of six valves in bridge connection. Calculate the percentage change required in the voltage of the transformer, which was originally at 90 kV, to produce a transmitted current of 800 A, other controls being unchanged. Comment on the reactive power requirements of the converters.
(Answer: 1.55 kA; 207 MW; 6.5 per cent)

**9.9** A thyristor rectifier supplies direct current to a load. On no-load the voltage across the load is 1.5 kV and the firing or delay angles 35°. The delay is controlled by a feedback system which holds the load voltage constant with changes in current. With this control the transformer reactance causes the ratio (voltage drop/no-load voltage) to be 0.08 p.u.

By considering only the fundamental components of the three-phase supply currents, plot a curve of volt–amperes reactive as a function of load current. Neglect the commutation angle.

**9.10** Draw a schematic diagram showing the main components of an h.v.d.c. link connecting two a.c. systems. Explain briefly the role of each component and how inversion into the receiving end a.c. system is achieved.

Discuss two of the main technical reasons for using h.v.d.c. in preference to a.c. transmission and list any disadvantages.

*(From E.C. Examination, 1996)*

# 10
# *Overvoltages and Insulation Requirements*

## 10.1 Introduction

An area of critical importance in the design of power systems is the consideration of the insulation requirements for lines, cables, and stations. At first glance this may appear to be a simple matter once the operating voltage of the system is decided, but unfortunately this is far from so. As well as the normal operating voltages, transients occur, causing overvoltages in the system as a result of switching, lightning strokes, and other causes; the peak values of these can be much in excess of the operating voltage. Because of this, devices must be provided to protect items of plant. The term extra-high voltage (E.H.V.) has generally been accepted as describing systems of 230 kV up to 765 kV, and for voltages above 765 kV the term ultra-high voltage (U.H.V.) is applied; below 230 kV, high voltage (H.V.) is in use.

Until recent years, lightning has largely determined the insulation requirements, i.e. size of bushings, number of insulators per string, and tower clearances of the system, and the insulation of equipment has been tested with voltages of a waveform approximately that of a lightning surge. With the much higher operating voltages now in use, and projected, the voltage transients or surges due to switching, i.e. the opening and closing of circuit breakers, have become the major consideration. In this context it is of interest to note that voltages of 500 kV and 765 kV are now in use, and active discussion and research are taking place to decide the next voltage level, likely to be in the range 1000–1500 kV if rights of way can be found.

A factor of major importance is the *contamination* of insulator surfaces caused by atmospheric pollution. This considerably modifies the performance of insulation, which becomes difficult to assess precisely. The presence of dirt, salt, etc., on the insulator discs or bushing surfaces results in these surfaces becoming slightly conducting, and hence flashover occurs.

A few terms frequently used in high-voltage technology need definition. They are as follows.

1. *Basic impulse insulation level or basic insulation level* (BIL)—reference levels expressed in impulse crest (peak) voltage with a standard wave not longer than a $1.2 \times 50 \,\mu s$ wave (see Figure 10.1). Apparatus insulation as demonstrated by suitable tests shall be equal or greater than the BIL. The two standard tests are the power frequency and 1.2/50 impulse-wave withstand tests. The *withstand voltage* is the level that the equipment will withstand for a given length of time or number of applications without disruptive discharge occurring, i.e. a failure of insulation resulting in a collapse of voltage and passage of current (sometimes termed 'sparkover' or 'flashover' when the discharge is on the external surface). Normally, several tests are performed and the number of flashovers noted. The BIL is usually expressed as a per unit of the peak (crest) value of the normal operating voltage to earth; e.g. for a maximum operating voltage of 362 kV,

$$1 \text{ p.u.} = \sqrt{2} \times \frac{362}{\sqrt{3}} = 300 \text{ kV}$$

so that a BIL of 2.7 p.u. $= 810 \text{ kV}$.

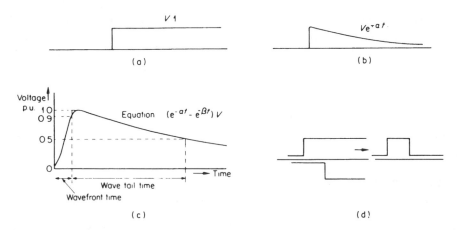

**Figure 10.1**  Basic impulse waveforms; (a) and (b) frequently used in calculations. (c) Shape of lightning and switching surges; the former have a rise time of, say, $1.2 \,\mu s$ and a fall time of half maximum value of $50 \,\mu s$ (hence 1.2/50 wave). Switching surges are much longer, the duration times varying with situation; a typical wave is 175/3000 $\mu s$. Equation for 1/50 wave is $v = 1.036 \, (e^{-0.0146t} - e^{-2.56t})$. (d) Use of two unit functions to form a wave of finite duration

2. *Critical flashover voltage* (CFO)—the peak voltage for a 50 per cent probability of flashover or disruptive discharge (sometimes denoted by $V_{50}$).

3. *Impulse ratio* (for flashover or puncture of insulation)—impulse peak voltage divided by the crest value of power-frequency voltage to cause flashover or puncture.

Impulse tests are normally performed with a voltage wave which rises in 1.2 $\mu$s and falls to half the peak value in 50 $\mu$s (see Figure 10.1); this is known as a 1.2/50 $\mu$s wave and typifies the lightning surge. In most impulse generators the shape and duration of the wave may be modified. Basic impulse waves are shown in Figure 10.1. Switching surges consist of damped oscillatory waves, the frequency of which is determined by the system configuration and parameters; they are normally of amplitude 2–2.8 p.u., although they can exceed 4 p.u. (per-unit values based on peak line-to-earth operating voltage as before).

## 10.2   Generation of Overvoltages

### 10.2.1   Lightning surges

A thundercloud is bipolar, often with positive charges at the top and negative at the bottom, usually separated by several kilometres. When the electric field strength exceeds the breakdown value a lightning discharge is initiated. The first discharge proceeds to the earth in steps (stepped leader stroke). When close to the earth a faster and luminous return stroke travels along the initial channel, and several such leader and return strokes constitute a flash. The ratio of negative to positive strokes is about 5 : 1 in temperate regions. The magnitude of the return stroke can be as high as 200 kA, although an average value is of the order of 20 kA.

Following the initial stroke, after a very short interval, a second stroke to earth occurs, usually in the ionized path formed by the original. Again, a return stroke follows. Usually, several such subsequent strokes (known as dart leaders) occur, the average being between three and four. The complete sequence is known as a multiple-stroke lightning flash and a representation of the strokes at different time intervals is shown in Figure 10.2. Normally, only the heavy current flowing over the first 50 $\mu$s is of importance and the current–time relationship has been shown to be of the form $i = i_{\text{peak}}(e^{-\alpha t} - e^{-\beta t})$.

When a stroke arrives on an overhead conductor, equal current surges of the above waveform are propagated in both directions away from the point of impact. The magnitude of each voltage surge set up is therefore $\frac{1}{2} \cdot Z_0 \cdot i_{\text{peak}}$ ($e^{-\alpha t} - e^{-\beta t}$), where $Z_0$ is the conductor surge impedance. For a current peak of 20 kA and a $Z_0$ of 350 $\Omega$ the voltage surges will have a peak value of $(350/2) \times 20 \times 10^3$, i.e. 3500 kV.

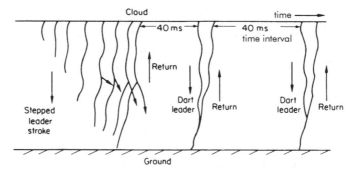

**Figure 10.2**   Sequence of strokes in a multiple lightning stroke

When a ground or earth wire exists over the overhead line, a stroke arriving on a tower or on the wire itself sets up surges flowing in both directions along the wire. On reaching neighbouring towers they are partially reflected and transmitted further. This process continues over the length of the line as towers are encountered. If the towers are 300 m apart the travel time between towers and back to the original tower is $(2 \times 300)/(3 \times 10^8)$, i.e. $2\,\mu s$, where the speed of propagation is $3 \times 10^8$ m/s. The voltage distribution may be obtained by means of the Bewley lattice diagram, to be described in section 10.6.

If an indirect stroke strikes the earth near a line, the induced current, which is normally of positive polarity, creates a voltage surge of the same waveshape which has an amplitude dependent on the distance from the ground. With a direct stroke the full lightning current flows into the line producing a surge that travels away from the point of impact in all directions. A direct stroke to a tower can cause a *back flashover* due to the voltage set up across the tower inductance and footing resistance by the rapidly changing lightning current (typically $10\,kA/\mu s$); this appears as an overvoltage between the top of the tower and the conductors (which are at a lower voltage).

## 10.2.2   Switching surges—interruption of short circuits and switching operations

When the arc between the circuit-breaker contacts breaks, the full system voltage (recovery voltage) suddenly appears across the open gap and hence across the $R\text{--}L\text{--}C$ circuit comprising the system. The simplest form of single-phase equivalent circuit is shown in Figure 10.3(a) and (b). The resultant voltage appearing across the circuit is shown in Figure 10.3(c). It consists of a high-frequency component superimposed on the normal system voltage, the total being known as the restriking voltage and constituting a switching surge. The equivalent circuit of Figure 10.3 may be analysed by means of the Laplace transform. When the circuit breaker opens, the cessation of current may be

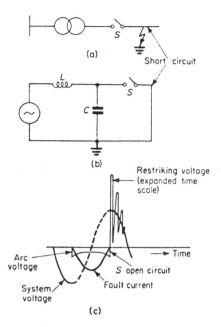

**Figure 10.3** Restriking voltage set up on fault interruption. (a) System diagram. (b) Equivalent circuit. $L$ = transformer leakage inductance; $C$ = transformer and connections capacitance to ground. (c) Current and voltage waveforms

simulated by the injection of an equal and opposite current into the system at time zero. Although this current is sinusoidal in form it may be approximated by a ramp function as shown in Figure 10.4. Let the fault current, $i = \sqrt{2}I \sin \omega t$.

The arc is extinguished at a current pause or zero (see Chapter 11), at which

$$\frac{\mathrm{d}i}{\mathrm{d}t} = (\sqrt{2}I\omega \cos \omega t)_{t=0}$$

$$= \sqrt{2}I\omega$$

The short-circuit current $= I = V/\omega L$ (r.m.s. value). Hence the equation for the ramp function of injected current is

$$\left(\frac{\mathrm{d}i}{\mathrm{d}t}\right)_{t=0} \cdot t = \sqrt{2}\left(\frac{V}{L}\right)t \tag{10.1}$$

The transform of the circuit after breaker opening is shown in Figure 10.5. Transform of the ramp function

$$= i(\mathrm{s}) = \frac{\sqrt{2}V}{Ls^2} \tag{10.2}$$

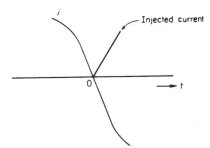

**Figure 10.4** Application of Laplace transform to circuit of Figure 10.3. Ramp function of current

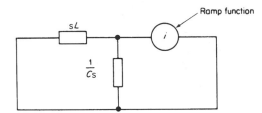

**Figure 10.5** Equivalent circuit with ramp applied

Voltage across the breaker contacts, i.e. across $C$ in Figure 10.3(b) is

$$v(s) = \frac{\sqrt{2}V}{Ls^2} \cdot \frac{L/C}{Ls + 1/Cs}$$

$$= \sqrt{2}V \frac{\omega_0^2}{s(s^2 + \omega_0^2)}$$

where $\omega_0^2 = 1/LC$.

$$\therefore v(s) = \sqrt{2}V\left(\frac{A}{s} + \frac{Bs + D}{s^2 + \omega_0^2}\right)$$

$$= \sqrt{2}V\left(\frac{1}{s} - \frac{s}{s^2 + \omega_0^2}\right)$$

since $A = 1$, $B = -1$, $D = 0$ from initial conditions.

From a table of transforms,

$$v(t) = \sqrt{2}V[1 - \cos(\omega_0 t)] \tag{10.3}$$

Equation (10.3) assumes that the full-system voltage $V$ exists across the open switch; strictly the voltage across $C$ is

$$V\left(\frac{X_c}{X_c - X_L}\right)$$

where $X_c$ and $X_L$ are the capacitive and inductive reactances at system angular frequency $\omega$ so that equation (10.3) becomes

$$\frac{\sqrt{2}V}{[1 - (X_L/X_c)]}(1 - \cos\omega_0 t) \tag{10.4}$$

The difference, in practice, between these two expressions is small.

When series resistance ($R\,\Omega$) is significant, equation (10.1) becomes,

$$\sqrt{2}\left(\frac{V\omega}{R + j\omega L}\right)t$$

and the ramp expression

$$i(s) = \frac{\sqrt{2}\omega V}{(R + j\omega L)s^2} \tag{10.5}$$

and $v(s) = i_s Z(s)$.

The analysis of this case yields the expression,

$$v(t) = \omega IL(1 - e^{-\alpha t}\cos\omega_0 t) \tag{10.6}$$

where $\omega$ is the system angular frequency, $\omega_0$ is the natural angular frequency of the circuit, and $I$ is the short-circuit current prior to the breaker opening,

$$\alpha = \frac{R}{2L}, \qquad (\alpha^2 + \omega_0^2) = \frac{1}{LC}$$

and

$$V \approx I\omega L$$

From equation (10.6)

$$(\mathrm{d}v/\mathrm{d}t)_{\text{maximum}} \approx \omega I\sqrt{\left(\frac{L}{C}\right)}e^{-\alpha t} \tag{10.7}$$

The restriking voltage ($v$) can rise to a maximum value of $2V$, where $V$ is the peak value of the system recovery voltage. A similar expression is obtained when a line is suddenly energized by the system voltage.

If a resistance $R_s$ be connected across the contacts of the circuit breaker, the surge will be critically damped when $R_s = \frac{1}{2}\sqrt{(L/C)}$, and this offers an important method of reducing the severity of the transient. The initial rate of rise of the surge is very important as this determines whether the contact gap, which is highly polluted with arc products, breaks down again after the initial open circuit occurs. In the system shown in Figure 10.6(a), a double-frequency transient (Figure 10.6(b)) is set up, the two frequencies being determined by the circuit on each side of the switch, i.e.

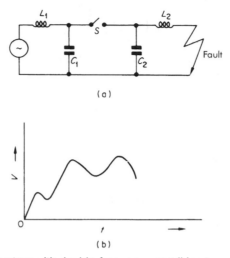

(a)

(b)

**Figure 10.6** (a) System with double-frequency restriking transient. (b) Typical waveform of double frequency, although initial amplitude is low, rate of rise is high

$$\omega_1 = \frac{1}{\sqrt{(L_1 C_1)}} \quad \text{and} \quad \omega_2 = \frac{1}{\sqrt{(L_2 C_2)}}$$

## *Example 10.1*

Determine the relative attenuation occurring in 5 cycles in the overvoltage surge set up on a 66 kV cable fed through an air-blast circuit breaker when the breaker opens on a system short circuit. The breaker incorporates resistance switching, i.e. an optimum resistance switched in across the contact gap on opening. The network parameters are as follows:

$$R = 7.8\,\Omega \qquad L = 6.4\,\text{mH} \qquad C = 0.0495\,\mu\text{F}$$

## *Solution*

$$\text{Switching resistance} = \tfrac{1}{2}\sqrt{(L/C)} = 180\,\Omega$$

$$\alpha = \frac{1}{2}\cdot\frac{R}{L} = 610$$

$$\omega = \frac{1}{\sqrt{LC}} = 5.61 \times 10^4\,\text{rad/s}$$

i.e. transient frequency = 8.93 kHz.

Hence

$$\text{time for 5 cycles} = \frac{1}{8.93 \times 10^3} \times 5\,\text{s}$$

and

$$e^{-\alpha t} = \exp\left(-610 \times \frac{5}{8.93 \times 10^3}\right) = 0.712$$

The maximum theoretical voltage set up with the circuit breaker opening on a short-circuit fault would be

$$2 \times \frac{66\,000}{\sqrt{3}} \times \sqrt{2}$$

i.e. 107.5 kV, and in 5 cycles (i.e. 560 $\mu$s) this becomes $107.5 \times (1 - 0.712)$, i.e. 31 kV.

(Note that resistance switching lowers the current to be broken and raises the power factor so that the voltage is not at peak value at opening.)

### 10.2.3 Switching surges—interruption of capacitive circuits

The interruption of capacitive circuits is shown in Figure 10.7. At the instant of arc interruption, the capacitance is left charged to value $v_m$ but half a cycle later the system voltage is $-v_m$, giving a gap voltage of $2v_m$. If the gap breaks down, an oscillatory transient is set up (as previously discussed), which can increase the gap voltage still further, as shown in Figure 10.7.

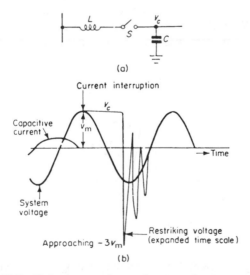

Figure 10.7 Voltage waveform when opening a capacitive circuit

## Current chopping

Current chopping arises with air-blast or puffer-assisted circuit breakers which operate on the same gas pressure and velocity for all values of interrupted current. Hence, on low-current interruption the breaker tends to open the circuit before the current natural-zero, and the electromagnetic energy present is rapidly converted to electrostatic energy, i.e.

$$\tfrac{1}{2}Li_0^2 = \tfrac{1}{2}Cv^2 \qquad \text{and} \qquad v = i_0\sqrt{\frac{L}{C}} \tag{10.8}$$

The voltage waveform is shown in Figure 10.8.

An extension of equation (10.8) to include resistance and time yields

$$v_r = i_0\sqrt{\frac{L}{C}}e^{-\alpha t}\sin\omega_0 t \tag{10.9}$$

where $\omega_0 = 1\sqrt{LC}$ and $i_0$ is the value of the current at the instant of chopping. High transient voltages may be set up on opening a highly inductive circuit such as a transformer on no-load.

## Faults

Overvoltages may be produced by certain types of asymmetrical fault, mainly on systems with unground neutrals. The voltages set up are of normal operating frequency. Consider the circuit with a three-phase earth fault as shown in Figure 10.9. If the circuit is not grounded the voltage across the first gap to open is $1.5\,V_{\text{phase}}$. With the system grounded the gap voltage is limited to the phase voltage. This has been discussed more fully in Chapter 7.

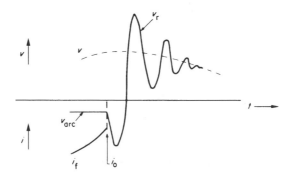

**Figure 10.8**  Voltage transient due to current chopping: $i_f$ = fault current; $v$ = system voltage, $i_0$ = current magnitude at chop, $v_r$ = restriking voltage

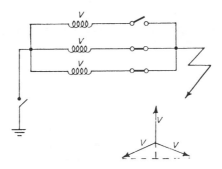

**Figure 10.9** Three-phase system with neutral earthing (grounding)

## *Resonance*

It is well known that in resonant circuits severe overvoltages occur, dependent on the resistance present; the voltage at resonance across the capacitance can be high. Although it is unlikely that resonance in a supply network can be obtained at normal supply frequencies, it is possible to have this condition at harmonic frequencies. Resonance is normally associated with the capacitance to earth of items of plant and is often brought about by an opened phase caused by a broken conductor or a fuse operating.

In circuits containing windings with iron cores, e.g. transformers, a condition due to the shape of the magnetization curve, known as *ferroresonance*, is possible. This can produce resonance with overvoltages and also sudden changes from one condition to another.

A summary of important switching operations is given in Table 10.1.

# 10.3  Protection Against Overvoltages

## *10.3.1  Modification of transients*

When considering the protection of a power system against overvoltages, the transients may be modified or even eliminated before reaching the substations, or, if this is not possible, the lines and substation equipment may be protected, by various means, from flashover or insulation damage. By the use of overhead earth (ground) wires, phase conductors may be shielded from direct lightning strokes and the effects of induced surges from indirect strokes lessened. The shielding is not complete, except perhaps for a phase conductor immediately below the earth wire. The effective amount of shielding is often described by an angle $\phi$, as shown in Figure 10.10; a value of 35° appears to agree with practical experience. Obviously, two earth wires horizontally separated provide much

**Table 10.1**   Summary of the more important switching operations

| Switching operation | System | Voltage across contacts |
|---|---|---|
| 1. Terminal short circuit | | |
| 2. Short line fault | | |
| 3. Two out-of-phase systems— voltage depends on grounding conditions in systems | | |
| 4. Small inductive currents, current chopped (unloaded transformer) | Transformer | |
| 5. Interrupting capacitive currents— capacitor banks, lines, and cables on no-load | | See Figure 10.7 |

(*Permission of* Brown Boveri Review, *December 1970.*)

better shielding. Often, for reasons of economy, earth wires are installed over the last kilometre or so of line immediately before it enters a substation. It has already been seen that the switching-in of resistance across circuit-breaker contacts reduces the high overvoltages produced on opening, especially on capacitive or low-current inductive circuits.

An aspect of vital importance, quite apart from the prevention of damage, is the maintenance of supply, especially as most flashovers cause no permanent damage and therefore a complete and lasting removal of the circuit from operation is not required. This may be achieved by the use of *autoreclosing* circuit breakers.

A *surge modifier* may be produced by the connection of a shunt capacitor between the line and earth or an inductor in series with the line; oscillatory effects may be reduced by the use of fast-acting voltage or current injections. It is uneconomical to attempt to modify or eliminate most overvoltages, and means are required to protect the various items of power systems. Surge divertors are connected in shunt across the equipment and divert the transient to earth; surge modifiers are connected in series with the line entering the substation and attempt to reduce the steepness of the wavefront—hence reduce its severity.

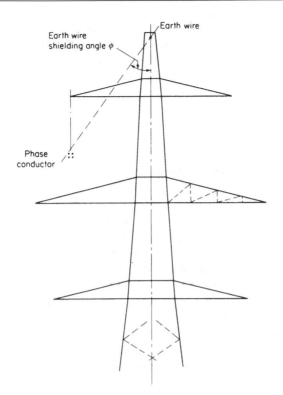

**Figure 10.10**   Single earth wire protection; shielding angle $\phi$ normally 35°

## 10.3.2  Surge diverters

The basic requirements for diverters are that they should pass no current at normal voltage, interrupt the power frequency follow-on current after a flash-over, and break down as quickly as possible after the abnormal voltage arrives. The simplest form is the *rod gap* connected across a bushing or insulator (as shown in Figure 10.48). This may also take the form of rings (arcing rings) around the top and bottom of an insulator string. The breakdown voltage for a given gap is polarity dependent to some extent. It does not interrupt the post-flashover follow-on current (i.e. the power-frequency current which flows in the path created by the flashover), and hence the circuit protection must operate, but it is by far the cheapest device for plant protection against surges. It is usually recommended that a rod gap be set to a breakdown voltage not less than 30 per cent below the voltage-withstand level of the protected equipment. For a given gap, the time for breakdown varies roughly inversely with the applied voltage; there exists, however, some dispersion of values. The times for positive voltages are lower than those for negative ones.

Typical curves relating the critical flashover voltage and time to breakdown for rod gaps of different spacings are shown in Figure 10.11. In Figure 10.12, the flashover voltage of vertically mounted insulator strings is shown. The flashover voltage is dependent on the length of the lower clearance to ground. For low values of this length there is a difference between positive (lower values) and negative flashover voltages.

## Expulsion gaps or tubes (protector tubes)

A disadvantage of the plain rod gap is the power-frequency current (follow-on current) that flows after breakdown which can only be extinguished by circuit-breaker operation. An improvement is the expulsion tube which consists of a spark gap in a fibre tube, as shown in Figure 10.13. When a sparkover occurs between the electrodes, the follow-on current arc is contained in the relatively small fibre tube. The high temperature of the arc vaporizes some of the organic material of the tube wall, causing a high gas pressure to build up in the tube. This gas possesses considerable turbulence and it extinguishes the arc. The hot gas rapidly leaves the tube, which is open at the ends. Very-high currents have been interrupted in such tubes. The breakdown voltage is slightly lower than for plain rod gaps for the same spacing.

An improved but more expensive surge diverter is the *lightning arrester*. A porcelain bushing contains a number of spark gaps in series with silicon carbide discs, the latter possessing low resistance to high currents and high resis-

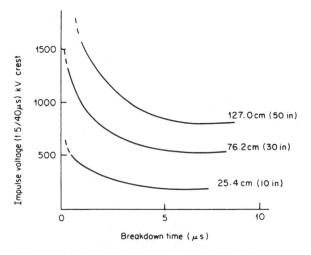

**Figure 10.11**   Breakdown characteristics of rod gaps

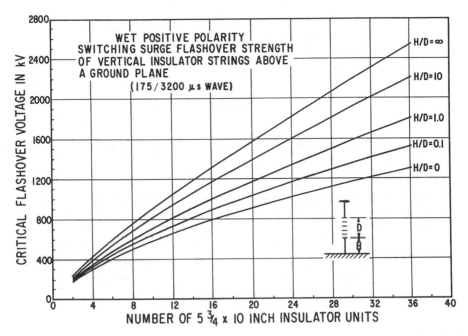

**Figure 10.12** Wet positive-polarity switching-surge flashover strength of vertical insulator strings above a ground plane *(Permission of the Edison Electric Institute)*

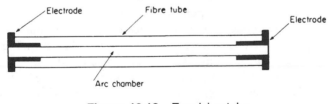

**Figure 10.13** Expulsion tube

tance to low currents, i.e. it obeys a law of the form $V = aI^{(0.2)}$, where $a$ depends on the material and its size. The overvoltage breaks down the gaps and then the power-frequency current is determined by the discs and limited to such a value that the gaps can quickly interrupt it at the first current zero. The voltage–time characteristic of a lightning arrester is shown in Figure 10.14; the gaps break down at S and the characteristic after this is determined by the current and the discs; the maximum voltage at R should be in the same order as at S. For high voltages a stack of several such units is used. The basic arrangement of discs and gaps, which are housed inside a porcelain bushing, is shown in Figure 10.15(a).

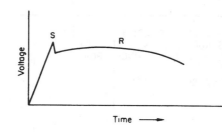

**Figure 10.14**   Characteristic of lightning arrester

Although with multiple spark gaps, diverters can withstand high rates of rise of recovery voltage (RRRV), the non-uniform voltage distribution between the gaps presents a problem. To overcome this, capacitors and non-linear resistors are connected in parallel across each gap. With the high-speed surge, the voltage is mainly controlled by the gap capacitance and hence capacitive grading is used. At power-frequencies a non-linear resistor provides effective voltage grading. The equivalent circuit is shown in Figure 10.15(b).

### Metal oxide (MO) arresters

Developments in materials have produced an arrester which does not require gaps to reduce the current after operation to a low value. Zinc oxide material is employed in stacked cylindrical blocks encased in a ceramic weatherproof insulator. A typical current–voltage characteristic is shown in Figure 10.16, where it can be seen that with a designed steady-state voltage of 4.25 kV, the residual current is below $10^{-2}$ A, depending upon temperature. At an effective rating of 3 kV the current is negligible. Consequently, no action is required to extinguish the current after operation, as may be required with a gapped arrester.

Metal oxide arresters up to the highest transmission voltages are now employed almost to the exclusion of gapped devices, although many of these latter still remain in operation.

## 10.4   Insulation Coordination

The equipment used in a power system comprises items having different breakdown or withstand voltages and different voltage–time characteristics. In order that all items of the system be adequately protected there is a need to consider the situation as a whole and not items of plant in isolation, i.e. the insulation

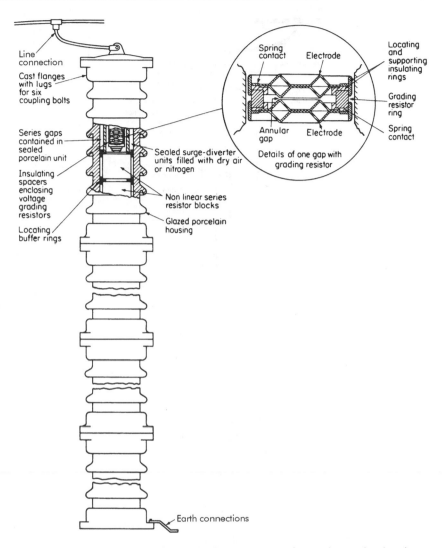

**Figure 10.15** (a) Diagram of a lightning arrester—four-unit stack showing part section of one unit (*Permission of the Electricity Council*)

protection must be coordinated. To assist this process, standard insulation levels are recommended and these are summarized in Tables 10.2 and 10.3. Reduced basic insulation impulse levels are used when considering switching surges and these are also summarized in Table 10.3.

Coordination is rendered difficult by the different voltage–time characteristics of plant and protective devices, e.g. a gap may have an impulse ratio of 2 for a 20 $\mu$s front wave and 3 for a 5 $\mu$s wave. At the higher frequencies (shorter wavefronts) corona cannot form in time to relieve the stress concentration

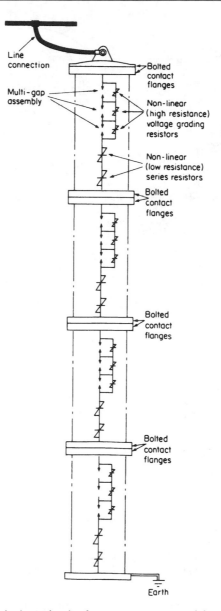

**Figure 10.15** (b)   Equivalent circuit of components comprising a single-phase four-unit surge-diverter stack (*Permission of the Electricity Council*)

on the gap electrodes. With a lightning surge a higher voltage can be withstood because a discharge requires a certain discrete amount of energy as well as a minimum voltage, and the applied voltage increases until the energy reaches this value. Figure 10.17 shows the voltage–time characteristics for the system

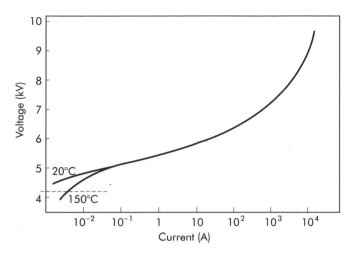

**Figure 10.16** Current–voltage characteristic of standard ZnO blocks, 80 mm in diameter and 35 mm thick. The reference voltage of the block is 4.25 kV d.c. and its continuous rating is 3kV (effective value) (*Permission of the Electricity Association*)

**Table 10.2** British substation practice (now I.E.C. standard)

| Nominal voltage (kV) | Impulse (peak kV) | Power frequency withstand (peak kV) | Minimum clearance to ground (m) | Minimum clearance between phases (m) | Coordinated gap setting (m) |
|---|---|---|---|---|---|
| 11 | 100 | 29 | 0.2 | 0.25 | 2 × 0.03 |
| 132 | 550 | 300 | 1.1 | 1.25 | 0.66 |
| 400 | 1425 | 675 | 3.05 | 3.55 | 1.5–1.8 |

**Table 10.3** Recommended BILs at various operating voltages (U.S. practice)

| Voltage class (kV) | 15 | 23 | 34.5 | 46 | 69 | 92 | 115 |
|---|---|---|---|---|---|---|---|
| BIL (kV) | 110 | 150 | 200 | 250 | 350 | 450 | 550 ⎫ |
| Reduced BIL (kV) | | | 125 | | | | 450 ⎬ |
| | | | | | | | 350 ⎭ |

| Voltage class (kV) | 138 | 161 | 196 | 230 | 287 | 345 | 500 |
|---|---|---|---|---|---|---|---|
| BIL (kV) | 650 | 750 | 900 | 1050 | 1300 | 1550 | 1800 |
| Reduced BIL (kV) | 550 ⎱ | 650 ⎱ | | 900 ⎫ | 1175 ⎫ | 1425 ⎫ | 1675 ⎫ |
| | 450 ⎰ | 550 ⎰ | | 825 ⎬ | 1050 ⎬ | 1300 ⎬ | 1550 ⎬ |
| | | | | 750 ⎭ | 900 ⎭ | 1050 ⎭ | 1300 ⎭ |

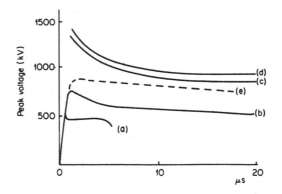

**Figure 10.17** Insulation coordination in an H.V. substation. Voltage–time character-istics of plant for a 1.5/40 $\mu$s wave. (a) Characteristics of lightning arrester. (b) Transformer. (c) Line insulator string. (d) Busbar insulation. (e) Maximum surge applied waveform

elements comprising a substation, and the protection of the weakest items by the arrester is illustrated.

Up to an operating voltage of 345 kV the insulation level is determined by lightning, and the standard impulse tests suffice, along with normal frequency tests. Above this value, however, the overvoltages resulting from switching are higher in magnitude and therefore they decide the insulation. The character-istics of air gaps and some solid insulations are different for switching surges than for the standard impulse waves, and closer coordination of insulation is required because of the lower attenuation of switching surges, although their amplification by reflection is less than with lightning. Recent work indicates that, for transformers, the switching impulse strength is of the order of 0.95 of the standard value while for oil-filled cables it is 0.7 to 0.8.

The design withstand level is selected by specifying the risk of flashover, e.g. for 550 kV towers a 0.13 per cent probability has been used. At 345 kV, design is carried out by accepting a switching impulse level of 2.7 p.u., which corre-sponds to the lightning level. At 500 kV, however, a 2.7 p.u. switching impulse would require 40 per cent more tower insulation than that governed by light-ning. The tendency is therefore for the design switching impulse level to be forced lower with increasing system operating voltage and for control of the surges to be made by the more widespread use of resistance switching in the circuit breakers or use of surge diverters. For example, for the 500 kV network the level is 2 p.u. and at 765 kV it is reduced to 1.7 p.u.; if further increases in system voltage occur, it is hoped to decrease the level to 1.5 p.u.

The problem of switching surges is illustrated in Figure 10.18, in which flashover probability is plotted against peak (crest) voltage; *critical flashover voltage* (CFO) is the peak voltage for a particular tower design for which there is a 50 per cent probability of flashover. For lightning, the probability of

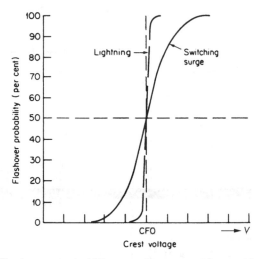

**Figure 10.18** Flashover probability—peak surge voltage. Lightning flashover is close-grouped near the critical flashover voltage (CFO), whereas switching-surge probability is more widely dispersed and follows normal Gaussian distribution (*Permission of the Westinghouse Electrical Corporation, East Pittsburgh, Pennsylvania, U.S.A.*)

flashover below the CFO is slight, but with switching surges having much longer fronts the probability is higher, with the curve following the normal Gaussian distribution; the tower clearances must be designed for a CFO much higher than the maximum transient expected.

An example of the application of lightning-arrester characteristics to system requirements is illustrated in Figure 10.19. In this figure, variation of crest voltage for lightning and switching surges with time is shown for an assumed BIL of the equipment to be protected. Over the impulse range (up to $10\,\mu s$) the fronts of wave and chopped-wave peaks are indicated. Over the switching surge region the insulation withstand strength is assumed to be 0.83 of the BIL (based on transformer requirements). The arrester maximum sparkover voltage is shown along with the maximum voltage set up in the arrester by the follow-on current (10 or 20 per cent). A protective margin of 15 per cent between the equipment withstand strength and the maximum arrester sparkover is assumed.

## 10.5  Propagation of Surges

The basic differential equations for voltage and current in a distributed-constant line are as follows:

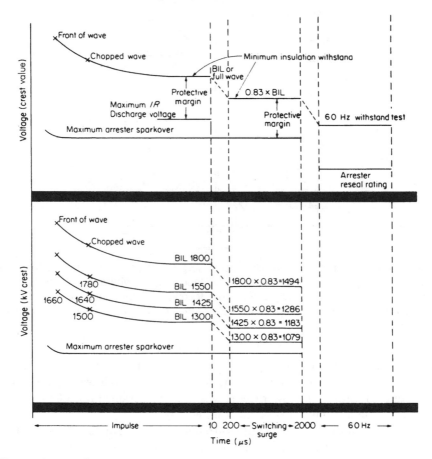

**Figure 10.19** Coordination diagrams relating lightning-arrester characteristics to system requirements Top: Explanation of characteristics. Bottom: Typical values for 500k V system (300 kV to ground) (*Permission of the Ohio Brass Co.*)

$$\frac{\partial^2 v}{\partial x^2} = LC\frac{\partial^2 v}{\partial t^2} \quad \text{and} \quad \frac{\partial^2 i}{\partial x^2} = LC\frac{\partial^2 i}{\mathrm{d}t^2}$$

and these equations represent travelling waves. The solution for the voltage may be expressed in the form,

$$v = F_1(t - x\sqrt{LC}) + F_2(t + x\sqrt{LC})$$

i.e. one wave travels in the positive direction of $x$ and the other in the negative. Also, it may be shown that because $\partial v/\partial x = -L(\partial i/\partial t)$, the solution for current is

$$i = \sqrt{\frac{C}{L}}[F_1(t - x\sqrt{LC}) - F_2(t + x\sqrt{LC}) \quad \text{noting that } \sqrt{\frac{C}{L}} = \frac{1}{Z_0}$$

In more physical terms, if a voltage is injected into a line (Figure 10.20) a corresponding current $i$ will flow, and if conditions over a length $dx$ are considered, the flux set up between the go and return wires is equal to $iL\,dx$, where $L$ is the inductance per unit length. The induced back e.m.f. is $-d\Phi/dt$, i.e. $-Li(dx/dt)$ or $-iLU$, where $U$ is the wave velocity. The applied voltage $v$ must equal $iLU$. Also, charge is stored in the capacitance over $dx$, i.e. $Q = idt = vC\,dx$ and $i = vCU$. Hence, $vi = viLCU^2$ and $U = 1/\sqrt{LC}$. Also, $i = v\sqrt{C/L} = v/Z_0$, where $Z_0$ is the characteristic or surge impedance. For single-circuit three-phase overhead lines (conductors not bundled) $Z_0$ lies in the range 400–600 $\Omega$. For overhead lines, $U = 3 \times 10^8$ m/s, i.e. the speed of light, and for cables

$$U = \frac{3 \times 10^8}{\sqrt{\varepsilon_r \mu_r}} \text{ m/s}$$

where $\varepsilon_r$ is usually from 3 to 3.5, and $\mu_r = 1$.

From the above relations,

$$\tfrac{1}{2}Li^2 = \tfrac{1}{2}(iLU)\left(\frac{i}{U}\right)$$

$$= \tfrac{1}{2}\left(\frac{vi}{U}\right) = \tfrac{1}{2}Cv^2$$

The incident travelling waves of $v_i$ and $i_i$, when they arrive at a junction or discontinuity, produce a reflected current $i_r$ and a reflected voltage $v_r$ which travel back along the line. The incident and reflected components of voltage and current are governed by the surge impedance $Z_0$, so that

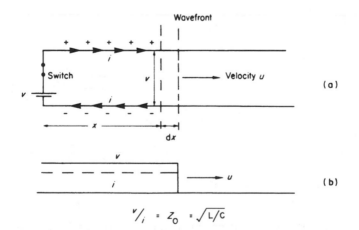

**Figure 10.20** Distribution of charge and current as wave progresses along a previously unenergized line. (a) Physical arrangement. (b) Symbolic representation

$$v_i = Z_0 i_i \qquad \text{and} \qquad v_r = -Z_0 i_r$$

In the general case of a line of surge impedance $Z_0$ terminated in $Z$ (Figure 10.21), the total voltage at $Z$ is $v = v_i + v_r$ and the total current is $i = i_i + i_r$. Also,

$$(v_i + v_r) = Z(i_r + i_i)$$
$$Z_0(i_i - i_r) = Z(i_r + i_i)$$

and hence

$$i_r = \left(\frac{Z_0 - Z}{Z_0 + Z}\right) i_i \qquad\qquad (10.10)$$

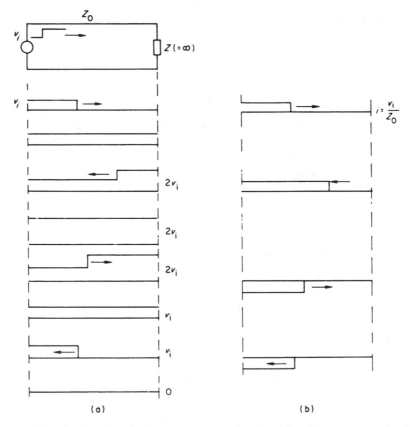

(a)                                             (b)

**Figure 10.21**  Application of voltage to unenergized loss-free line on open circuit at far end. (a) Distribution of voltage. (b) Distribution of current. Voltage source is an effective short circuit

Again,

$$v_i + v_r = Z(i_i + i_r)$$

but, as $v_r = -Z_0 i_r$, we have

$$v_i + v_r = Z\left(\frac{v_i - v_r}{Z_0}\right)$$

or,

$$v_r = \left(\frac{Z - Z_0}{Z + Z_0}\right)v_i = \alpha v_i \qquad (10.11)$$

where $\alpha$ is the *reflection coefficient* equal to

$$\left(\frac{Z - Z_0}{Z + Z_0}\right)$$

Hence,

$$v = \left(\frac{2Z}{Z + Z_0}\right)v_i \qquad (10.12)$$

and

$$i = \left(\frac{2Z_0}{Z + Z_0}\right)i_i \qquad (10.13)$$

From the above, if $Z \to \infty$, $v = 2v_i$ and $i = 0$. Also, if $Z = Z_0$ (matched line), $\alpha = 0$, i.e. no reflection. If $Z > Z_0$, then $v_r$ is positive and $i_r$ is negative, but if $Z < Z_0$, $v_r$ is negative and $i_r$ is positive. The reflected waves will travel back and forth along the line, setting up, in turn, further reflected waves at the ends, and this process will continue indefinitely unless the waves are attenuated because of resistance and corona.

Summarizing, at an open circuit the reflected voltage is equal to the incident voltage and this wave, along with a wave $(-i_i)$, travels back along the line; note that at the open circuit the total current is zero. Conversely, at a short circuit the reflected voltage wave is $(-v_i)$ in magnitude and the current reflected is $(i_i)$, giving a total voltage at the short circuit of zero and a total current of $2i_i$. For other termination arrangements, Thevenin's theorem may be applied to analyse the circuit. The voltage across the termination when it is open-circuited is seen to be $2v_i$ and the equivalent impedance looking in from the open-circuited termination is $Z_0$; the termination is then connected across the terminals of the Thevenin equivalent circuit (Figure 10.22).

Consider two lines of different surge impedance in series. It is required to determine the voltage across the junction between them (Figure 10.23):

$$v_{AB} = \left(\frac{2v_i}{Z_0 + Z_1}\right)Z_1 = \beta v_i \qquad (10.14)$$

**Figure 10.22** Analysis of travelling waves—use of Thevenin equivalent circuit. (a) System. (b) Equivalent circuit

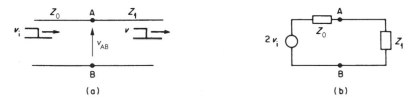

**Figure 10.23** Analysis of conditions at the junction of two lines or cables of different surge impedance

The wave entering the line $Z_1$ is the refracted wave and $\beta$ is the *refraction coefficient*, i.e. the proportion of the incident voltage proceeding along the second line $(Z_1)$.

$$v_r = v_i \alpha = v_i \left( \frac{Z_1 - Z_0}{Z_1 + Z_0} \right) \quad \text{and} \quad i = \frac{v_{AB}}{Z_1} = \frac{2v_1}{Z_1 + Z_0} = \text{refracted current}$$

When several lines are joined to the line on which the surge originates (Figure 10.24), the treatment is similar, e.g. if there are three lines having equal surge impedances $(Z_1)$, then

$$i_A = \frac{2v_i}{(Z_0 + Z_1/3)} \quad \text{and} \quad v_{AB} = \left( \frac{2v_i}{Z_0 + Z_1/3} \right)\left( \frac{Z_1}{3} \right) \tag{10.15}$$

An important practical case is that of the clearance of a fault at the junction of two lines and the surges produced. The equivalent circuits are shown in Figure 10.25; the fault clearance is simulated by the insertion of an equal and opposite current $(I)$ at the point of the fault. From the equivalent circuit, the magnitude of the resulting voltage surges $(v)$

$$= I\left( \frac{Z_1 Z_2}{Z_1 + Z_2} \right)$$

and the currents entering the lines are

$$I\left( \frac{Z_1}{Z_1 + Z_2} \right) \quad \text{and} \quad I\left( \frac{Z_2}{Z_1 + Z_2} \right)$$

The directions are as shown in Figure 10.25(c).

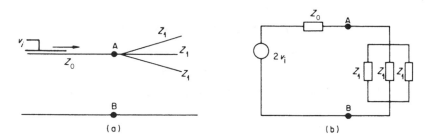

**Figure 10.24** Junction of several lines. (a) System. (b) Equivalent circuit

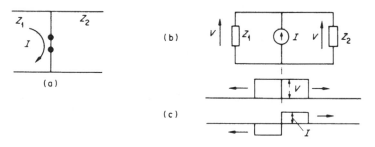

**Figure 10.25** Surge set up by fault clearance. (a) Equal and opposite current ($I$) injected in fault path. (b) Equivalent circuit. (c) Voltage and current waves set up at a point of fault with direction of travel

### 10.5.1  Termination in inductance and capacitance

*Shunt capacitance*

Using the Thevenin equivalent circuit as shown in Figure 10.26, the voltage rise across the capacitor C is $v_c = 2v_i(1 - e^{-t/Z_0 C})$, where $t$ is the time commencing with the arrival of the wave at C. The current through C is given by,

$$\left(\frac{2v_i}{Z_0}\right)e^{-t/Z_0 C}$$

The reflected wave,

$$v_r = v_c - v_i = 2v_i(1 - e^{-t/Z_0 C}) - v_i = v_i(1 - 2e^{-t/Z_0 C}) \tag{10.16}$$

As to be expected, the capacitor acts initially as a short circuit and finally as an open circuit.

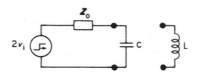

**Figure 10.26** Termination of line (surge impedance $Z_0$) in a capacitor C or inductance L

## Shunt inductance

Again, from the equivalent circuit the voltage across the inductance is

$$v_L = 2v_i\, e^{-(Z_0/L)t}$$

and

$$v_r = v_L - v_i = v_i[2\,e^{-(Z_0/L)t} - 1] \tag{10.17}$$

Here, the inductance acts initially as an open circuit and finally as a short circuit.

## Capacitance and resistance in parallel (Figure 10.27(a))

Open-circuit voltage across AB (Figure 10.27(b))

$$= \left(\frac{2v_i}{Z_0 + R}\right) R$$

Equivalent Thevenin resistance

$$= \frac{RZ_0}{R + Z_0}$$

Voltage across $R$ and $C$ is

$$v = \frac{2v_i}{Z_0 + R}(1 - e^{-[t(R+Z_0)]/RZ_0C}) \tag{10.18}$$

This is the solution to the practical system shown in Figure 10.27(c), where $C$ is used to modify the surge. The reflected wave is given by $(v - v_i)$.

## Example 10.2

An overhead line of surge impedance 500 Ω is connected to a cable of surge impedance 50 Ω through a series resistor (Figure 10.28(a)). Determine the magnitude of the resistor

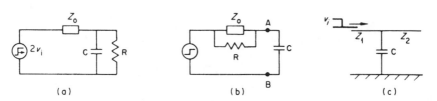

**Figure 10.27**  Two lines surge impedances **Z₁** and **Z₂** grounded at their junction through a capacitor C. (a) and (b) Equivalent circuits. (c) System diagram

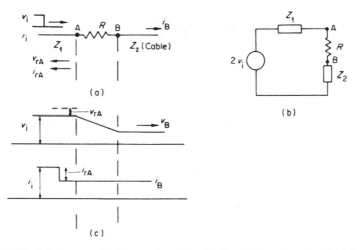

**Figure 10.28**  (a) System for Example 10.2. (b) Equivalent circuit. (c) Voltage and current surges

such that it absorbs maximum energy from a surge originating on the overhead line and travelling into the cable. Calculate:

- the voltage and current transients reflected back into the line, and
- those transmitted into the cable, in terms of the incident surge voltage;
- the energies reflected back into the line and absorbed by the resistor.

Let the incident voltage and current be $v_i$ and $i_i$, respectively. From the equivalent circuit (Figure 10.28(b)),

$$v_B = \frac{2v_i \cdot Z_2}{Z_1 + Z_2 + R}$$

and the reflected voltage at A

$$v_{rA} = \frac{2v_i(Z_2 + R)}{Z_1 + Z_2 + R} - v_i = \frac{Z_2 + R - Z_1}{Z_1 + Z_2 + R} \cdot v_i$$

As

$$v_i = Z_1 i_i$$

$$i_B = \frac{2v_i}{Z_1 + Z_2 + R} = \frac{2Z_1 i_i}{Z_1 + Z_2 + R}$$

and

$$i_{rA} = i_i \frac{(Z_1 - Z_2 - R)}{Z_1 + Z_2 + R}$$

Power absorbed by the resistance

$$= R\left(i_i \frac{2Z_1}{Z_1 + Z_2 + R}\right)^2 \quad \text{(W)}$$

This power is a maximum when

$$\frac{\mathrm{d}}{\mathrm{d}R}\left[\frac{R}{(Z_1 + Z_2 + R)^2}\right] = 0$$

if $i_i$ is given and constant; from which $R = (Z_1 + Z_2)$. With this resistance the maximum energy is absorbed from the surge.

Hence $R$ should be $500 + 50 = 550\,\Omega$. With this value of $R$,

$$i_B = \left(\frac{2 \times 500}{500 + 50 + 550}\right) \cdot i_i = 0.91 i_i$$

$$i_{rA} = -0.091 i_i$$

$$v_B = 0.091 v_i \quad \text{and} \quad v_{rA} = 0.091 v_i$$

Also, the surge energy entering $Z_2$

$$= v_B i_B = 0.082 v_i i_i$$

The energy absorbed by $R = (0.91 i_i)^2 \times 550$

$$= 455\left(\frac{v_i}{Z_1}\right) i_i = 0.91 v_i i_i$$

and the energy reflected $= v_i i_i (1 - 0.082 - 0.91)$

$$= 0.008 v_i i_i$$

The waveforms are shown in Figure 10.28(c).

## 10.6  Determination of System Voltages Produced by Travelling Surges

In the previous section the basic laws of surge behaviour were discussed. The calculation of the voltages set up at any node or busbar in a system at a given instant in time is, however, much more complex than the previous section would suggest. When any surge reaches a discontinuity its reflected waves travel back and are, in turn, reflected so that each generation of waves sets up further waves which coexist with them in the system.

To describe completely the events at any node involves, therefore, an involved book-keeping exercise. Although many mathematical techniques are available and, in fact, used, the graphical method due to Bewley (1961) indicates clearly the physical changes occurring in time, and this method will be explained in some detail.

## 10.6.1  Bewley lattice diagram

This is a graphical method of determining the voltages at any point in a transmission system and is an effective way of illustrating the multiple reflections which take place. Two axes are established: a horizontal one scaled in distance along the system, and a vertical one scaled in time. Lines indicating the passage of surges are drawn such that their slopes give the times corresponding to distances travelled. At each point of change in impedance the reflected and transmitted waves are obtained by multiplying the incidence-wave magnitude by the appropriate reflection and refraction coefficients $\alpha$ and $\beta$.

The method is best illustrated by an example.

### Example 10.3

A loss-free system comprising a long overhead line $(Z_1)$ in series with a cable $(Z_2)$ will be considered. Typically, $Z_1$ is $500\,\Omega$ and $Z_2$ is $50\,\Omega$.

### Solution

Referring to Figure 10.29, the following coefficients apply:

Line-to-cable reflection coefficient,

$$\alpha_1 = \frac{50 - 500}{50 + 500} = -0.818 \qquad \text{(note order of values in numerator)}$$

Line-to-cable refraction coefficient,

$$\beta_1 = \frac{2 \times 50}{50 + 500} = 0.182$$

Cable-to-line reflection coefficient,

$$\alpha_2 = \frac{500 - 50}{500 + 50} = 0.818$$

Cable-to-line refraction coefficient,

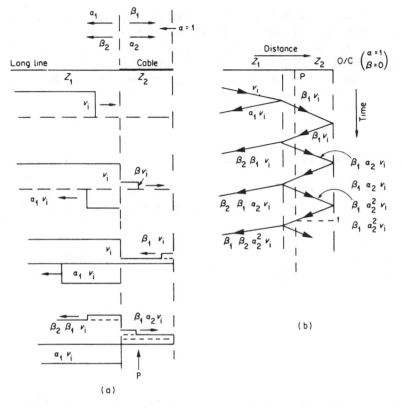

**Figure 10.29** Bewley lattice diagram—analysis of long overhead line and cable in series. (a) Position of voltage surges at various instants over first complete cycle of events, i.e. up to second reflected wave travelling back along line. (b) Lattice diagram

$$\beta_2 = \frac{2 \times 500}{500 + 50} = 1.818$$

As the line is long, reflections at its remote (sending) end will be neglected. The remote end of the cable is considered to be open-circuited, giving an $\alpha$ of 1 and a $\beta$ of zero at that point.

When the incident wave $v_i$ (see Figure 10.29) originating in the line reaches the junction, a reflected component travels back along the line ($\alpha_1 v_i$), and the refracted or transmitted wave ($\beta_1 v_i$) traverses the cable and is reflected from the open-circuited end back to the junction ($1 \times \beta_1 v_i$). This wave then produces a reflected wave back through the cable ($1 \times \beta_1 \alpha_2 v_i$) and a transmitted wave ($1 \times \beta_2 \beta_1 v_i$) through the line. The process continues and the waves multiply as indicated in Figure 10.29(b). The total voltage at a point P in the cable at a given time ($t$) will be the sum of the voltages at P up to time $t$, i.e. $v_i \beta_1 (2 + 2\alpha_2 + 2\alpha_2^2)$, and the voltage at infinite time will be $2v_i \beta_1 (1 + \alpha_2 + \alpha_2^2 + \alpha_2^3 + \alpha_2^4 + \cdots)$.

The voltages at other points are similarly obtained. The time scale may be determined from a knowledge of length and surge velocity; for the line the latter is of the order of 300 m/$\mu$s and for the cable 150 m/$\mu$s. For a surge 50 $\mu$s in duration and a cable 300 m in length there will be 25 cable lengths traversed and the terminal voltage will approach $2v_i$. If the graph of voltage at the cable open-circuited end is plotted against time, an exponential rise curve will be obtained similar to that obtained for a capacitor.

The above treatment applies to a rectangular surge waveform, but may be modified readily to account for a waveform of the type illustrated in Figure 10.1(b) or (c). In this case the voltage change with time must also be allowed for and the process is more complicated.

## 10.6.2 Short-line faults

A particular problem for circuit breakers on h.v. systems is interrupting current if a solid short circuit occurs a few kilometres from the circuit breaker. This situation is illustrated in Figure 10.30.

It will be recalled that in Figure 10.3, fault interruption gives rise to a restriking transient on the supply side of the breaker. The sudden current interruption sends a step surge down the short line to the fault, which is reflected at the short-circuit point back to the breaker. In a manner similar to the Bewley lattice calculation of Figure 10.29, a surge is propagated back and forth in $Z_0$ as shown in Figure 10.30(b). The resulting combination of restriking voltage on the supply side and the continuously reflected surge on the line side of the breaker produces a steeply rising initial restriking voltage, greater than the restriking transient of Figure 10.3. Since this occurs just following current zero, it provides the most onerous condition of all for the circuit breaker to deal with, and must be included in the series of proving tests on any new design.

The first peak of the line-side component of the transient recovery voltage may be obtained from a knowledge of the fault current $I_f$ and the value of $Z_0$. Assuming ideal interruption at current zero, the rate of rise of transient recovery voltage is

$$\frac{dv}{dt} = Z_0 \frac{di}{dt} = \sqrt{2}\omega Z_0 I_f \times 10^{-6}\, \text{V}/\mu\text{s}$$

The amplitude of the first peak of reflected wave is

$$V = t \frac{dv}{dt}\, \text{(V)}$$

where $t$ is twice the transit time of the surge through $Z_0$ and $\omega$ is the system angular frequency. To this value should be added any change of $v_s$ on the supply side, calculated as given by equation (10.6).

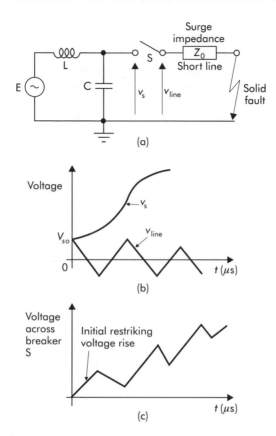

**Figure 10.30** Short-line fault. (a) System diagram. (b) Supply-side and line-side transients. (c) Transient voltage across circuit breaker contacts

### 10.6.3 Effects of line loss

Attenuation of travelling waves is caused mainly by corona which reduces the steepness of the wavefronts considerably as the waves travel along the line. Attenuation is also caused by series resistance and leakage resistance and these quantities are considerably larger than the power-frequency values. The determination of attenuation is usually empirical and use is made of the expression $v_x = v_i \mathrm{e}^{-\gamma x}$, where $v_x$ is the magnitude of the surge at a distance $x$ from the point of origination. If a value for $\gamma$ is assumed, then the wave magnitude of the voltage may be modified to include attenuation for various positions in the lattice diagram. For example, in Figure 10.29(b), if $\mathrm{e}^{-\gamma x}$ is equal to $a_{\mathrm{L}}$ for the length of line traversed and $a_{\mathrm{c}}$ for the cable, then the magnitude of the first reflection from the open circuit is $a_{\mathrm{L}} v_i a_{\mathrm{c}} \beta_1$ and the voltages at subsequent times will be similarly modified.

Considering the power and losses over a length $dx$ of a line of resistance and shunt conductance per unit length $R\,(\Omega)$ and $G\,(\Omega^{-1})$, the power loss

$$dp = i^2 R\,dx + v^2 G\,dx \quad \text{(W)}$$

also,

$$p = vi = i^2 Z_0 \qquad \text{and} \qquad dp = 2iZ_0\,di.$$

As $dp$ is a loss it is considered negative and

$$-2iZ_0\,di = (i^2 R + v^2 G)\,dx$$

hence,

$$\frac{di}{i} = -\tfrac{1}{2}\left(\frac{R + Z_0^2 G}{Z_0}\right)dx$$

giving

$$i = i_i \exp\left[-\frac{1}{2}\left(\frac{R}{Z_0} + GZ_0\right)x\right] \tag{10.19}$$

where $i_i =$ surge amplitude (A).
   Also, it may be shown that

$$v = v_i \exp\left[-\frac{1}{2}\left(\frac{R}{Z_0} + GZ_0\right)x\right] \tag{10.20}$$

and the power at $x$ is

$$v_x i_x = vi = v_i i_i\, e^{-[(R/Z_0)+GZ_0]x} \tag{10.21}$$

If $R$ and $G$ are realistically assessed (including corona effect), attenuation may be included in the travelling-wave analysis.

### 10.6.4 Digital methods

The lattice diagram becomes very cumbersome for large systems and digital methods are usually applied. Digital methods may use strictly mathematical methods, i.e. the solution of the differential equations or the use of Fourier or Laplace transforms. These methods are capable of high accuracy, but require large amounts of data and long computation times. The general principles of the graphical approach described above may also be used to develop a computer program which is very applicable to large systems. Such a method will now be discussed in more detail and an example considered.
   A rectangular wave of infinite duration is used. The theory developed in the previous sections is applicable. The major role of the program is to scan the nodes of the system at each time interval and compute the voltages. In Figure 10.31 a

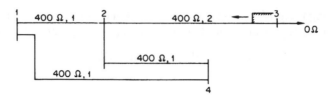

**Figure 10.31** Application of digital method. Each line is labelled with surge impedance and surge travel time (multiples of basic unit), e.g. 400 Ω 1

particular system (single-phase representation) is shown and will be used to illustrate the method.

The relevant data describing the system are given in tabular form (see Table 10.4). Branches are listed both ways in ascending order of the first nodal number and are referred to under the name BRANCH. The time taken for a wave to travel along a branch is recorded in terms of a positive integer (referred to as PERIOD) which converts the basic time unit into actual travel time. Reflection coefficients are stored and referred to as REFLECT and the corresponding refraction coefficients are obtained, i.e. $(1 + \alpha_{ij})$. Elements of time as multiples of the basic integer are also shown in Table 10.4.

The method is illustrated by examining the system after the arrival of the rectangular wave at node 3 at TIME (0). This voltage (magnitude 1 p.u.) is entered in the BRANCH (3, 2), TIME (0) element of the BRANCH–TIME matrix. On arrival at node 2 at time equal to zero plus PERIOD (3, 2), two waves are generated, on BRANCH (2, 1) and BRANCH (2, 4), both of magnitude $1(1 + \alpha_{32})$,

**Table 10.4**  Data for lattice calculations

| BRANCH $i$ | 1 | 1 | 2 | 2 | 2 | 3 | 4 | 4 |
| $j$ | 2 | 4 | 1 | 3 | 4 | 2 | 1 | 2 |
| PERIOD | 1 | 1 | 1 | 2 | 1 | 2 | 1 | 1 |
| REFLECT $\alpha_{ij}$ | $-\frac{1}{3}$ | 0 | 0 | $-1$ | 0 | $-\frac{1}{3}$ | 0 | $-\frac{1}{3}$ |
| $1 + \alpha_{ij}$ | $\frac{2}{3}$ | 1 | 1 | 0 | 1 | $\frac{2}{3}$ | 1 | $\frac{2}{3}$ |
| TIME 0 | | | | | | 1 | | |
| 1 | | | | | | | | |
| 2 | | | | $\frac{2}{3}$ | $-\frac{1}{3}$ | $\frac{2}{3}$ | | |
| 3 | 0 | $\frac{2}{3}$ | | | | | | |
| 4 | | | | | | | | |

i.e. 2/3. A reflected wave is also generated on BRANCH (2, 3), of magnitude $1 \times \alpha_{32}$, i.e. $-1/3$. These voltages are entered in the appropriate BRANCH in the TIME (2) row of Table 10.4. On reaching node 1, TIME (3), a refracted wave of magnitude $\frac{2}{3}(1 + \alpha_{21})$, i.e. 2/3, is generated on BRANCH (1, 4) and a reflected wave $\frac{2}{3} \times \alpha_{21}$, i.e. 0, is generated on BRANCH (1, 2). This process is continued until a specified time is reached. All transmitted waves for a given node are placed in a separate node–time array; a transmitted wave is considered only once even though it could be entered into several BRANCH–TIME elements. Current waves are obtained by dividing the voltage by the surge impedance of the particular branch. The flow diagram for the digital solution is shown in Figure 10.32.

It is necessary for the programs to cater for semi-infinite lines (i.e. lines so long that waves reflected from the remote end may be neglected) and also inductive/capacitive terminations. *Semi-infinite* lines require the use of artificial nodes labelled (say) 0. For example, in Figure 10.33, node 2 is open-circuited and this is accounted for by introducing a line of infinite surge impedance between nodes 2 and 0, and hence, if

$$Z \to \infty \qquad \alpha_{12} = \frac{Z - Z_0}{Z + Z_0} \to 1$$

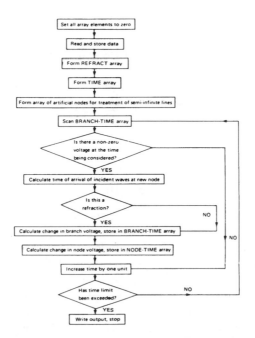

**Figure 10.32** Flow diagram of digital method for travelling-wave analysis

**Figure 10.33** Treatment of terminations—use of artificial node 0 and line of infinite impedance to represent open circuit at node 2

Although the scanning will not find a refraction through node 2, it is necessary for the computer to consider there being one to calculate the voltage at node 2. Lines with short-circuited nodes may be treated in a similar fashion with an artificial line of zero impedance.

Inductive/capacitive terminations may be simulated by *stub lines*. An inductance $L$ (henrys) is represented by a stub transmission line short-circuited at the far end and of surge impedance $Z_L = L/t$, where $t$ is the travel time of the stub. Similarly, a capacitance $C$ (farads) is represented by a line open-circuited and of surge impedance $Z_c = t/C$. For the representation to be exact, $t$ must be small and it is found necessary for $Z_L$ to be of the order of 10 times and $Z_c$ to be one-tenth of the combined surge impedance of the other lines connected to the node.

For example, consider the termination shown in Figure 10.34(a). The equivalent stub-line circuit is shown in Figure 10.34(b). Stub travel times are chosen to be short compared with a quarter cycle of the natural frequency, i.e.

$$\frac{2\pi\sqrt{LC}}{4} = \frac{2\pi}{4}\sqrt{0.01 \times 4 \times 10^{-8}} = 0.315 \times 10^{-4}\,\text{s}$$

Let $t = 5 \times 10^{-6}$ s for both $L$ and $C$ stubs (corresponding to the total stub length of 1524 m). Hence,

$$Z_c = \frac{t}{C} = \frac{2.5 \times 10^{-6}}{4 \times 10^{-8}} = 67.5\,\Omega$$

and

$$Z_L = L/t = \frac{0.01}{2.5 \times 10^{-6}} = 4000\,\Omega$$

The configuration in a form acceptable for the computer program is shown in Figure 10.34(c). Refinements to the program to incorporate attenuation, waveshapes, and non-linear resistors may be made without changing its basic form.

A typical application is the analysis of the nodal voltages for the system shown in Figure 10.31. This system has previously been analysed by a similar method by Barthold and Carter (1961) and good agreement found. The printout of nodal voltages for the first 20 $\mu$s is shown in Table 10.5, and in Figure 10.35 the voltage plot for nodes 1 and 4 is compared with a transient analyser solution obtained by Barthold and Carter.

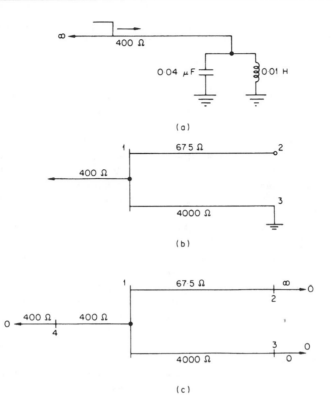

**Figure 10.34** Representation of line terminated by *L–C* circuit by means of stub lines. (a) Original system. (b) Equivalent stub lines. (c) Use of artificial nodes to represent open- and short-circuited ends of stub lines

### 10.6.5 Three-phase analysis

The single-phase analysis of a system as presented in this chapter neglects the mutual effects which exist between the three phases of a line, transformer, etc. The transient voltages due to energization may be further increased by this mutual coupling and also by the three contacts of a circuit breaker not closing at the same instant. The difference resulting between the use of three-phase and single-phase representation has been discussed by Bickford and Doepel (1967).

## 10.7 Electromagnetic Transient Program (EMTP)

The digital method previously described is very limited in scope. A much more powerful method has been developed by the Bonneville Power Administration and is known as EMTP. This is widely used, especially in the U.S.A.

**Table 10.5** Digital computer printout: node voltages (system of Figure 10.31)

| Time ($\mu$s) | Node | | | |
|---|---|---|---|---|
| | 1 | 2 | 3 | 4 |
| 0 | 0.0000E–01 | 0.0000E-01 | 1.0000E 00 | 0.0000E–01 |
| 1 | 0.0000E–01 | 0.0000E-01 | 1.0000E 00 | 0.0000E–01 |
| 2 | 0.0000E–01 | 6.6670E-01 | 1.0000E 00 | 0.0000E–01 |
| 3 | 6.6670E–01 | 6.6670E-01 | 1.0000E 00 | 6.6670E–01 |
| 4 | 1.3334E 00 | 6.6670E-01 | 1.0000E 00 | 1.3334E 00 |
| 5 | 1.3334E 00 | 1.5557E 00 | 1.0000E 00 | 1.3334E 00 |
| 6 | 1.5557E 00 | 1.7779E 00 | 1.0000E 00 | 1.5557E 00 |
| 7 | 2.0002E 00 | 1.7779E 00 | 1.0000E 00 | 2.0002E 00 |
| 8 | 2.2224E 00 | 2.0743E 00 | 1.0000E 00 | 2.2224E 00 |
| 9 | 2.2965E 00 | 1.7779E 00 | 1.0000E 00 | 2.2965E 00 |
| 10 | 1.8520E 00 | 1.8520E 00 | 1.0000E 00 | 1.8520E 00 |
| 11 | 1.4075E 00 | 1.9508E 00 | 1.0000E 00 | 1.4075E 00 |
| 12 | 1.5062E 00 | 1.0617E 00 | 1.0000E 00 | 1.5062E 00 |
| 13 | 1.1604E 00 | 7.6534E–01 | 1.0000E 00 | 1.1604E 00 |
| 14 | 4.1944E–01 | 8.2297E–01 | 1.0000E 00 | 4.1955E–01 |
| 15 | 8.2086E–02 | 2.6307E–01 | 1.0000E 00 | 8.2086E–02 |
| 16 | −7.4396E–02 | 1.6435E–01 | 1.0000E 00 | −7.4396E–02 |
| 17 | 7.8720E–03 | 1.0732E–02 | 1.0000e 00 | 7.8720E–03 |
| 18 | 9.3000E–02 | −2.5556E–01 | 1.0000E 00 | 9.3000E–02 |
| 19 | −1.7043E–01 | 4.1410E–01 | 1.0000E 00 | −1.7043E–01 |
| 20 | 1.5067E–01 | 6.2634E–01 | 1.0000E 00 | 1.5067E–01 |

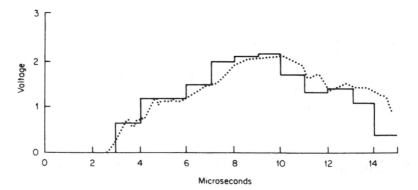

**Figure 10.35** Voltage–time relationship at nodes 1 and 4 of system in Figure 10.31. ——, computer solution; . . . ., transient analyser
(*Permission of the Institute of Electrical and Electronic Engineers*)

It is assumed that the variables of interest are known at the previous time step $t - \Delta t$, where $\Delta t$ is the time step. The value of $\Delta t$ must be small enough to give reasonable accuracy with a finite difference method.

## 10.7.1 Lumped element modelling

Consider an inductance $L$ (Figure 10.36(a)) and the voltage–time curve shown in 10.36(b).

$$v = L \, di/dt$$

and

$$i = \int \frac{v}{L} \, dt$$

$$\tfrac{1}{2} \frac{\Delta t}{L} [v(t) - v(t - \Delta t)]$$

Hence the current at time interval $(t)$ is

$$i(t) = i(t - \Delta t) + \frac{\Delta t [v(t) - v(t - \Delta t)]}{2L}$$

or

$$i(t) = v(t) \frac{\Delta t}{2L} + i(t - \Delta t) + v(t - \Delta t) \frac{\Delta t}{2L} = \frac{v(t)}{R} + I \qquad (10.22)$$

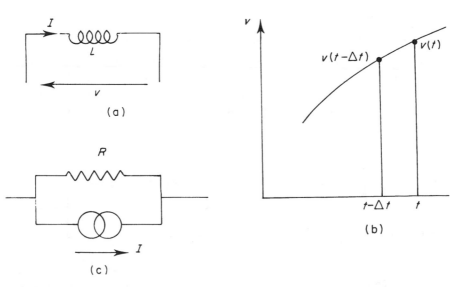

**Figure 10.36** (a) Lumped inductance. (b) Voltage–time curve for inductance. (c) Equivalent circuit

where

$$R = \frac{2L}{\Delta t} \quad \text{and} \quad I = i(t - \Delta t) + v(t - \Delta t)/R$$

Here, $R$ is constant and $I$ varies with time. The equivalent circuit is shown in Figure 10.36(c).

A similar treatment applies to capacitance ($C$) (see Figure 10.37(a)). Here,

$$v = \int \frac{i}{C} \, dt$$

and, over the interval $\Delta t$,

$$i(t) = i(t - \Delta t) + \frac{2C}{\Delta t} [v(t) - v(t - \Delta t)]$$

Again, if

$$R = \Delta t/2C$$

then

$$I = -i(t - \Delta t) - \frac{v(t - \Delta t)}{R} \tag{10.23}$$

giving the equivalent circuit of Figure 10.37(b).

Resistance is represented directly by a piecewise linear curve, as in Figure 10.39(a).

The procedure is as follows:

1. From initial conditions, determine $i(t - \Delta t) = i(0)$ and $v(t - \Delta t) = v(0)$.

2. Solve for $i(t)$ and $v(t)$. Increase time step by $\Delta t$ and calculate new values of $i$ and $v$, and so on. The analysis is carried out by use of the nodal admittance matrix and Gaussian elimination. If mutual coupling between elements exists then the representation becomes very complex.

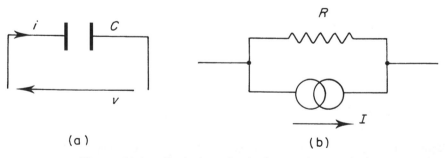

(a)                                    (b)

**Figure 10.37**   Equivalent circuits for transient analysis

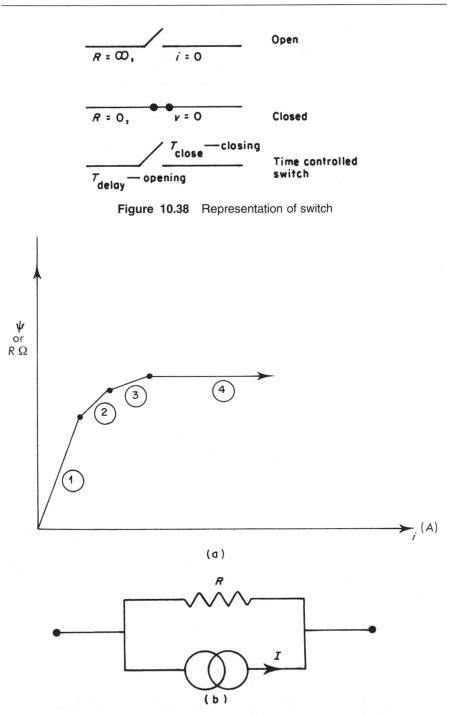

**Figure 10.38** Representation of switch

**Figure 10.39** (a) Non-linear characteristic; (b) circuit representation

## 10.7.2  Switching

The various representations are shown in Figure 10.38. A switching operation changes the topology of the network and hence the $[Y]$ matrix. If $[Y]$ is formed with all the switches open, then the closure of a switch is obtained by the adding together of the two rows and columns of $[Y]$ and the associated rows of $[i]$.

Another area where switching is used is to account for non-linear $\psi - i$ characteristics of transformers, reactors, etc. The representation is shown in Figure 10.39(a) and (b), in which

$$R = 2b_k/\Delta t$$

and

$$I = v(t - \Delta t)/R + i(t - \Delta t) \tag{10.24}$$

where $b_k$ = incremental inductance.

If $\psi$ is outside of the limits of segment $K$, the operation is switched to either $k - 1$ or $k + 1$. This changes the $[Y]$ matrix.

Because of the random nature of certain events, e.g. switching time or lightning incidence, Monte Carlo (statistical) methods are sometimes used. Further information can be obtained from the references at the end of this book.

## 10.7.3  Travelling-wave approach

Lines and cables would require a large number of $\pi$ circuits for accurate representation. An alternative would be the use of the travelling-wave theory.

Consider Figure 10.40,

$$i(x, t) = f_1(x - Ut) + f_2(x + Ut)$$
$$v(x, t) = z_0 f_1(x - Ut) + z_0 f_2(x + Ut)$$

where s = speed of propagation and $z_0$ = characteristic impedance. At node $k$,

$$i_k(t) = \frac{v_k(t)}{z_0} + I_k \tag{10.25}$$

where

$$I_k = -i_m(t - \tau) - v_m(t - \tau)/z_0$$

and

$$\tau = \frac{d}{U} = \text{travel time}$$

The equivalent circuit is shown in Figure 10.41. An advantage of this method is that the two ends of the line are decoupled. The value of $I_k$ depends

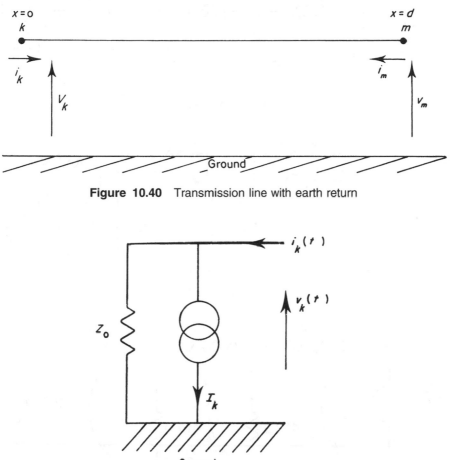

**Figure 10.40**  Transmission line with earth return

**Figure 10.41**  Equivalent circuit for single-phase line

on the current and voltage at the other end of the line $\tau$ seconds previously, e.g. if $\tau = 0.36$ ms and $\Delta t = 100\,\mu$s, storage of four previous times is required.

Detailed models for synchronous machines and h.v.d.c. converter systems are given in the references.

## Example 10.4

The equivalent circuit of a network is shown in Figure 10.42. Determine the network which simulates this network for transients using the EMTP method after the first time step of the transient of $5\,\mu$s.

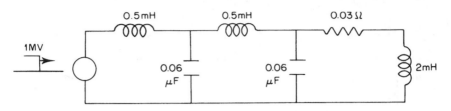

**Figure 10.42**(a)

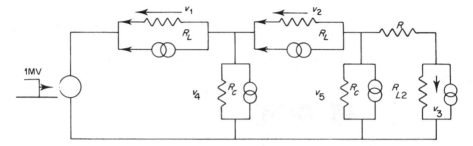

**Figure 10.42**(b)

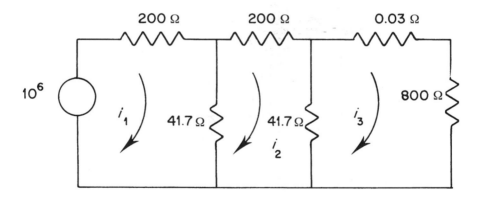

**Figure 10.42**(c)

*Solution*

$$R_{\mathrm{L}} = \frac{2 \times 0.0005}{5 \times 10^{-6}} = 200\,\Omega$$

$$R_{\mathrm{c}} = \frac{5 \times 10^{-6}}{2 \times 0.06 \times 10^{-6}} = 41.7\,\Omega$$

$$R_{\mathrm{L2}} = \frac{2 \times 0.002}{5 \times 10^{-6}} = 800\,\Omega$$

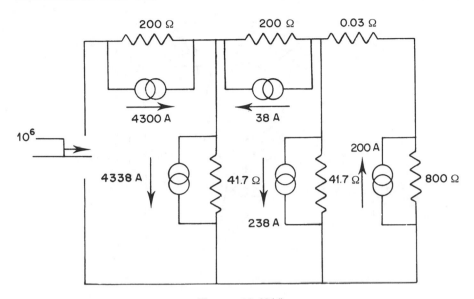

**Figure 10.42(d)**

At $t = 0$, note current sources are zero.

$$\begin{bmatrix} 10^6 \\ 0 \\ 0 \end{bmatrix} = \begin{bmatrix} 241.7 & -41.7 & 0 \\ -41.7 & 241.7 & -41.7 \\ 0 & -41.7 & 800 \end{bmatrix} \begin{bmatrix} i_1 \\ i_2 \\ i_3 \end{bmatrix}$$

Invert

$$\begin{bmatrix} i_1 \\ i_2 \\ i_3 \end{bmatrix} = \begin{bmatrix} 0.0043 & 0.0043 & 0.0007 \\ 3.8 \times 10^{-5} & 0.0007 & 0.0043 \\ 0.0002 & 3.8 \times 10^{-5} & 0.0002 \end{bmatrix} \begin{bmatrix} 10^6 \\ 0 \\ 0 \end{bmatrix}$$

$$i_1 = 4300 \, \text{A}$$
$$i_2 = -38, \qquad i_3 = -200$$

The equivalent circuit after $5 \, \mu s$ is shown in Figure 10.42(d).

The process is then repeated for the next $5 \, \mu s$ step using Figure 10.42(d) as the starting condition.

## 10.8 Ultra-High-Voltage Transmission

In this section, arrangements for extra-high voltage (E.H.V.) lines will be reviewed and the factors influencing designs for U.H.V. discussed. At the present time, for E.H.V. lines, 'V'-type insulator strings are used to support bundle conductors in horizontal formation on steel lattice towers with two earth (ground) wires; a typical structure is shown in Figure 10.43. In this figure, dimensions of interest are labelled and some critical distances are shown. Span

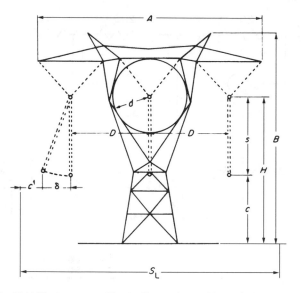

**Figure 10.43**  U.H.V. tower—critical dimensions (*Permission of the Institute of Electrical and Electronics Engineers*)

lengths (i.e. distances between towers) are of the order of 400–500 m at present and are expected to continue in this range. The number of subconductors in each bundle is determined by radio-interference and audible noise levels stipulated along the route.

Tower dimensions are determined by insulation requirements and the importance of the switching surge in this connection has already been discussed. The tower surge withstand voltage is set below the critical flashover (i.e. 50 per cent probability) voltage to give a reasonable value of flashover probability. When designing a tower, the insulation string length, insulator surface creep distance, tower grounding, and strike distances must be decided upon. Tower strike distance is the distance from the conductor to the tower structure. These dimensions are chosen to make the insulation strength such that the applied surge results in an acceptable surge flashover rate.

An example of laboratory assessment of the strength of tower insulation (V-string insulators) is shown in Figure 10.44. It has been shown (Figure 10.18) that for a given insulation the flashover voltage follows a Gaussian cumulative distribution curve to at least four standard deviations ($\sigma_f$) below the CFO; $\sigma_f$ is about 4.6–5 per cent of the CRO. The procedure is to equate the switching-surge voltage to the withstand strength of the insulation where withstand is defined as that voltage which results in a 0.13 per cent flashover probability, i.e. $3\sigma_f$ below the CFO. The CFO is determined from a knowledge of $\sigma_f$ and the withstand voltage, and from it the strike distances are obtained. This process is illustrated in Figure 10.45 for CFOs under wet and dry conditions.

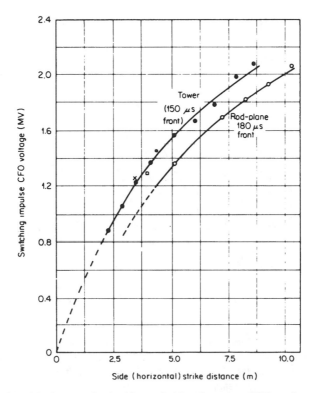

**Figure 10.44** Maximum obtainable switching impulse CFO voltage for a specific side strike distance (*Permission of the Westinghouse Electric Corporation*)

*Contamination* requirements are expressed in terms of the shortest distance along the insulator disc from cap to pin (creep distance); the creep distance required per kilovolt for flashover-free operation with normal practice is, at present, between 0.83 in (2.1 cm) and 1 in (2.54 cm) per kilovolt. The various factors involving flashover and insulation requirements are summarized in Figure 10.46, in which system operating voltage is related to the minimum distance for flashover between the phase conductor and tower sides (strike distance) for a 'V' formation of string insulators. The switching-surge curves refer to various per-unit peak values of switching surge, control over which may be exercised by circuit-breaker resistance switching. It is seen that increases in system voltages require progressively larger increases in the strike distance (conductor-to-tower minimum distance), and the tower dimensions would become intolerable from both economic and appearance standpoints unless the per-unit value of the surge is reduced.

The method illustrated in this section has been questioned for the U.H.V. region. It does not produce an estimate of the switching-surge flashover rate and only matches the withstand voltage to the maximum switching surge. In

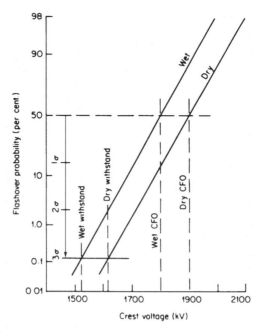

**Figure 10.45**   Selection of tower surge withstand level. Design withstand level for a tower is established by specifying some acceptable risk of flashover, in this case three standard deviations below the 50 per cent probability CFO voltage (*Permission of the Westinghouse Electric Corporation*)

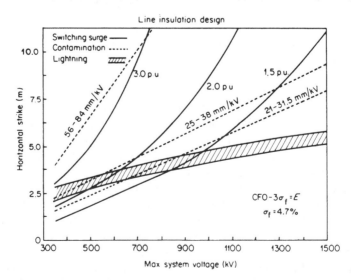

**Figure 10.46**   Estimates   of   E.H.V.–U.H.V.   tower   insulation   requirements (*Permission of the Westinghouse Electric Corporation*)

fact, two probability distributions are involved; one for surge magnitude, the other for insulation strength. The exact form of the surge magnitude probability curves for systems is not as yet known to an acceptable degree of accuracy for the calculation of flashover probability. This complete probability approach is being studied and revised versions of the curves shown in Figure 10.46 obtained.

Paris (1969) has suggested various criteria for the design of U.H.V. towers, e.g. dimensions are chosen in such a way as to make the cost of the inactive components a constant ratio (about 0.8) of the active components (i.e. the conductors). The tower size is defined by two parameters: the 'line size' giving the width of the 'right of way' (width of the strip of ground required underneath the line) and the 'tower size' defined as the product of tower width, height, and number of towers per kilometre. As the height of the conductors will not increase proportionally with the voltage, the magnitude of the electric field at the ground surface becomes a critical factor. Paris's analysis suggests that towers for voltages up to 1500 kV are feasible and some of his suggested towers are shown in Figure 10.47, based on line characteristics given in Table 10.6 and the criteria mentioned above. The guyed structures already in use lead to the possible use of towers able to withstand vertical and transverse forces only, with just a small number of special towers able to withstand the longitudinal forces as well, say one special tower to five light two-dimensional towers. Other possibilities for reducing size at high voltages include the use of insulated cross-arms of the form shown in Figure 10.48.

## 10.9   Design of Insulation by Digital Computer

The design of the complete system involves many parameters and variables and the design-analysis procedure is complex. Some factors, e.g. loss, radio interference (RI), and the swinging of insulator strings and conductors, depend on the weather conditions. Hence, some form of statistical analysis dependent on the geographical position of the line is important to obtain an economical design. Although the presentation here is of an introductory nature, flow diagrams of typical programs will be presented mainly to indicate the general manner in which design studies are developing at the present time. In particular, such programs endeavour to give a realistic account of weather conditions and perform computations to cover a long time-period, i.e. 10–20 years, to obtain reasonable statistical distributions. As well as weather conditions, it is also necessary statistically to account for switching-surge magnitudes and flashover occurrences. A general critical-path diagram covering the steps in the design of an E.H.V. line is shown in Figure 10.49.

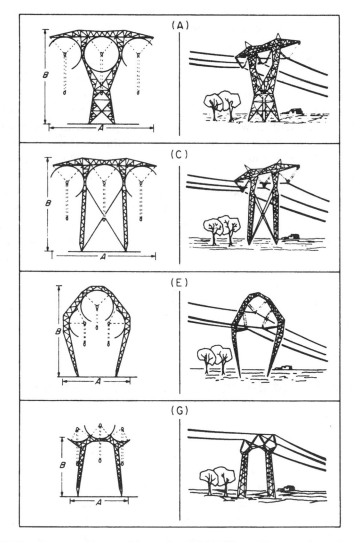

**Figure 10.47**   Proposed types of tower for 1500 kV lines. Proposed new tower types, C–H. Traditional tower types, A and B (*Permission of the Institute of Electrical and Electronics Engineers*)

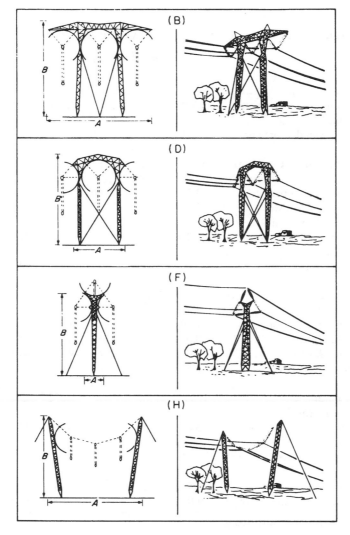

**Figure 10.47** (*continued*)

**Table 10.6**    Characteristics of lines at various system voltages

| | Highest system voltage $(V_m)$(kV) | | | | | |
|---|---|---|---|---|---|---|
| | 420 | 525 | 765 | 1000 | 1300 | 1500 |
| Overall aluminium section per phase, $S$ (mm$^2$) | 1240 | 1660 | 2680 | 3780 | 5250 | 6300 |
| Number of subconductors per phase, $n$ | 2 | 3 | 4 | 6 | 8 | 8 |
| Subconductor diameter, $\phi$ (mm) | 34.5 | 32.4 | 35.8 | 34.7 | 35.5 | 38.8 |
| Conductor-tower clearance, $d$ (m) | 3.00 | 3.90 | 5.60 | 7.20 | 8.50 | 9.40 |
| Switching impulse 50 per cent discharge voltage of tower insulation, $V_{50\%}$ (per unit) | 3.2 | 2.95 | 2.60 | 2.25 | 1.95 | 1.80 |
| Conductor-ground clearance at midspan, $C$ (m) | 7.2 | 8.45 | 10.8 | 13.1 | 15.0 | 16.2 |
| Span length, $L$ (m) | 400 | 420 | 445 | 475 | 500 | 515 |
| Midspan sag, $s$ (m) | 12 | 13.5 | 15 | 17 | 19 | 20 |
| Conductor height at the tower, $H$ (m) | 19.2 | 21.7 | 25.8 | 30.1 | 34.0 | 36.2 |
| Interphase distance, $D$ (m) | 7.30 | 9.20 | 12.8 | 16.1 | 19.0 | 20.8 |
| Tower width, $A$ (m) | 20.0 | 25.4 | 35.6 | 45.2 | 53.3 | 58.4 |
| Tower height, $B$ (m) | 24.6 | 28.2 | 35.5 | 42.25 | 47.9 | 51.5 |
| Line-size parameter (right of way), $S_L$ (m) | 35.5 | 42.3 | 52.0 | 62.5 | 72.0 | 76.5 |
| Tower-size parameter, $S_T = 1000AB/L$ (m$^2$/km) | 1230 | 1700 | 2840 | 4020 | 5110 | 5840 |
| RI limit gradient of lateral phase conductor, (kV/m) | 15.8 | 15.7 | 15.35 | 15.5 | 15.25 | 14.85 |
| Voltage gradient at ground, $G$ (kV/m) | 7.35 | 9.50 | 11.4 | 13.1 | 16.55 | 17.55 |
| Surge impedance, $Z_s$($\Omega$) | 284 | 268 | 264 | 249 | 240 | 245 |
| Surge impedance loading, $P_s$ (MW) | 560 | 925 | 1970 | 3615 | 6335 | 8265 |

(*Permission of I.E.E.E.—Paris, 1969*)

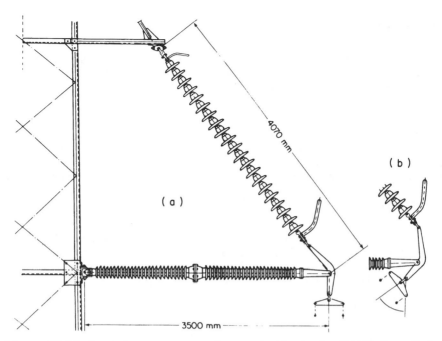

**Figure 10.48** Insulating cross-arm for a double-circuit 420 kV line (Italian). (a) Normal conditions. (b) Windy conditions

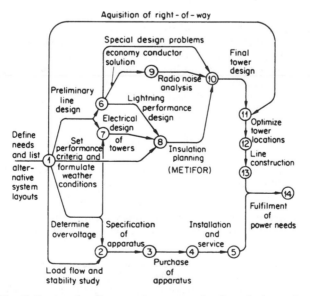

**Figure 10.49** Critical-path diagram for steps in the design of an E.H.V. line (*Permission of the Edison Electric Institute*)

# Problems

**10.1**   A 345 kV, 60 Hz system has a fault current of 40 kA. The capacitance of a busbar to which a circuit breaker is connected is 25 000 pF. Calculate the surge impedance of the busbar and the frequency of the restriking (recovery) voltage on opening.
(Answer: 674 $\Omega$, 875 Hz)

**10.2**   A highly capacitive circuit of capacitance per phase 100 $\mu$F is disconnected by circuit breaker, the source inductance being 1 mH. The breaker gap breaks down when the voltage across it reaches twice the system peak line-to-neutral voltage of 38 kV. Calculate the current flowing with the breakdown, and its frequency, and compare it with the normal charging current of the circuit.
(Answer: 34 kA, 503 Hz; note $\hat{I} = 2 V_p/Z_0$)

**10.3**   A 10 kV, 64.5 mm$^2$ cable has a fault 9.6 km from a circuit breaker on the supply side of it. Calculate the frequency of the restriking voltage and the maximum voltage of the surge after 2 cycles of the transient. The cable parameters are (per km), capacitance per phase = 1.14 $\mu$F, resistance = 5.37 $\Omega$, inductance per phase = 1.72 mH. The fault resistance is 6 $\Omega$.
(Answer: 374 Hz; 16 kV)

**10.4**   The effective inductance and capacitance of a faulted system as viewed by the contacts of a circuit breaker are 2 mH and 500 $\mu$F, respectively. The circuit breaker chops the fault current when it  has an instantaneous value of 100 A. Calculate the restriking voltage set up across the circuit breaker. Neglect resistance.
(Answer: 200 kV)

**10.5**   A 132 kV circuit breaker interrupts the fault current flowing into a symmetrical three-phase-to-earth fault at current zero. The fault infeed is 2500 MVA and the shunt capacitance, $C$, on the source side is 0.03 $\mu$F. The system frequency is 50 Hz. Calculate the maximum voltage across the circuit breaker and the restriking-voltage frequency.
   If the fault current is prematurely chopped at 50 A, estimate the maximum voltage across the circuit breaker on the first current chop.
(Answer: 215.5 kV; 6.17 kHz; 45 kV)

**10.6**   Repeat Example 10.2 but with the surge travelling from the cable into the overhead line.
(Answer: Current into line = 0.091× incident surge current; current reflected back into cable = 0.91× incident current, reflected energy = 0.83 × incident surge energy)

**10.7**   Repeat Example 10.2 but with zero resistance between the line and cable.
(Answer: Energy reflected back to line = 0.67× incident surge energy)

**10.8**   A cable of inductance 0.188 mH per phase and capacitance per phase of 0.4 $\mu$F is connected to a line of inductance of 0.94 mH per phase and capacitance 0.0075 $\mu$F per phase. All quantities are per km. A surge of 1 p.u. magnitude travels along the cable towards the line. Determine the voltage set up at the junction of the line and cable.
(Answer: 1.85 p.u.)

**10.9** A long overhead line has a surge impedance of $500\,\Omega$ and an effective resistance at the frequency of the surge of $7\,\Omega/\mathrm{km}$. If a surge of magnitude $500\,\mathrm{kV}$ enters the line at a certain point, calculate the magnitude of this surge after it has traversed $100\,\mathrm{km}$ and calculate the resistive power loss of the wave over this distance. The wave velocity is $3 \times 10^5\,\mathrm{km/s}$.
(Answer: $250\,\mathrm{kV}$; $375\,\mathrm{MW}$)

**10.10** A rectangular surge of $2\,\mu\mathrm{s}$ duration and magnitude $2\,\mathrm{p.u.}$ travels along a line of surge impedance of $350\,\Omega$. The latter is connected to another line of equal impedance through an inductor of $800\,\mu\mathrm{H}$. Calculate the value of the surge transmitted to the second line.
(Answer: $v = v_i[1 - \mathrm{e}^{-(2Z_0/L)t}]$ (i.e. $1.67\,\mathrm{p.u.}$))

**10.11** A lightning arrester employs a thyrite material possessing a resistance characteristic described by $R = (72 \times 10^3)/(I^{0.75})$. An overhead line of surge impedance $500\,\Omega$ is terminated by the arrester. Determine the voltage across the end of the line when a rectangular travelling surge of magnitude $500\,\mathrm{kV}$ travels along the line and arrives at the termination. (A graphical method using the voltage–current characteristics is useful.)
(Answer: $375\,\mathrm{kV}$)

**10.12** A rectangular surge of $1\,\mathrm{p.u.}$ magnitude strikes an earth (ground) wire at the centre of its span between two towers of effective resistance to ground of $200\,\Omega$ and $50\,\Omega$. The ground wire has a surge impedance of $500\,\Omega$. Determine the voltages transmitted beyond the towers to the earth wires outside the span.
(Answer: $0.44v_i$ from $200\,\Omega$ tower and $0.18v_i$ from $50\,\Omega$ tower)

**10.13** A system consists of the following elements in series: a long line of surge impedance $500\,\Omega$, a cable ($Z_0$ of $50\,\Omega$), a short line ($Z_0$ of $500\,\Omega$), a cable ($Z_0$ of $50\,\Omega$), a long line ($Z_0$ of $500\,\Omega$). A surge takes $1\,\mu\mathrm{s}$ to traverse each cable (they are of equal length) and $0.5\,\mu\mathrm{s}$ to traverse the short line connecting the cables. The short line is half the length of each cable. Determine, by means of a lattice diagram, the p.u. voltage of the junction of the cable and the long line if the surge orginates in the remote long line.
(Answer: see Figure 10.50)

**10.14** A 3 p.h., 50 Hz, 11 kV star-connected generator, with its star point earthed, is connected via a circuit breaker to a busbar. There is no load connected to the busbar. The capacitance to earth on the generator side terminals of the circuit breaker is $0.007\,\mu\mathrm{F}$ per phase. A 3 phase-to-earth short circuit occurs at the busbar with a symmetrical subtransient fault current of $5000\,\mathrm{A}$. The fault is then cleared by the circuit breaker. Assume interruption at current zero.

(a) Sketch the voltage across the circuit breaker terminals of the first phase to clear.
(b) Neglecting damping, calculate the peak value of the transient recovery voltage of this phase.
(c) Determine the time to this peak voltage and hence the average rate of rise of recovery voltage.

(Answer: (b) $17.96\,\mathrm{kV}$; (c) $76.7\,\mu\mathrm{s}$, $1.075\,\mathrm{kV}/\mu\mathrm{s}$)
(*From E.C. Examination, 1996*)

**10.15** A very long transmission line AB is joined to an underground cable BC of length $5\,\mathrm{km}$. At end C, the cable is connected to a transmission line CD of $15\,\mathrm{km}$ length. The

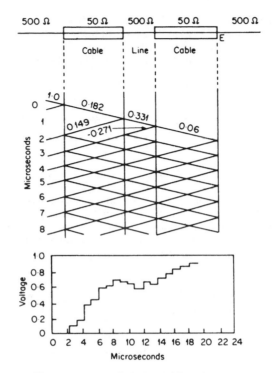

**Figure 10.50**   Solution of Problem 10.13

transmission line is open-circuit at D.

The cable has a surge impedance of 50 Ω and the velocity of wave propagation in the cable is $150 \times 10^6$ m/s. The transmission lines each have a surge impedance of 500 Ω. A voltage step of magnitude 500 kV is applied at A and travels along AB to the junction B with the cable.

Use a lattice diagram to determine the voltage at:

(a) D shortly after the surge has reached D;
(b) D at a time 210 μs after the surge first reaches B;
(c) B at a time 210 μs after the surge first reaches B.

Sketch the voltage at B over these 210 μs.
(Answer: (a) 331.2 kV)
    (*From E.C. Examination, 1995*)

# 11

# Substations and Protection

## 11.1 Introduction

In Chapter 7 attention was confined to the analysis of various types of faults which may occur in a power system. Although the design of electrical plant is influenced by a knowledge of fault conditions, the major use of fault analysis is in the specification of switchgear and protective gear, both being housed in 'substations'. Circuit-breaker ratings are determined by the fault MVA at their particular locations. The maximum circuit-breaker rating is of the order of 50 000–60 000 MVA, and this is achieved by the use of several interrupter heads in series per phase. Not only has the circuit breaker to extinguish the fault-current arc, with the substation connections it has also to withstand the considerable forces set up by short-circuit currents, which can be very high.

A knowledge of the currents resulting from various types of fault at a location is essential for the effective operation of what is known as 'system protection'. If faults occur on the system, the control engineers, noting the presence of the fault, can operate the appropriate circuit breakers to remove the faulty line or plant from the network. This, however, takes considerable time and experience. Faults on a power system resulting in high currents and also possible loss of sychronism must be removed in the minimum of time. Automatic means, therefore, are required to detect abnormal currents and voltages and, when detected, to open the appropriate circuit breakers. It is the object of protection to accomplish this. In a large interconnected network, considerable design knowledge and skill is required to remove the faulty part from the network and leave the healthy remainder working intact.

There are many varieties of automatic protective systems, ranging from simple overcurrent electromechanical relays to sophisticated electronic systems transmitting high-frequency signals along the power lines. The simplest but extremely effective form of protection is the electromechanical relay, which closes contacts and hence energizes the circuit-breaker opening mechanisms when currents larger than specified pass through the equipment.

The protection used in a network can be looked upon as a form of insurance in which a percentage of the total capital cost (about 5 per cent) is used to safeguard apparatus and ensure continued operation when faults occur. In a highly industrialized community the maintenance of an uninterrupted supply to consumers is of paramount importance and the adequate provision of protection systems is essential.

Summarizing, protection and the automatic tripping (opening) of associated circuit breakers has two main functions: (1) to isolate faulty equipment so that the remainder of the system can continue to operate successfully; and (2) to limit damage to equipment caused by overheating, mechanical forces, etc.

## 11.2  Switchgear

Some of the functions of the switches or circuit breakers are obvious and apply to any type of circuit, others are peculiar to high-voltage equipment. For maintenance to be carried out on plant, it must be isolated from the rest of the network and hence switches must be provided on each side. If these switches are not required to open under working conditions, i.e. with fault or load current and normal voltage, a cheaper form of switch known as an *isolator* can be used; this can close a live circuit but not open one. Owing to the high cost of circuit breakers, much thought is given, in practice, to obtaining the largest degree of flexibility in connecting circuits with the minimum number of switches. Popular arrangements of switches are shown in Figure 11.1 and a typical feeder bay layout in Figure 11.2.

High-voltage circuit breakers, 11 kV and above, take five basic forms: oil immersed, air blast, small oil volume, sulphur hexafluoride ($SF_6$), and vacuum. In addition, air circuit breakers with long contact gaps and large arc-splitter plates have been used for up to 6.6 kV, but they are bulky and expensive compared with the five types listed above. These five types will   now be described in more detail.

### 11.2.1  The bulk-oil circuit breaker

A cross-section of an oil circuit breaker with all three phases in one tank is shown in Figure 11.3(a). There are two sets of contacts per phase. The lower and moving contacts are usually cylindrical copper rods and make contact with the upper fixed contacts. The fixed contacts consist of spring-loaded copper

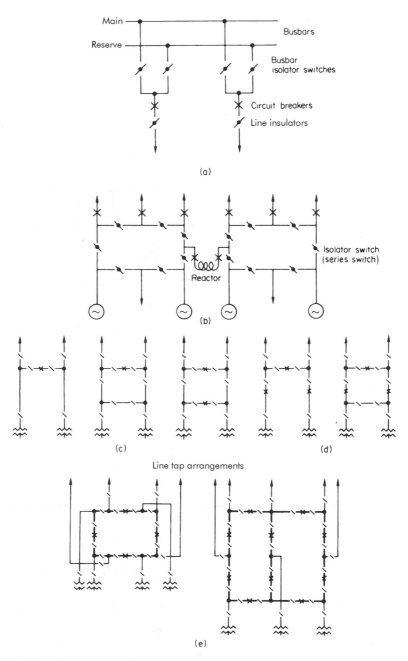

**Figure 11.1** Possible switchgear arrangements. (a) Double-busbar selection arrangements. (b) Double-ring busbars with connecting reactor busbars can be isolated for maintenance, but circuits cannot be transferred from one side of the reactor to the other. (c) Open-mesh switching stations, transformers not switched. (d) Open-mesh switching stations, transformers switched. (e) Closed-mesh switching stations. (Isolators are sometimes called series switches.) Arrangements can be indoors or outdoors

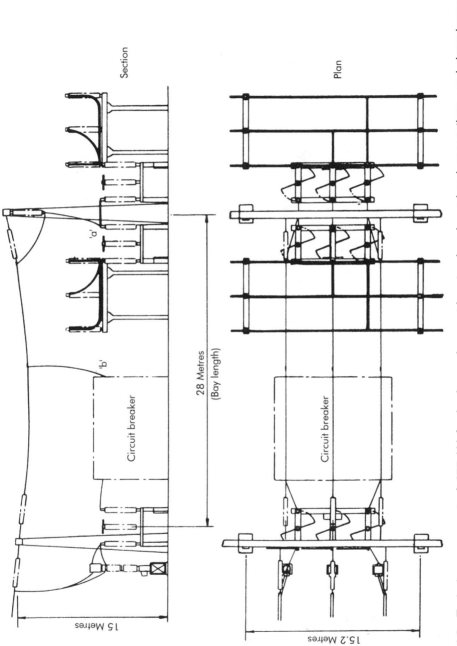

**Figure 11.2** Typical arrangement of 275 kV feeder bay using gantries, tensioned overhead connections, and downdroppers (*Permission of I.E.E.*)

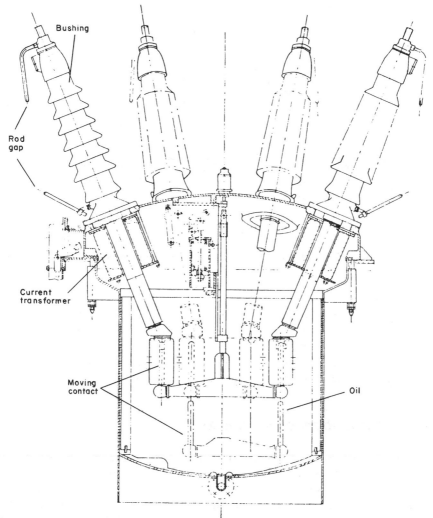

**Figure 11.3**(a)  Cross-section of a 66 kV bulk-oil circuit breaker. Three phases in one tank (Now obsolete) (*Permission of English Electric Co. Ltd*)

segments which exert pressure on the lower contact rod, when closed, to form a good electrical contact. On opening, the lower contacts move rapidly downwards and draw an arc. When the circuit breaker opens under fault conditions many thousands of amperes pass through the contacts and the extinction of the arc (and hence the effective open-circuiting of the switch) are major engineering problems. Effective opening is only possible because the instantaneous voltage and current per phase reduces to zero during each alternating current cycle. The arc heat causes the evolution of a hydrogen bubble in the oil and this high-

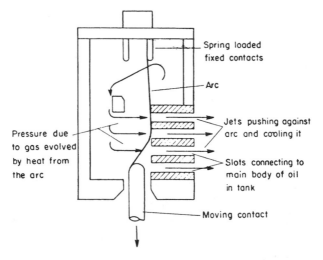

**Figure 11.3**(b)    Cross-jet explosion pot for arc extinction in bulk-oil circuit breakers

pressure gas pushes the arc against special vents in a device surrounding the contacts and called a 'turbulator' (Figure 11.3(b)). As the lowest contact moves downwards the arc stretches and is cooled and distorted by the gas and so eventually breaks. The gas also sweeps the arc products from the gap so that the arc does not re-ignite when the voltage rises to its full open-circuit value. Various patentable forms of oil-immersed arc-extinction devices are in use.

## 11.2.2  *The air-blast circuit breaker*

For voltages above 120 kV the air-blast breaker has been popular because of the feasibility of having several contact gaps in series per phase. Schematic diagrams of two types of air-blast head are shown in Figure 11.4. Air normally stored at $1.38 \, \text{MN/m}^2$ is released and directed at the arc at high velocities, thus extinguishing it. The air also actuates the mechanism of the movable contact. Figure 11.5 shows a 132 kV air-blast breaker with two breaks (interrupters) per phase and its associated isolator or series switch. Although air-blast circuit breakers have been developed and installed up to the highest voltages they have now been largely superseded by the $SF_6$ gas circuit breaker (see later).

## 11.2.3  *Small- or low-oil-volume circuit breakers*

Here, the quenching mechanisms are enclosed in vertical porcelain insulation compartments and the arc is extinguished by a jet of oil issuing from the moving (lower contact) as it opens. The volume of oil is much smaller than

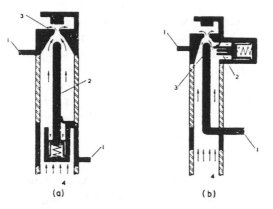

**Figure 11.4** Schematic diagrams of two types of air-blast head. (a) Axial flow with axially moving contact. (b) Axial flow with side-moving contact.
1 = Terminal, 2 = moving contact, 3 = fixed contact, and 4 = blast pipe
(*Permission of A. Reyrolle & Company Ltd*)

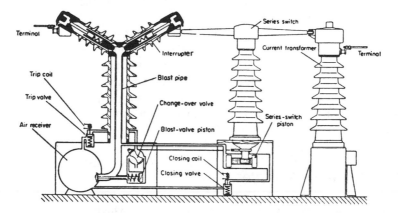

**Figure 11.5** Schematic arrangement for typical air-blast circuit breaker

in the bulk-oil type, thereby reducing hazards from explosion and fire. Although many oil circuit breakers are in use, particularly at distribution voltages, modern practice is to use $SF_6$ or vacuum breakers to avoid the presence of flammable liquids for circuit-interruption purposes.

## 11.2.4 Sulphur hexafluoride (SF₆) gas

The advantages of using sulphur hexafluoride ($SF_6$) as an insulating and interrupting medium in circuit breakers arise from its high electric strength and outstanding arc-quenching characteristics. $SF_6$ circuit breakers are much smaller than air breakers of the same rating, the electric strength of $SF_6$ at atmospheric pressure being roughly equal to that of air at the pressure of 10 atm. Temperatures of the order of 30 000 K are likely to be experienced in arcs in $SF_6$ and these are, of course, well above the dissociation temperature of the gas (about 2000 K); however, nearly all the decomposition products are electronegative so that the electric strength of the gas recovers quickly after the arc has been extinguished. Filters are provided to render the decomposition products harmless and only a small amount of fluorine reacts with metallic parts of the breaker. A sectional view of an $SF_6$, 420/525 kV switchgear unit is shown in Figure 11.6.

The arrangement of equipment in a 380/110 kV substation $SF_6$ is illustrated in Figure 11.7. The switchgear also contains all necessary measuring and other facilities as follows: $SF_6$-insulated, toroidal current transformers and voltage transformers, cable terminations, gas storage cylinders, cable isolator, grounding switches, bus isolator, and bus system. Such substations are of immense value in urban areas because of their greatly reduced size compared with air blast. $SF_6$ circuit breakers rated at 45 GVA are available and designs for 1300 kV have been produced.

$SF_6$ switchgear is now widely used at lower voltages from 6.6 kV to 132 kV for distribution systems. A typical interrupter, shown in Figure 11.8, is enclosed in a sealed porcelain cylinder with $SF_6$ under about 5 atm pressure. The movement of the contact forces gas into the opening contact by a 'puffer' action, thereby forcing extinction. $SF_6$ interrupters are very compact and robust; they require very little maintenance when mounted in metal cabinets to form switchboards.

### 11.2.5  Vacuum interrupter

A pair of contacts opening in vacuum draws an arc which burns in the vaporized contact material. Consequently, the contact material and its arcing root shape are crucial to the design of a commercial interrupter. A typical design is illustrated in Figure 11.9.

The main advantages of a vacuum interrupter are: (1) the very small damage normally caused to the contacts on operation, so that a life of 30 years can be expected without maintenance; (2) the small mechanical energy required for tripping; and (3) the low noise caused on operation. The nature of the vacuum arc depends on the current; at low currents the arc is diffuse and can readily be interrupted, but at high currents the arc tends to be constricted. Electrode contour geometries have been produced to give diffuse arcs with current densities of $10^6$–$10^8$ A/cm$^2$; electron velocities of $10^8$ cm/s are experienced in the arc and ion velocities of $10^6$ cm/s.

Considerable progress has been made in increasing the current-breaking capacity of vacuum interrupters, but not their operating voltage. Today,

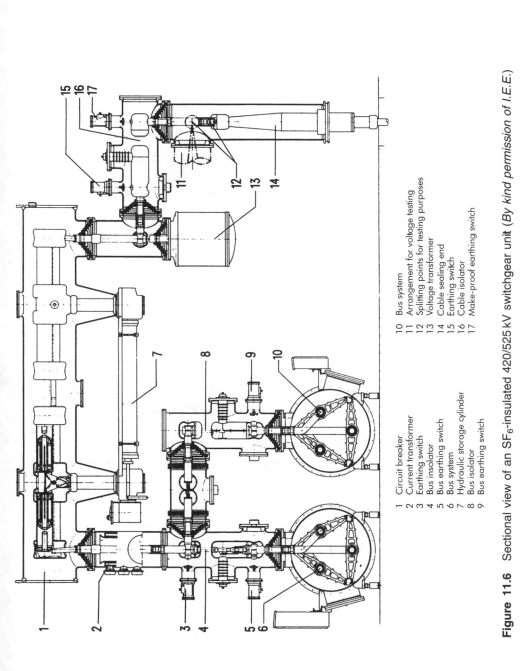

1  Circuit breaker
2  Current transformer
3  Earthing switch
4  Bus insolator
5  Bus earthing switch
6  Bus system
7  Hydraulic storage cylinder
8  Bus isolator
9  Bus earthing switch

10  Bus system
11  Arrangement for voltage testing
12  Splitting points for testing purposes
13  Voltage transformer
14  Cable sealing end
15  Earthing switch
16  Cable isolator
17  Make-proof earthing switch

**Figure 11.6**  Sectional view of an SF$_6$-insulated 420/525 kV switchgear unit (*By kind permission of I.E.E.*)

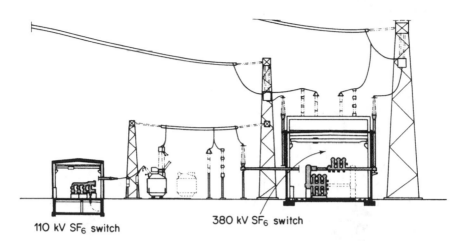

110 kV SF₆ switch

380 kV SF₆ switch

**Figure 11.7** Arrangement of 380 kV/110 kV substation using SF₆ switchgear (*Permission of Brown Boveri*)

they are available with ratings of 500 MVA at 30 kV and are extensively used in 11, 20, and 25 kV switchboards. A three-phase vacuum breaker with horizontally mounted interrupters is shown in Figure 11.10 with SF₆ insulation surrounding the interrupters.

## 11.2.6  Summary of circuit-breaker requirements

A circuit breaker must fulfil the following conditions:

1.  open and close in the shortest possible time under any network condition;

2.  conduct rated current without exceeding rated design temperature;

3.  withstand, thermally and mechanically, any short-circuit currents;

4.  maintain its voltage to earth and across the open contacts under both clean and polluted conditions;

5.  not create any large overvoltage during opening and closing;

6.  be easily maintained;

7.  be not too expensive.

Although air has now been largely overtaken by SF₆ as an interrupting medium at high voltages, it has given good performance over the years as it was easily able to achieve 2-cycle ($\approx$ 40 ms) interruption after receipt of a tripping signal to the operating coil. Increasing need, at voltages up to 725 kV, to reduce interruption times to $1\frac{1}{2}$ cycles to maintain stability and to

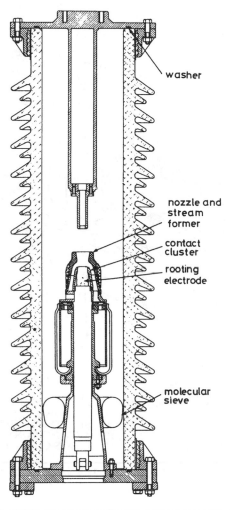

washer

nozzle and stream former

contact cluster

rooting electrode

molecular sieve

**Figure 11.8**  An $SF_6$ single-pressure puffer-type interrupter (*By kind permission of A. Reyrolle & Co. Ltd*)

reduce fault damage has meant that $SF_6$ switchgear has been required. A 60 kA rating has been possible with resistive switching to avoid unnecessary over-voltages, and a much lower noise level on operation has further confirmed $SF_6$ as the preferred medium. Being factory-sealed and fully enclosed has considerably reduced radio-interference and audible noise due to corona discharge.

The increasing performance and low maintenance has led to the development of on-line monitoring of the health of switchgear and the associated connections and components (known as 'condition monitoring'). This is particularly required for $SF_6$ enclosed units as even small dust particles can cause breakdown. Fast protection gear is essential to match these developments.

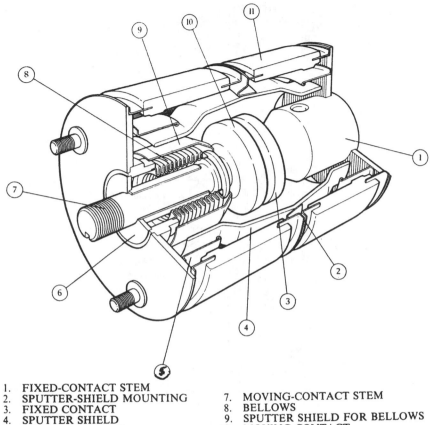

1. FIXED-CONTACT STEM
2. SPUTTER-SHIELD MOUNTING
3. FIXED CONTACT
4. SPUTTER SHIELD
5. GRADING SHIELD
6. MOVING-CONTACT GUIDE
7. MOVING-CONTACT STEM
8. BELLOWS
9. SPUTTER SHIELD FOR BELLOWS
10. MOVING CONTACT
11. GLASS-CERAMIC BODY

**Figure 11.9** Constructional features of an 11 kV vacuum interrupter (*Courtesy of A. Reyrolle & Co. Ltd*)

## 11.3 Qualities Required of Protection

A few terms often used to describe the effectiveness of protective gear will now be described.

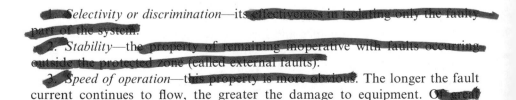

1. *Selectivity or discrimination*—its effectiveness in isolating only the faulty part of the system.
2. *Stability*—the property of remaining inoperative with faults occurring outside the protected zone (called external faults).
3. *Speed of operation*—this property is more obvious. The longer the fault current continues to flow, the greater the damage to equipment. Of great

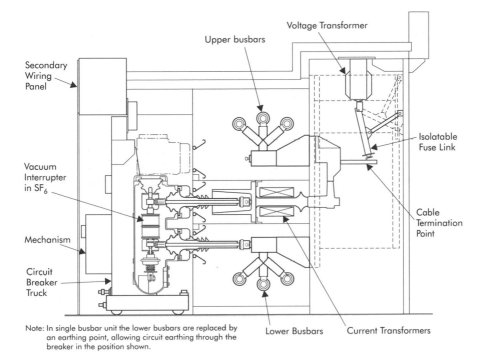

Note: In single busbar unit the lower busbars are replaced by
an earthing point, allowing circuit earthing through the
breaker in the position shown.

**Figure 11.10** HMX double-busbar configuration (*Reproduced by permission of G.E.C. Alsthom T&D Distribution Switchgear Ltd.*)

importance is the necessity to open faulty sections before the connected synchronous generators lose synchronism with the rest of the system. This aspect is dealt with in detail in Chapter 8. A typical fault clearance time in H.V. systems is 80 ms, and this requires very high-speed relaying as well as breaking.

4. *Sensitivity*—this is the level of magnitude of fault current at which operation occurs, which may be expressed in current in the actual network (primary current) or as a percentage of the current-transformer secondary current.

5. *Economic consideration*—in distribution systems the economic aspect almost overrides the technical one, owing to the large number of feeders, transformers, etc., provided that basic safety requirements are met. In transmission systems the technical aspects are more important. The protection is relatively expensive, but so is the system or equipment protected, and security of supply is vital. In transmission two separate protective systems are used, one main (or primary) and one back-up. In some instances, two out of three systems must operate before circuit-breaker tripping is initiated.

6. *Reliability*—this property is self-evident. A major cause of circuit 'outages' is mal-operation of the protection itself. On average, in the British

system (not including faults on generators), nearly 10 per cent of outages are due to this cause.

## 11.3.1 Back-up protection

Back-up protection, as the name implies, is a completely separate arrangement which operates to remove the faulty part should the main protection fail to operate. The back-up system should be as independent of the main protection as possible, possessing its own current transformers and relays. Often, only the circuit-breaker tripping and voltage transformers are common.

Each main protective scheme protects a defined area or *zone* of the power system. It is possible that between adjacent zones a small region, e.g. between the current transformers and circuit breakers, may be unprotected, in which case the back-up scheme (known as remote back-up), will afford protection because it overlaps the main zones, as shown in Figure 11.11. In distribution the application of back-up is not as widespread as in transmission systems; often it is sufficient to apply it at strategic points only. Remote back-up is slow and usually disconnects more of the supply system than is necessary to remove the faulty part.

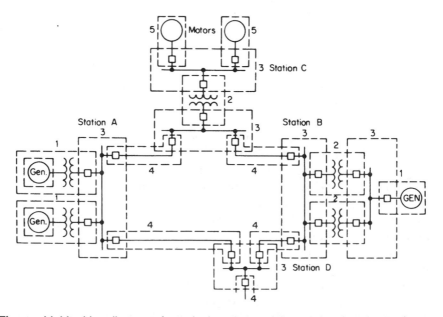

**Figure 11.11**   Line diagram of a typical system and the overlapping zones of protection (*Permission of Westinghouse Electrical Corporation*)

# 11.4    Components of Protective Schemes

## 11.4.1    Current transformers (CTs)

In order to obtain currents which are proportional to the system (primary) currents, and which can be used in control circuits, current transformers are used. Often the primary conductor itself, e.g. an overhead line, forms a single primary turn (bar primary). Whereas instrument current transformers have to remain accurate only up to slight overcurrents, protection current transformers must retain proportionality up to at least 20 times normal full load.

A major problem exists when two current transformers are used which should retain identical characteristics up to the highest fault current, e.g. in pilot wire schemes. Because of saturation in the silicon steel used and the possible existence of a direct component in the fault current, the exact matching of such current transformers is difficult. The nominal secondary current rating of current transformers is now usually 1 A, but 5 A has been used in the past.

### Linear couplers

The problems associated with current transformers have resulted in the development of devices called linear couplers, which serve the same purpose but, having air cores, remain linear at the highest currents. These are also known as Rogowski coils and are particularly suited to digital schemes.

## 11.4.2    Voltage (or potential) transformers (VTs or PTs)

These provide a voltage which is much lower than the system voltage, the nominal secondary voltage being 110 V. There are two basic types: the wound (electromagnetic), virtually a small power transformer, and the capacitor type. In the latter a tapping is made on a capacitor bushing (usually of the order of 12 kV) and the voltage from this tapping is stepped down by a small voltage transformer. The arrangement is shown in Figure 11.12; the reactor (X) and the capacitor (C) constitute a tuned circuit which corrects the phase-angle error of the secondary voltage. In H.V. systems, the capacitor divider is a separate unit mounted within an insulator. It can also be used as a line coupler for high frequency signalling (power line carrier, see Section 11.10).

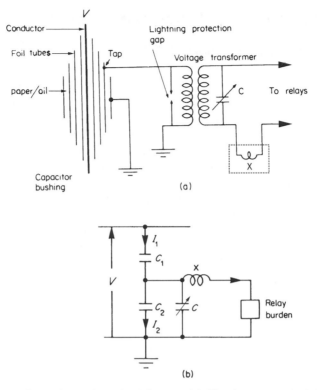

**Figure 11.12**   Capacitor voltage transformer. (a) Circuit arrangement. (b) Equivalent circuit—burden = impedance of transformer and load referred to primary winding

### 11.4.3   Relays

A relay is a device which, when energized by appropriate system quantities, indicates an abnormal condition. When the relay contacts close, the associated circuit-breaker trip-circuits are energized and the breaker contacts open, isolating the faulty part from the system. There are two main forms of relay: electromagnetic and semiconductor, including digital (numerical). For some purposes (e.g. overload protection) a bimetallic-strip thermal action is used.

The basic forms of the electromagnetic type comprise induction disc, induction cup, hinged armature, and plunger action. The hinged-armature and plunger-type devices (Figure 11.13) are the simplest and rely on the attraction of an armature or plunger due to an electromagnet which may be energized by a.c. or d.c.

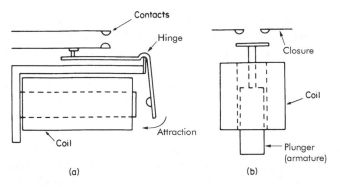

**Figure 11.13** (a) Hinged-armature relay. (b) Plunger-type relay

## Induction-disc relay

Here, a copper or aluminium disc is free to rotate between the poles of an electromagnet which produces two alternating magnetic fields displaced in phase and space. The eddy currents due to one flux and the remaining flux interact to produce a torque on the disc. In early relays the flux displacement was produced by a copper band around part of the magnet pole (shading ring) which displaced the flux contained by it. Modern relays employ a wattmetric principle in which two electromagnets are employed, as shown in Figure 11.14.

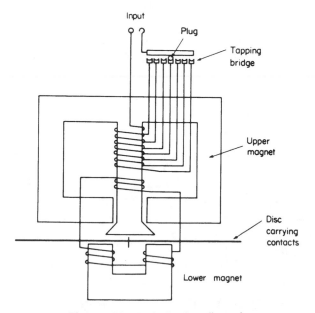

**Figure 11.14** Induction-disc relay

The current in the lower electromagnet is induced by transformer action from the upper winding and sufficient displacement between the two fluxes results. This, however, may be adjusted by means of a reactor in parallel with the secondary winding.

The basic mode of operation of the induction disc is indicated in the phasor diagram of Figure 11.15. The torques produced are proportional to $\Phi_2 i_1 \sin \alpha$ and $\Phi_1 i_2 \sin \alpha$, so that the total torque is proportional to $\Phi_1 \Phi_2 \sin \alpha$ as $\Phi_1$ is proportional to $i_1$ and $\Phi_2$ to $i_2$.

This type of relay is fed from a current transformer (CT) and the sensitivity may be varied by the plug arrangement shown in Figure 11.14. The operating characteristics are shown in Figure 11.16. To enable a single characteristic curve to be used for all the relay sensitivities (plug settings) a quantity known as the current (or plug) setting multiplier is used as the abscissa instead of current magnitude, as shown in Figure 11.16. To illustrate the use of this curve (usually shown on the relay casing) the following example is given.

### Example 11.1

Determine the time of operation of a 1 A, 3 s overcurrent relay having a plug setting of 125 per cent and a time multiplier of 0.6. The supplying CT is rated 400 : 1 A and the fault current is 4000 A.

### Solution

The relay coil current $= (4000/400) \times 1 = 10$ A. The normal relay coil current is $1 \times (125/100) = 1.25$ A. Therefore the relay fault current as a multiple of the plug

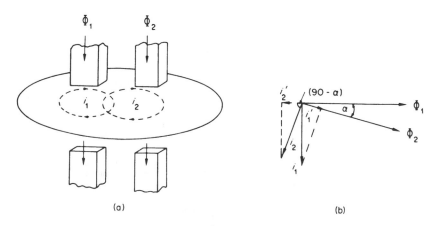

(a)                    (b)

**Figure 11.15** Operation of disc-type electromagnetic relay. (a) Fluxes. (b) Phasor diagram. $i_1$ and $i_2$ are induced currents in disc

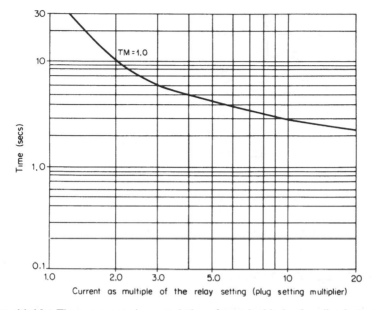

**Figure 11.16** Time–current characteristics of a typical induction disc in terms of the plug-setting multiplier. TM = time multiplier. (Note: Time is along the vertical axis)

setting $= (10/1.25) = 8$. From the relay curve (Figure 11.16), the time of operation is 3.3 s for a time setting of 1. The time multiplier (TM) controls the time of operation by changing the angle through which the disc moves to close the contacts. The actual operating time $= 3.3 \times 0.6 = 2.0$ s.

Induction-disc relays may be made responsive to power flow by feeding the upper magnet winding in Figure 11.14 from a voltage via a potential transformer and the lower winding from the corresponding current. As the upper coil will consist of a large number of turns, the current in it lags the applied voltage by 90°, whereas in the lower (small number of turns) coil they are almost in phase. Hence, $\Phi_1$ is proportional to **V**, and $\Phi_2$ is proportional to **I**, and torque is proportional to $\Phi_1 \Phi_2 \sin \alpha$, i.e. to $\Phi_1 \Phi_2 \sin(90 - \phi)$, or $VI \cos \phi$ (where $\phi$ is the angle between **V** and **I**).

The direction of the torque depends on the power direction and hence the relay is directional. A power relay may be used in conjunction with a current-operated relay to provide a directional property.

### Induction-cup relay

The operation is similar to the induction disc; here, two fluxes at right angles induce eddy currents in a bell-shaped cup which rotates and carries the moving contacts. A four-pole relay is shown in Figure 11.17.

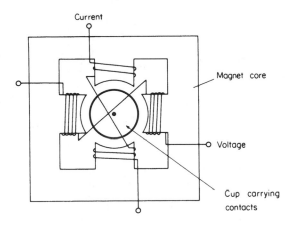

**Figure 11.17**    Four-pole induction-cup relay

## Permanent-magnet moving coil

The action in one type is similar to a moving-coil indicating instrument with the moving-coil assembly carrying the contacts. In a second type the action is basically that of the loudspeaker in which the coil moves axially in the gap of a permanent magnet. The time–current characteristic is inverse with a definite minimum time.

## Balanced beam

The basic form of this relay is shown in Figure 11.18. The armatures at the ends of the beam are attracted by electromagnets which are operated by the appropriate parameters, usually voltage and current. A slight mechanical bias is incorporated to keep the contacts open, except when operation is required.

The pulls on the armatures by the electromagnets are equal to $K_1 V^2$ and $K_2 I^2$, where $K_1$ and $K_2$ are constants, and for operation (i.e. contacts to close), i.e.

$$K_1 V^2 > K_2 I^2$$

then

$$\frac{V}{I} < \sqrt{\frac{K_2}{K_1}} \quad \text{or} \quad Z < \sqrt{\frac{K_2}{K_1}}$$

This shows that the relay operates when the impedance it 'sees' is less than a predetermined value. The characteristic of this relay, when drawn on $R$ and $jX$ axes, is a circle, as shown in Figure 11.19.

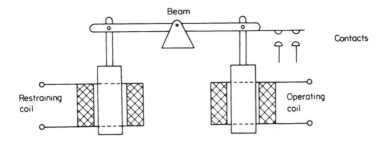

**Figure 11.18**   Schematic diagram of balanced-beam relay

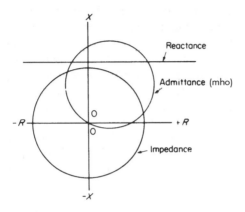

**Figure 11.19**   Characteristics of impedance, reactance, and admittance (mho) relays shown on the *R–X* diagram

## Distance relays

The balanced-beam relay, because it measures the impedance of the protected line, effectively measures distance. Two other relay forms also may be used for this purpose:

1.  the reactance relay which operates when $(V/I)\sin\phi \leqslant$ constant, having the characteristic shown in Figure 11.19;

2.  the mho or admittance relay, the characteristic of which is also shown in Figure 11.19.

The above relays operate with any impedance phasor lying inside the characteristic circle or below the reactance line. The mho characteristic may be obtained by devices balancing two torque-producing elements and an induction cup could be used. A rectifier bridge supplying a moving-coil relay is also often used.

## Negative-phase sequence

This is used in generator protection and is sensitive to the presence of negative-sequence currents. The protection comprises a bridge circuit supplying a current-operated relay.

## Solid-state (passive) devices

These relays are extremely fast in operation, having no moving parts and are very reliable. Detection involving phase angles and current and voltage magnitudes are made with appropriate circuits. Most required current–time characteristics may be readily obtained and solid-state devices are now firmly established. Inverse-characteristic, overcurrent, and earth-fault relays have a minimum time lag and the operating time is inversely related to some power of the input (e.g. current). In practical static relays it is advantageous to choose a circuit which can accommodate a wide range of alternative inverse time characteristics, precise minimum operating levels, and definite minimum times.

Electromechanical relays are vulnerable to corrosion, shock vibration, and contact bounce and welding. They require regular maintenance by skilled personnel. Not surprisingly, with the advent of the microprocessor and integrated circuits, digital (numerical) protection devices are now the norm.

## Digital (numerical) relaying

Having monitored currents and voltages through primary transducers (CTs and VTs), these analogue quantities can be sampled and converted to digital form for numerical manipulation, analysis, display, and recording. This process provides a flexible and very reliable relaying function, thereby enabling the same basic hardware units to be used for almost any kind of relaying scheme. With the continuous reduction in digital-circuit costs and increases in their functionality, considerable cost–benefit improvement ensues. It is now often the case that the cost of the relay housing, connections, and EMC protection dominates the hardware but, as is usual in digital systems, the software development and proving procedure are the most expensive items in the overall scheme.

Since digital relays can store data, they are particularly suited to post-fault analysis and can therefore be used in a self-adaptive mode, which is impossible with conventional devices. Additionally, they are capable of self-monitoring and communication with hierarchical controllers. By these means, not only can fast and selective fault clearance be obtained, but also fault location can be flagged to mobile repair crews. Minor (non-vital) protection-system faults can also be indicated for maintenance attention. With the incorporation of a satellite-timing signal receiver using Global Positioning Satellite (GPS) to give a 1 $\mu$s synchronized signal, faults on overhead circuits can be located to within 300 m.

The basic hardware elements of a digital relay are indicated in Figure 11.20 and a flow chart for the software is outlined in Figure 11.21. A digital relay unit is shown in Figure 11.22. For the advanced student, many good references to digital or numerical relaying are now available (see the references at the end of this book).

### Summation transformer

In some relaying schemes it is necessary to transmit the secondary currents of the current transformers considerable distances in order to compare them with currents elsewhere. To avoid the use of wires from each of the three CTs in a three-phase system, a summation transformer is used which gives a single-phase output, the magnitude of which depends on the nature of the fault. The arrangement is shown in Figure 11.23, in which the ratios of the turns are indicated. On balanced through-faults there is no current in the winding between c and n. The phase (a) current energizes the 1 p.u. turns between a and b and the phasor sum of $I_a$ and $I_b$ flows in the 1 p.u. turns between b and c.

The arrangement gives a much greater sensitivity to earth faults than to phase faults. When used in phase-comparison systems, however, the actual value of output current is not important and the transformer usually saturates on high-fault currents, so protecting the secondary circuits against high voltages.

## 11.5 Protection Systems

The application of the various relays and other equipment to form adequate schemes of protection forms a large and complex subject. Also, the various schemes are largely dependent on the methods of individual manufacturers. The main intention here is to present a survey of general practice and to outline the principles of the methods used. Some schemes are discriminative to fault location and involve several parameters, e.g. time, direction, current, distance, current balance, and phase comparison. Others discriminate according to the type of fault, e.g. negative-sequence relays, and some use a combination of location and type of fault.

A convenient classification is the division of the systems into *unit* and *non-unit* types. Unit protection signifies that an item of equipment or zone is being uniquely protected independently of the adjoining parts of the system. Non-unit schemes are those in which several relays and associated equipment are

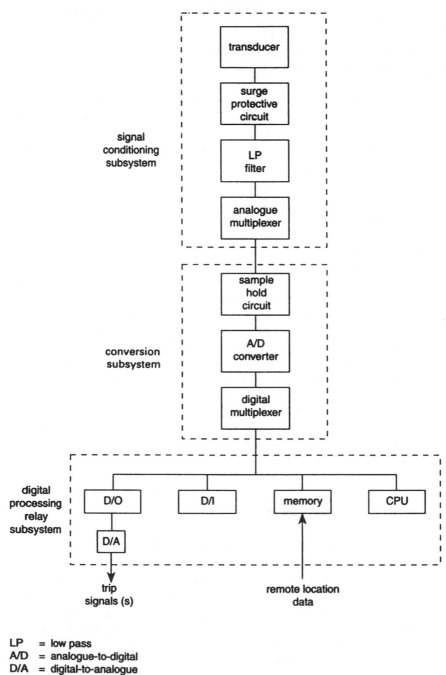

LP  = low pass
A/D = analogue-to-digital
D/A = digital-to-analogue
CPU = central processor unit
D/I = data input
D/O = data output

**Figure 11.20** Basic components of a digital relay (*Permission of Peter Peregrinus, I.E.E.*)

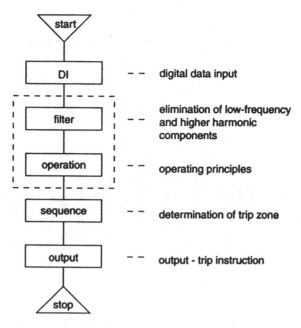

**Figure 11.21** Flow chart for the software of a digital protective relay (*Permission of Peter Peregrinus, I.E.E.*)

used to provide protection covering more than one zone. Examples of both types are classified in Figure 11.24. Non-unit schemes represent the most widely used and cheapest forms of protection and these will be discussed first.

### 11.5.1 Overcurrent protection

This basic method is widely used in distribution networks and as a back-up in transmission systems. It is applied to generators, transformers, and feeders. The arrangement of the components is shown in Figure 11.25. The relay normally employed is the induction-disc type with two electromagnets, as shown in Figure 11.14, but numerical overcurrent relays are becoming more frequently used.

The application to feeders is illustrated in Figure 11.26. Along the radial feeder the relaying points and circuit breakers are shown. The operating times are graded to ensure that only that portion of the feeder remote from the infeed side of a fault is disconnected. When determining selectivity, allowance must be made for the operating time of the circuit breakers. Assume Figure 11.26 to be a distribution network with slow-acting breakers operating in 0.3 s and that the relays have true inverse-law characteristics. Selectivity is obtained with a through-fault of 200 per cent full load, with the fault between D and E as

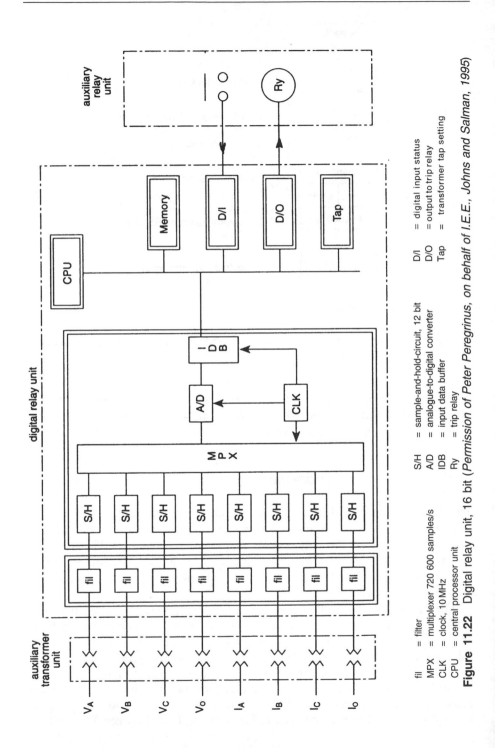

fil = filter
MPX = multiplexer 720 600 samples/s
CLK = clock, 10 MHz
CPU = central processor unit

S/H = sample-and-hold-circuit, 12 bit
A/D = analogue-to-digital converter
IDB = input data buffer
Ry = trip relay

D/I = digital input status
D/O = output to trip relay
Tap = transformer tap setting

**Figure 11.22** Digital relay unit, 16 bit (*Permission of Peter Peregrinus, on behalf of I.E.E., Johns and Salman, 1995*)

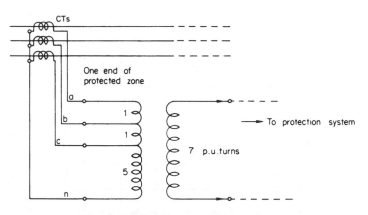

**Figure 11.23** Summation transformer

illustrated because the time difference between relay operations is greater than 0.3 s. Relay D operates in 0.5 s and its circuit breaker trips in 0.8 s. The fault current ceases to flow (normal-load current is ignored for simplicity) and the remaining relays do not close their contacts. Consider, however, the situation when the fault current is 800 per cent of full load. The relay operating times are now: A, 0.5 s [i.e. $2 \times (200/800)$]; B, 0.375 s; C, 0.25 s; D, 0.125 s; and the time for the breaker at D to open is $0.125 + 0.3 = 0.425$ s. By this time, relays B and C will have operated and selectivity is not obtained. This illustrates the fundamental drawback of this system, i.e. that for correct discrimination to be obtained the times of operation close to the supply point become large.

### 11.5.2 *Overcurrent and directional*

To obtain discrimination in a loop or networked system, relays with an added directional property are required. For the system shown in Figure 11.27, directional and non-directional overcurrent relays have time lags for a given fault current as shown. Current feeds into fault at the location indicated from both directions, and the first relay to operate is at B (0.6 s). The fault is now fed along route ACB only, and next the relay at C (1 s) operates and completely isolates the fault from the system. Assuming a circuit-breaker clearance time of 0.3 s, complete selectivity is obtained at any fault position. Note, however, that directional relays require a voltage input, non-directional ones do not.

## 11.6 Distance Protection

The shortcomings of graded overcurrent relays have led to the widespread use of distance protection. The distance between any point in the feeder and the

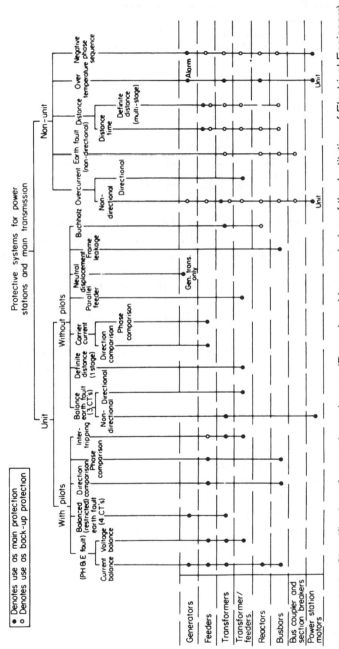

**Figure 11.24**  Classification of protection schemes (Reproduced by permission of the *Institution of Electrical Engineers*)

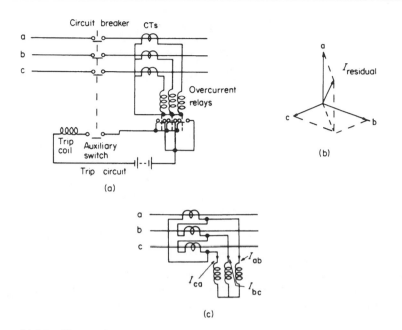

Figure 11.25   Circuit diagram of simple overcurrent protection scheme. (a) CTs in star connection. (b) Phasor diagram of relay currents, star connection. (c) CTs in delta connection

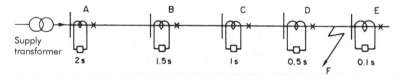

Figure 11.26   Application of overcurrent relays to feeder protection

fault is proportional to the ratio (voltage/current) at that point and relays responsive to impedance, admittance (mho), or reactance may be used. Although a variety of time–distance characteristics are available for providing correct selectivity, the most popular one is the stepped characteristic shown in Figure 11.28. Here, A, B, C, and D are distance relays with directional properties and A and C only measure distance when the fault current flows in the indicated direction. Relay A trips its associated breaker if a fault occurs within the first 80 per cent of the length of feeder 1. For faults in the remaining 20 per cent of feeder 1 and the initial 30 per cent of feeder 2 (called the stage 2 zone), relay A initiates tripping after a short time delay. A further delay in relay A is introduced for faults further along feeder 2 (stage 3 zone). Relays B and D have similar characteristics when the fault current flows in the opposite direction.

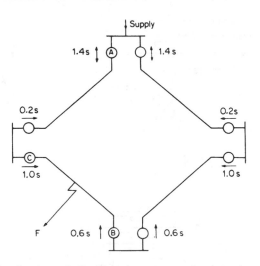

**Figure 11.27** Application of directional overcurrent relays to a loop network. ↔ Relay responsive to current flow in both directions; →, relay responsive to current flow in direction of arrow

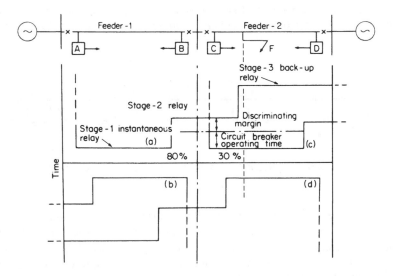

**Figure 11.28** Characteristic of three-stage distance protection

The selective properties of this scheme can be understood by considering a fault such as at F in feeder 2, when fault current flows from A to the fault. For this fault, relay A starts to operate, but before the tripping circuit can be completed, relay C trips its circuit breaker and the fault is cleared. Relay A then resets and feeder 1 remains in service. The margin of selectivity provided is indicated by the vertical intercept between the two characteristics for relays A and C at the position F, less the circuit-breaker operating time.

It will be noted that the stage 1 zones are arranged to extend over only 80 per cent of a feeder from each end. The main reason for this is because practical distance relays and their associated equipment have errors, and a margin of safety has to be allowed if incorrect tripping for faults which occur just inside the next feeder is to be avoided. Similarly, the stage 2 zone is extended well into the next feeder to ensure definite protection for that part of the feeder not covered by stage 1. The object of the stage 3 zone is to provide general back-up protection for the rest of the adjacent feeders.

The characteristics shown in Figure 11.28 require three basic features: namely, response to direction, response to impedance, and timing. These features need not necessarily be provided by three separate relay elements, but they are fundamental to all distance protective systems. As far as the directional and measuring relays are concerned, the number required in any scheme is governed by the consideration that three-phase, phase-to-phase, phase-to-earth, and two-phase-to-earth faults must be catered for. For the relays to measure the same distance for all types of faults, the applied voltages and currents must be different. It is common practice, therefore, to provide two separate sets of relays, one set for phase faults and the other for earth faults, and either of these caters for three-phase faults and double-earth faults. Each set of relays is, in practice, usually further divided into three, since phase faults may concern any pair of phases, and, similarly, any phase can be faulted to earth. With digital relays, a pre-selection of relaying quantities allows just one processor to deal with all types of fault.

# 11.7   Unit Protection Schemes

With the ever-increasing complexity of modern power systems the methods of protection so far described may not be adequate to afford proper discrimination, especially when the fault current flows in parallel paths. In unit schemes, protection is limited to one distinct part or element of the system that is disconnected if any internal fault occurs. On the other hand, the protected part should remain connected with the passage of current flowing into an external fault.

## 11.7.1   Differential relaying

At the extremities of the zone to be protected the currents are continuously compared and balanced by suitable relays. Provided that the currents are equal in magnitude and phase, no relay operation will occur. If, however, an internal fault (inside the protected zone) occurs, this balance will be disturbed (see Figure 11.29) and the relay will operate. The current transformers at the ends of each phase should have identical characteristics to ensure perfect balance on through-faults. Unfortunately, this is difficult to achieve and a restraint, or bias, is applied (see Figure 11.30) that carries a current proportional to the full system current and restrains the relay operation on large through-fault currents. The corresponding characteristic is shown in Figure 11.30. This principle (circulating current) may be applied to generators, feeders, transformers, and busbars, and provides excellent selectivity. By suitable connections and current summation, teed or multi-ended circuits can be protected using the same principles.

## 11.8   Generator Protection

Large generators are invariably connected to their own step-up transformer and the protective scheme usually covers both items. A typical scheme is shown

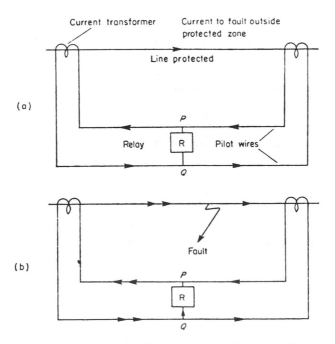

**Figure 11.29** Circulating current, differential protection (one phase only shown). (a) Current distribution with through-fault—no current in relay. (b) Fault on line, unequal currents from current transformers and current flows in relay coil. Relay contacts close and trip circuit breakers at each end of the line

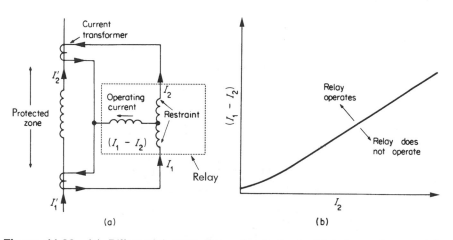

**Figure 11.30** (a) Differential protection—circuit connections (one phase only)—relay with bias. (b) Characteristic of bias relay in differential protection. Operating current plotted against circulating or restraint current

in Figure 11.31, in which separate differential circulating-current protections are used to cover the generator alone and the generator plus transformer. When differential protection is applied to a transformer the current transformer on each side of a winding must have ratios which give identical secondary currents. In many countries the generator neutral is often grounded through a distribution transformer. This energizes a relay which operates the generator main and field breakers when a ground fault occurs in the generator or transformer. The ground fault is usually limited to about 10 A by the distribution transformer or a resistor, although an inductor has some advantages. The field circuit of the generator must be opened when the differential protection operates in order to avoid the machine feeding the fault.

The relays of the differential protection on the stator windings (see Figure 11.32) are set to operate at about 10–15 per cent of the circulating current produced by full-load current in order to avoid current-transformer errors. If the phase e.m.f. generated by the winding is $E$, the minimum current for a ground fault at the star-point end, and hence with the whole winding in circuit, is $E/R$, where $R$ is the neutral effective resistance. For a fault at a fraction $x$ along the winding from the neutral (Figure 11.32(c)), the fault current is $xE/R$ and 10–15 per cent of the winding is unprotected. With the neutral grounded via the transformer, $R$ is high and earth faults are detected by a sensitive relay across the transformer secondary. With an interturn fault (turn-to-turn short circuit) on a phase of the stator winding (Figure 11.32(b)), current balance at the ends is retained and no operation of the differential relay takes place. the relays operate only with phase-to-phase and ground faults.

On unbalanced loads or faults the negative-sequence currents in the generator produce excessive heating on the rotor surface and generally ($I_2^2 t$)

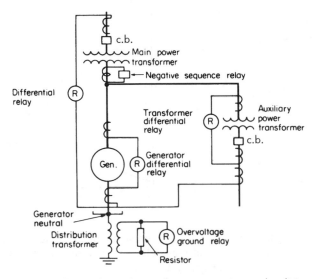

**Figure 11.31** Protection scheme for a generator and unit transformer

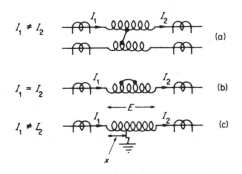

**Figure 11.32** Generator winding faults and differential protection. (a) Phase-to-phase fault. (b) Interturn fault. (c) Phase-to-earth fault

must be limited to a certain value for a given machine (between 3 and 4 $(p.u.)^2$ s for 500 MW machines), where $t$ is the duration of the fault in seconds. To ensure this happens, a relay is installed which detects negative-sequence current and trips the generator main breakers when a set threshold is exceeded. When loss of excitation occurs, reactive power ($Q$) flows into the machine, and if the system is able to supply this the machine will operate as an induction generator, still supplying power to the network. The generator output will oscillate slightly as it attempts to lock into synchronism. Relays are connected to isolate the machine when a loss of field occurs, which can be readily detected by a reactance relay.

# 11.9 Transformer Protection

A typical protection scheme is shown in Figure 11.33(a), in which the differential circulating-current arrangement is used. The specification and arrangement of the current transformers is complicated by the main transformer connections and ratio. Current-magnitude differences are corrected by adjusting the turns ratio of the current transformers to account for the voltage ratio at the transformer terminals. In a differential scheme the phase of the secondary currents in the pilot wires must also be accounted for with star–delta transformers. In Figure 11.33(a) the primary-side current transformers are connected in delta and the secondary ones in star. The corresponding currents are shown in Figure 11.33(b) and it is seen that the final (pilot) currents entering the connections between the current transformers are in phase for balanced-load conditions and hence there is no relay operation. The delta current–transformer connection on the main transformer star-winding also ensures stability with through earth-fault conditions which would not be obtained with both sets of current transformers in the star connection. The distribution of currents in a Y–Δ transformer is shown in Figure 11.34.

Troubles may arise because of the magnetizing current inrush when energization operates the relays, and often restraints sensitive to third-harmonic components of the current are incorporated in the relays. As the inrush current has a relatively high third-harmonic content the relay is restrained from operating.

Faults occurring inside the transformer tank due to various causes give rise to the generation of gas from the insulating oil or liquid. This may be used as a means of fault detection by the installation of a gas/oil-operated relay in the pipe between the tank and conservator. The relay normally comprises hinged floats and is known as the *Buchholz* relay (see Figure 11.35). With a small fault, bubbles rising to flow into the conservator are trapped in the relay chamber, disturbing the float which closes contacts and operates an alarm. On the other hand, a serious fault causes a violent movement of oil which moves the floats, making other contacts which trip the main circuit breakers.

# 11.10 Feeder Protection

### 11.10.1 Differential pilot wire

The differential system already described can be applied to feeder protection. The current transformers situated at the ends of the feeder are connected by insulated wires known as pilot wires. In Figure 11.29, *P* and *Q* must be at the electrical midpoints of the pilot wires, and often resistors are added to obtain a geographically convenient midpoint. By reversing the current-transformer

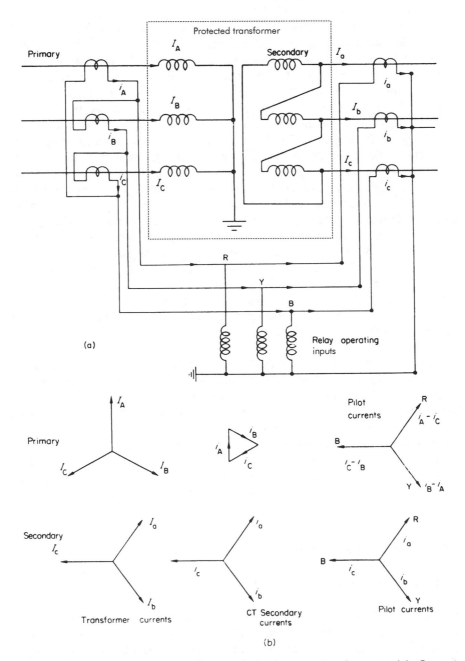

**Figure 11.33** Differential protection applied to Y–Δ transformers. (a) Current-transformer connections. (b) Phasor diagrams of currents in current transformers

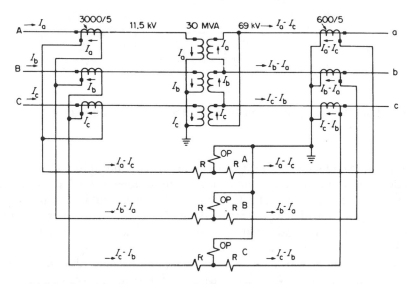

**Figure 11.34**  Currents in Y–Δ transformer differential protection. OP = operate input; R = restraint input

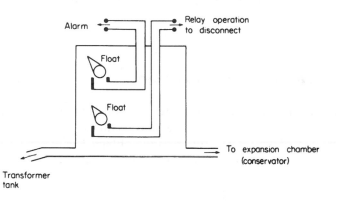

**Figure 11.35**  Schematic diagram of Buchholz relay arrangement

connections (Figure 11.36) the current-transformer e.m.f.s oppose and no current flows in the wires on normal or through-fault conditions. This is known as the *opposed voltage* method. As, under these conditions, there are no back ampere-turns in the current-transformer secondaries, on heavy through-faults the flux is high and saturation occurs. Also, the voltages across the pilot wires may be high under this condition and unbalance may occur due to capacitance currents between the pilot wires. To avoid this, sheathed pilots are used.

Pilot wires may be installed underground or strung on towers. In the latter method, care must be taken to cater for the induced voltages from

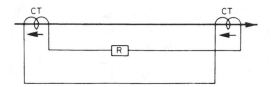

**Figure 11.36**    Pilot-wire differential feeder protection—opposed voltage connections

the power-line conductors. Sometimes it is more economical to rent wires from the telephone companies, although special precautions to limit pilot voltages are then required. A typical scheme using circulating current is shown in Figure 11.37, in which a mixing or summation device is used. With an internal fault at $F_2$ the current entering end A will be in phase with the current entering end B, as in H.V. networks the feeder will inevitably be part of a loop network and an internal fault will be fed from both ends. $V_A$ and $V_B$ become additive, causing a circulating current to flow, which causes relay operation. Thus, this scheme could be looked on as a phase-comparison method. If the pilot wires become short-circuited, current will flow and the relays can give a false trip. In view of this, the state of the wires is constantly monitored by the passage of a small d.c. current.

### 11.10.2   Carrier-current protection

Because of pilot capacitance, pilot wire relaying is limited to line lengths below 40 km (30 miles). Above this, distance protection may be used, although for discrimination of the same order as that obtained with pilot wires, carrier-current equipment may be used. In carrier-current schemes a high-frequency

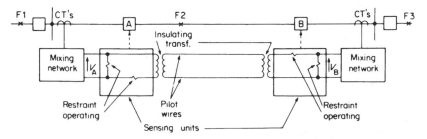

**Figure 11.37**   Differential pilot-wire practical scheme using mixing network (or summation transformer) and biased relays. $\mathbf{V}_A = \mathbf{V}_B$ for external faults, e.g. at F1 and F3; $V_A \neq V_B$ for internal faults, e.g. at F2

signal in the band 80–500 kHz and of low power level (1 or 2 W) is transmitted via the power-line conductors from each end of the line to the other. It is not convenient to superimpose signals proportioned to the magnitude of the line primary current, and usually the phases of the currents entering and leaving the protected zone are compared. Alternatively, directional and distance relays are used to start the transmission of a carrier signal to prevent the tripping of circuit breakers at the line ends on through faults or external faults. On internal faults other directional and distance relays stop the transmission of the carrier signal, the protection operates, and the breakers trip.

A further application, known as transferred tripping, uses the carrier signal to transmit tripping commands from one end of the line to the other. The tripping command signal may take account of, say, the operation or non-operation of a relay at the other end (permissive intertripping) or the signal may give a direct positive instruction to trip alone (intertripping).

Carrier-current equipment is complex and expensive. The high-frequency signal is injected on to the power line by coupling capacitors and may be coupled either to one phase conductor (phase-to-earth) or between two conductors (phase-to-phase), the latter being technically better but more expensive. A schematic diagram of a phase-comparison carrier-current system is shown in Figure 11.38. The wave or line trap is tuned to the carrier frequency and presents a high impedance to it, but a low impedance to power-frequency currents; it thus confines the carrier to the protected line. Information regarding the phase angles of the currents entering and leaving the line is transmitted from the ends by modulation of the carrier by the power current, i.e. by blocks of carrier signal corresponding to half-cycles of power current (Figure 11.39). With through faults or external faults the currents at the line ends are equal in

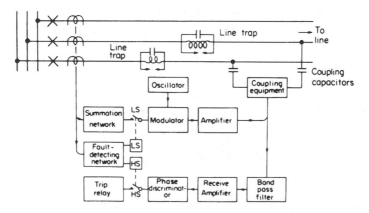

**Figure 11.38** Block diagram measuring and control equipment for carrier-current phase-comparison scheme. LS = low-set relay; HS = high-set relay

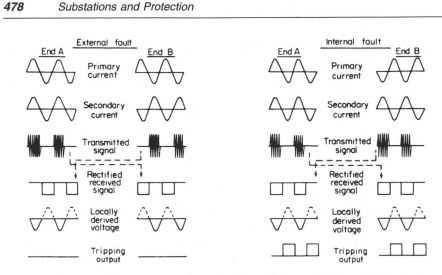

**Figure 11.39**  Waveforms of transmitted signals in carrier-type line protection

magnitude but 180° phase-displaced (i.e. relative to the busbars, it leaves one bus and enters the other). The blocks of carrier occur on alternate half-cycles of power current and hence add together to form a continuous signal, which is the condition for no relay operation. With internal faults the blocks occur in the same half-cycles and the signal comprises non-continuous blocks; this is processed to cause relay operation (Figure 11.39).

The currents from the current transformers are fed into a summation device which produces a single-phase output that is fed into a modulator (Figure 11.38). This combines the power frequency with the carrier to form a chopped 100 per cent modulated carrier signal, which is then amplified and passed to the line-coupling capacitors. The carrier signal is received via the coupling equipment, passed through a narrow-bandpass filter to remove any other carrier signals, amplified, and then fed to the phase discriminator which determines the relative phase between the local and remote signals and operates relays accordingly. The equipment is controlled by low-set and high-set relays that start the transmission of the carrier only when a relevant fault occurs. These relays are controlled from a starting network. Although expensive, this form of protection is very popular on overhead transmission lines.

### 11.10.3  Voice-frequency signalling

Increasingly, as the telephone network expands with the use of fibre-optic cabling and high-frequency multiplexing of communication channels, it is no longer possible to obtain a continuous metallic connection between the ends of unit-protection schemes. Consequently, differential protection, utilizing voice frequencies (600–4000 Hz), has been developed. In Figure 11.37 the pilot wires (after the insulating transformers) feed into a voltage-to-frequency (v.f.) converter with send and receive channels. The summated or derived 50 (or 60) Hz relaying signal is frequency-modulated onto the channel carrier and demodulated at the far end. The signal is then compared with the local signal and if a discrepancy is detected then relay operation occurs. It is important to ensure that both signal magnitude and phase are faithfully reproduced at each end after demodulation, independently of the channel characteristics which can change or distort during adverse transmission conditions. The usual methods built into the v.f. channel relays to ensure reliability utilize an automatic control of signal level and a regular measurement (say, every 100 ms) of channel delay so that phase correction can be applied to the demodulated signal.

# 11.11   System Monitoring and Control

## 11.11.1   Introduction

For a power system to be able to supply all its customers within normal voltage and frequency limits, it must be able to ride through unavoidable disturbances, some of which could be quite abnormal. Examples of abnormal, but nevertheless credible, disturbances are:

- shunt faults and consequent line outages;
- equipment failure with subsequent isolation, e.g. generators, transformers, busbars;
- switching surges and lightning strikes;
- mechanical damage, e.g. double-circuit line tower failure.

Some of these disturbances can be dealt with by protective devices, as discussed in previous sections, and the system restored to  normal within a few cycles. In these cases no further control action is needed. Others may cause transient oscillations which could last for several seconds, producing large oscillations in power flow, abnormal voltages and frequency, and subsequent tripping of plant items. If tripping occurs, then corrective control actions are required. For this purpose, an Energy Management System (EMS) is vital for any power system.

## 11.11.2   Energy Management System

An EMS enables engineers to operate and control the network in real time and includes facilities to capture the current state of the system and to instruct generating plant and other controllable system components such that all consumers are supplied, at least-cost, with security. Considerable back-up facilities are necessary, including special software programs, displays, and support staff. The hierarchy of a power system with EMS is shown in Figure 11.40. At level 0 is the power system with its isolators, switchgear, interconnections, transmission lines, cables, transformers, etc.; substation (local) controls are situated at level 1. These may include protection relays, tap-change controllers, and compensator controls, with operating channels to the level 0 units. Level 1 controls often comprise digital/electronic devices for voltage and current measurement, interlocking and facilities for receiving and sending data to the next level up (area concentrators). In many cases, level 1 consists of racks of electronics within cabinets called remote terminal units (RTUs).

At level 2 (area), man–machine interfacing and data concentrators enable control and maintenance to be exercised so that the whole system can be kept in reliable and efficient condition. At the top level (3), the Supervisory Control and Data Acquisition (SCADA) system resides, usually in a single control centre (variously called Pool, National, or System control). The SCADA system accepts data from the various level 1 collectors and displays it in a meaningful way to the control engineers or operators, usually by a one-line mimic diagram on a colour video screen. With powerful computer processing, the SCADA feeds into an Alarm Management subsystem to supplement automatic relay operation and to give warning about any system abnormality that is not able to be detected locally at levels 1 or 2.

The EMS processes SCADA data in various ways, including topology identification by using the dynamic data from switchgear, isolators, and

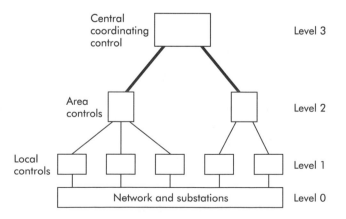

**Figure 11.40** Hierarchy of controls required for an Energy Management System (EMS)

other system connectors. This 'switch' data can also be combined, in both logical and estimated fashion, with current and voltage measurements to determine the system 'state'. Such a procedure, using mathematical methods, is called 'state estimation'. In a typical system, flows and connection status (i.e. circuit breaker and isolator position) data may involve many thousands of items on which surveillance is required. Bearing in mind that this data may change minute by minute, it is obvious that the most efficient way of receiving and recording it in the control centre is by on-line digital processors.

If this data is displayed to the operator, e.g. on a video display, then, because the operator has knowledge via other sources, such as the telephone, and he has experience of similar situations arising before, he can correlate items together. If a few data items are wrong or missing, no great harm results. On the other hand, if the incoming data is to be used automatically for further computer analysis, such as contingency checking, economic scheduling, and automatic frequency control (as required in modern computer-aided control systems), then the telemetered system data must be checked for consistency. Missing or erroneous items must be found *before* they will be worth using for further processing. If the data is checked by simple and, perhaps, ad hoc methods, such as limit checking and redundant measurement comparison (e.g. two busbar voltages should be of similar value if the busbars are connected together), then the process is called *data validation*. It can be performed within a single set of measurements from a substation at level 1 before being telemetered to the control centre, or it can be carried out at the control centre before being stored in the data base.

### 11.11.3  The basis of Power-System State Estimation (PSSE)

State estimation is used in modern control theory to enable processes to be controlled, mainly by an on-line digital computer. In a power system, the control engineer is primarily concerned with operation of the network in the steady state. Hence the process of 'static-state estimation' is normally applied. As computer methods and hardware continue to evolve (particularly powerful microprocessors), it may be possible to control the system in the dynamic or transient state, for which 'dynamic state estimation' could be employed.

Power-System State Estimation is defined as follows:

> PSSE is a process whereby data, telemetered from network measuring points to a central computer, can be formed into a set of reliable data (the 'data base') for control and recording purposes.

A *static-state estimator* is obtained from measurements taken within a short time interval (e.g. 5–10 s) and these values are the only ones used for the estimate (this is sometimes called the 'snapshot' estimator, for obvious reasons).

More measurements are actually taken than are required to identify the system at any given moment, and redundant measurements are available to 'observe' the system. In Chapter 6, load flows were calculated from given data which was just enough for a unique solution to be obtained. Common sense tells us that if we were to take a snapshot of our power system, a unique set of flows and voltages would be measured. And yet common practice is to measure, for example, power and reactive-power flows at both ends of an overhead line for which we know the impedance. Consequently, we strictly only require flow measurements at the *near* end of each line—we can then calculate the expected flows at the far end and compare them with the measured values.

Averaging of the two values provides an 'estimate' of the flows which, if our method is any good, can give us a better result than either of the original measurements. This, then, is our estimation process; namely, an attempt to minimize the errors in our measurements by statistical calculation. Note that we must assume that measurement errors are randomly distributed among our system measurements and that, if we take enough measurements, we can minimize the error through the estimator.

Once we have estimated all our measurements for one system 'snapshot', then the measurements can be stored in a data base and used for further calculation or even for updating the display to the operator. Control action required as a result of this latest information can then be made on the power system, thus completing the loop, as depicted in Figure 11.41.

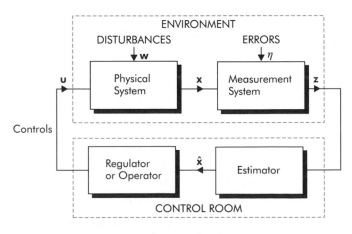

**Figure 11.41**  Outline of estimation system

### 11.11.4  *Mathematical description of the PSSE process*

The practical application of PSSE requires four main subroutines, as indicated by the flow diagram of Figure 11.42.

In PSSE the system states are usually denoted by the vector **x**. It is important to define these states and to make sure that there are enough, but only *just* enough, to uniquely describe the system. Now, in the load-flow problem, it is usual to calculate the nodal voltages from the given data and to use them subsequently to calculate line flows, and generator and load injections. We know that, provided we have knowledge of the network configuration and its nodal admittances, the complex nodal voltages define the state of the system. We therefore let

$$\mathbf{x}^t \equiv [a_1, a_2, a_3 \ldots a_n, b_2, b_3 \ldots b_n]$$

or

$$\mathbf{x}^t \equiv [x_1, x_2, x_3 \ldots x_n]$$

where voltage at node $k$ is $V_k = a_k + jb_k$ and there are $n$ nodes in the network. Note that we have omitted $b_1$ from $\mathbf{x}^t$ because node 1 is taken as the reference node, where $b_1 = 0$. The vector **x** contains $(2n - 1) = N$ elements for which, in a load-flow solution, we would require $(2n - 1)$ equations to determine. Of course, **x** can be expressed in polar form, if desired, in which case it will contain the same number of elements but there would be $n$ voltage magnitudes and $(n - 1)$ angles.

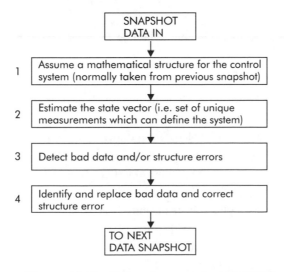

**Figure 11.42**  Subroutines required for PSSE

In PSSE the input-data is not necessarily in the form of busbar voltages because these are not always easy to measure with accuracy. In any case, measurement of phase angle between substations many miles apart is quite expensive. It is usually much cheaper and more direct to measure $P$ and $Q$ flows on many three-phase circuits, generator outputs, transformer feeds, etc., and to telemeter these to the control centre. Consequently, measurements could consists of any combination of $P$ and $Q$ flows or injections, busbar voltages, or perhaps even current flows where they are measured in preference to $P$ or $Q$, e.g. on cables or in distribution systems.

As well as a mixture of measurements, say $M$ in number, we know that, in practice, each measurement contains an error $\eta$, which we hope will be randomly distributed around a mean value. Consequently, we can state that our $M$ measurements (represented by $\mathbf{z}$) must be a non-linear function $h(\mathbf{x})$ of the state variable $\mathbf{x}$ with a 'noise' vector $\eta$ so that

$$\mathbf{z} = h(\mathbf{x}) + \eta \qquad (11.1)$$

This is our fundamental PSSE equation. It contains the load-flow equations developed in Chapter 6, using equations (6.9) and (6.10), from which flows $P_{kj}$ and $Q_{kj}$ from node $k$ to node $j$ can be derived.

Table 11.1 gives five ways of making up the vector $\mathbf{z}$ from sets of measurements. Case 1 corresponds to the load-flow calculation whilst case 5 assumes that all possible measurements are taken so that the redundancy of the measurements is a maximum. In practice, the redundancy will be somewhere between these two extremes, defined as

$$\text{Redundancy} = \frac{\text{number of measurements}}{(2n - 1)}$$

**Table 11.1**  Five basic measurement schemes for a system with $n$ buses and $m$ lines

| Case | Components of $\mathbf{z}$ | Dimension $\mathbf{z}$ |
|---|---|---|
| 1 | $P_k$ and $Q_k$ injections at all nodes (except $P_1$ at slack bus) <br> $\mathbf{z} = [P_2 \ldots P_k, Q_1 \ldots Q_k]$ | $2n - 1$ |
| 2 | Same as case 1, with voltage magnitudes $V_k$ at all nodes <br> $\mathbf{z} = [\ldots P_k, Q_k \ldots P_n, Q_n, V_1 \ldots V_k]$ | $3n - 1$ |
| 3 | Active $P_{kj}$ and reactive $Q_{kj}$ flows at both ends of each line <br> $\mathbf{z} = [\ldots P_{kj}, Q_{kj}]$ | $4m$ |
| 4 | Same as case 3 plus voltage magnitude at all nodes <br> $\mathbf{z} = [\ldots P_{kj}, Q_{kj} \ldots V_1 \ldots V_n]$ | $4m + n$ |
| 5 | Full-measurement system <br> $\mathbf{z} = [\ldots P_k, Q_k \ldots P_{kj}, Q_{kj} \ldots V_1 \ldots V_n, \theta_2 \ldots \theta_n]$ | $4n - 1 + 4m$ |

For any system, the economic case for PSSE depends upon the cost of redundant measurements weighed against the savings obtained due to better system control. Since, in most cases, the economic justification for control is difficult to establish, considerable off-line study needs to be done before a PSSE scheme is installed.

### 11.11.5  Minimization technique for PSSE

The fundamental PSSE equation has already been derived:

$$\mathbf{z} = h(\mathbf{x}) + \eta \tag{11.1}$$

and since this contains an error vector $\eta$, it cannot be 'solved' to give a unique solution as in the csse of the load-flow equations of Chapter 6. However, if it is possible to reduce the errors to a minimum, a good value of the state vector $\mathbf{x}$ can be determined. However, these will only be 'best' or 'optimal' estimates of $\mathbf{x}$ (denoted by $\hat{\mathbf{x}}$) and will never be the 'true' values, although we would hope that they could approach the true values if we could take enough measurements. In practice, the control engineer is happy if he/she feels that he/she knows their flows and voltages to between 1–2 per cent accuracy, but it must be remembered that if we are determining economic operating conditions, an error of 1 per cent in voltage could make a large difference in the calculation of losses on the system. Consequently, it is worthwhile spending some time and effort on obtaining the best error minimization for on-line application in a particular system. This method could well depend upon the system characteristics (closely meshed or long lines), as well as the type of measurements taken (injections, line flows, or a combination of both).

### 11.11.6  Least-squares estimation

The optimal estimate $\hat{\mathbf{x}}$ is given by that value of $\mathbf{x}$ for which the scalar sum $J$ of the squares of the errors has a minimum. This is given by

$$\text{min. } J = [\mathbf{z} - h(\mathbf{x})]^T R^{-1} [\mathbf{z} - h(\mathbf{x})] \tag{11.2}$$

where $R^{-1}$ is a factor or 'weight' which is inserted to obtain a better minimization depending upon the nature of the 'noise' or 'error' vector $\eta$.

To minimize $J$, we need to take its derivative and set it equal to zero, thus

$$\frac{\partial J}{\partial x} = -2\left[\frac{\partial h(\mathbf{x})}{\partial \mathbf{x}}\right]^T R^{-1}[\mathbf{z} - h(\mathbf{x})] = 0 \tag{11.3}$$

where

$$\left[\frac{\partial h(\mathbf{x})}{\partial \mathbf{x}}\right]$$

is the Jacobian matrix, as used in the Newton–Raphson load flow, consisting of elements as follows:

$$\left[\frac{\partial h(\mathbf{x})}{\partial \mathbf{x}}\right] = H(\mathbf{x})\Delta \begin{bmatrix} \dfrac{\partial h_1(x)}{\partial x_1} & \cdots & \dfrac{\partial h_1(x)}{\partial x_N} \\ \vdots & & \vdots \\ \dfrac{\partial h_M(x)}{\partial x_M} & & \dfrac{\partial h_m(x)}{\partial x_N} \end{bmatrix} \tag{11.4}$$

and is of dimensions $M$ (no. of measurements) $\times N$ ($= 2n - 1$).

Now we have to solve equation (11.3) by any means at our disposal in order to obtain $\mathbf{x}$ (our state variable defining the system), which will actually be $\hat{\mathbf{x}}$, i.e. our best estimate of the system states.

Many methods are available to solve equation (11.3) and these are reported in the literature, but it is worthwhile to note the following points.

1.  The dimensions of $\mathbf{z}$ and $\mathbf{x}$ can be quite large (e.g. 200 nodes, 500 measurements).

2.  The Jacobian $H(\mathbf{x})$ is sparse.

3.  Elements of $H(\mathbf{x})$ and $h(\mathbf{x})$ have similar properties so that algorithms written generally to obtain $h(\mathbf{x})$ can be used to obtain $H(\mathbf{x})$.

4.  Sparsity programming and optimal ordering can be used effectively to minimize storage and computation time. Yet again, it pays to study the system before implementing PSSE.

An efficient solution is as follows:

Use the iteration process of

$$\mathbf{G}^p[\mathbf{x}^{p+1} - \mathbf{x}^p = H(\mathbf{x})^p\ R^{-1}[z - h(\mathbf{x})^p] \tag{11.5}$$

where superscript $p$ denotes the iteration number and $\mathbf{G}$ is a 'gain' matrix which must be chosen, usually by trial and error, as in the case of an acceleration factor for a Gauss–Seidel load-flow solution. The unit matrix of full rank can be used if no experience of the measurement system is available beforehand.

An initial guess $x^0$ is required to start the process and, at convergence, $\mathbf{x}^{p+1} = \mathbf{x}^p = \hat{\mathbf{x}}$, although as usual, convergence is assumed if $|x^{p+1} - x^p| <$ tolerance. As noted in Chapter 6, since equation (11.5) is very sparse, sparsity techniques can be used to speed up solution time and reduce storage. Figure 11.43 gives a flow diagram for this computation.

It should be noted that simplifications of the algorithm for $\hat{\mathbf{x}}$ calculation can often be obtained if the 'decoupling' between $P$–$\theta$ and $Q$–$V$ is utilized. In some closely meshed networks, as in the NGC system for example, it may only be

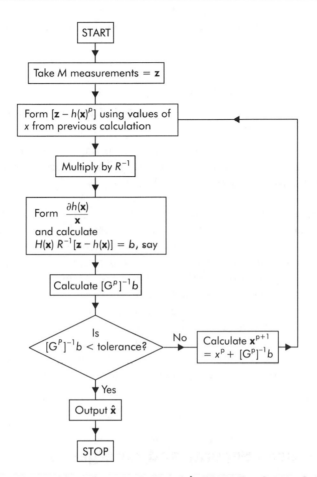

**Figure 11.43**  Flow diagram for calculation of $\hat{x}$. (Note: For Gauss–Seidel calculations, an inner loop, changing one element of **x** each time, is required)

necessary to estimate $P$ flows from a $P$–$\theta$ calculation using only $P$ measurements plus one or two voltages. This considerably eases the on-line computation problem, as well as reducing the amount of telemetered data required.

## 11.11.7  Errors and detection in PSSE

The criterion for a good estimator is not only that it is suitable for on-line calculation, but also that it can detect data errors, identify the cause, and apply a correction to the estimation process automatically. A full appreciation of the possibilities requires a knowledge of probability theory, but an indication of what can be done with practical estimators will be given in this section.

An estimate $\hat{x}$ of the state variables may be in error due to any or all of the following causes:

1.  metering errors, e.g. non-zero time constant of some instruments, non-linearities, noise on telemetering channels, etc.;

2.  bias error in the measurement—this should be constant and it could be accounted for in the data if it is known;

3.  bad data, e.g. observations during transients, major failures of meter/communication, errors in assumed system parameters $G$, $B$, etc., network structure errors unknown to the data processor.

Consequently, to *detect* errors, a statistical test is required during the process of determining the estimate of the system states. In practice, it is convenient to have a single 'residual' value which can be compared with a pre-set threshold value after each estimate $\hat{x}$ is completed. If this threshold is exceeded then a process of *identifying* which measurement or set of measurements is causing the error is started. Once the cause is identified, the estimator has to be *corrected*, usually by removing the affected measurement, after which the estimator can proceed as before, but now with a little less redundancy. Obviously, the control engineer must be informed of the removal of some measurements so that the cause of the discrepancy can be investigated and put right before that measurement is reinserted, by manual action, into the estimator.

# 11.12   System Security and Emergency Control

It has already been stated that the reason for designing and operating a system in a meshed form is to provide a path from every generator to every load, despite the possibility that one or two circuits could be outaged. A network configuration and loading state which enables any one circuit to be outaged without loss of supply to any load is called $n - 1$ secure. To determine the secure network configuration and state (i.e. loading) required for each hour of the day is a daunting task, even with large and powerful computers. Normally, secure network states are calculated for a few representative loading conditions (usually up to 6 per day), including the daily peak, night minimum, and intermediate subpeak conditions.

Nowadays, optimal power flows with many constraints are employed to determine the secure network configuration and economic loading, including any constrained-on or constrained-off plant. Security calculations are made at least 24 h ahead by many a.c. or d.c. load flows, taking out critical circuits one-by-one. Adjustments to the generator commitment and loading are made by

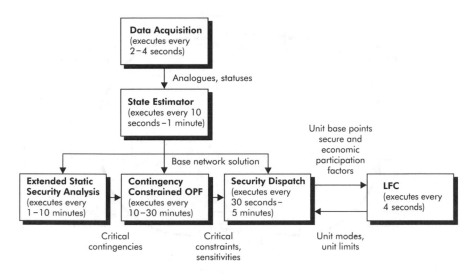

**Figure 11.44**  Optimal power flow (OPF) in secure optimal automatic generation control. LFC, load frequency control. (*Permission of Siemens Power System Controls, U.S.A.*)

feedback rules or by a 'contingency-constrained optimal power flow (OPF), which outputs a generator schedule, as illustrated in Figure 11.44.

Other constraints within which the operator must work are the requirements (1) to hold enough VAr reserves at strategic nodes in the system such that voltages can be maintained despite loss of a circuit, compensator, or generator; and (2) that across pre-defined boundaries, flows should not exceed certain values, otherwise a credible fault could plunge the system into instability. Both voltage-secure and stability-secure situations require considerable pre-event study and the use of sophisticated planning and operating software utilizing data from SCADA. Many programs to automatically derive secure dispatch and network configuration schedules are still under development and this provides a fruitful area for research.

### 11.12.1  *Emergency control* (Figure 11.45)

The following system-operating states and transition between states may be identified.

1.  *Normal* to *alert*—reductions of security level. This could be caused by unexpected load increases, loss of generating units, derating of plant due to environmental constraints, and rescheduled maintenance.

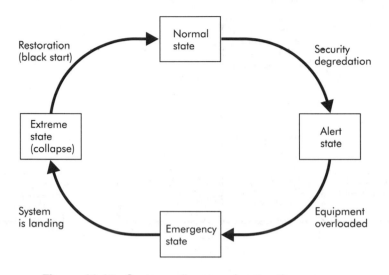

**Figure 11.45** System collapse and restoration process

2. *Alert* to *emergency*—inability of parts of system to meet requirements, e.g. lines (emergency ratings), voltage levels, frequency, machine and bus voltage angles. Caused by malfunction of protection, lightning, etc.

3. *Emergency* to *extreme condition* (collapse)—loss of integrity. Caused by loss of ties resulting in isolated generation islands which are unable to carry their internal loads. This is triggered by prolonged overloading of critical ties, malfunction of protection, and successive disturbances.

Corresponding control measures to meet the above situations are as follows.

1. *Alert*—This involves restoration of reserve margins, increased generation reserves, rescheduling of tie-lines, and voltage reduction.

2. *Emergency*—This involves fast valving on steam-turbines, dynamic braking of generators, load control, capacitor switching, and immediate action to clear equipment overloads.

3. *Extreme*—This involves all of the above plus load shedding and controlled operation of isolated power groups.

Following these emergencies action is taken to re-establish a viable system. This involves restarting and synchronization of generation units, load restoration, and resynchronization of all areas. Although action to prevent transient instability of individual generators has been a major factor in emergency control, such instabilities are not necessarily a major factor in system-extreme emergencies. Local immediate (reflexive) action may prevent damage to the equipment involved, but the resulting system security may be reduced to a dangerous level.

# Problems

**11.1**  A 132 kV supply feeds a line of reactance 13 Ω which is connected to a 100 MVA, 132/33 kV transformer of 0.1 p.u. reactance. The transformer feeds a 33 kV line of reactance 6 Ω which, in turn, is connected to an 80 MVA, 33/11 kV transformer of 0.1 p.u. reactance. This transformer supplies an 11 kV substation from which a local 11 kV feeder of 3 Ω reactance is supplied. This feeder energizes a protective overcurrent relay through 100/1 A current transformers. The relay has a true inverse-time characteristic and operates in 10 s with a coil current of 10 A.

If a three-phase fault oocurs at the load end of the 11 kV feeder, calculate the fault current and time of operation of the relay.
(Answer: 1575 A; 6.35 s)

**11.2**  A ring-main system consists of a number of substations designated A, B, C, D, and E, connected by transmission lines having the following impedances per phase (Ω): AB (1.5 + j2); BC (1.5 + j2); CD (1 + j1.5); DE (3 + j4); EA (1 + j1).

The system is fed at A at 33 kV from a source of negligible impedance. At each substation, except A, the circuit breakers are controlled by relays fed from 1500/5 A current transformers. At A, the current transformer ratio is 4000/5. The characteristics of the relays are as follows:

| Current (A): | | 7 | 9 | 11 | 15 | 20 |
|---|---|---|---|---|---|---|
| Operating | relays at A, D, and C: | 3.1 | 1.95 | 1.37 | 0.97 | 0.78 |
| time(s) | relays at B and E: | 4 | 2.55 | 1.8 | 1.27 | 1.02 |

Examine the sequence of operation of the protective gear for a three-phase symmetrical fault at the midpoint of line CD.

Assume that the primary current of the current transformer at A is the total fault current to the ring and that each circuit breaker opens 0.3 s after the closing of the trip-coil circuit. Comment on the disadvantages of this system.

**11.3**  The following currents were recorded under fault conditions in a three-phase system:

$$I_R = 1500 \angle 45° \text{ A}, \qquad I_Y = 2500 \angle 150° \text{ A}, \qquad I_B = 1000 \angle 300° \text{ A}.$$

If the phase sequence is R–Y–B, calculate the values of the positive, negative, and zero phase-sequence components for each line.
(Answer: $I_0 = (-200 + j480)$ A)

# 12

# *Basic Power-System Economics and Management*

## 12.1  Introduction

It is important that all power-system engineers understand the economic context within which the generation, transmission, distribution, and supply of electrical energy to customers operates. Figure 12.1 illustrates the  hierarchy that exists between the whole economy of a country (the *macroeconomy*) and the *electricity subsector*. Note that at the *intermediate level*, competition cannot exist between the different forms of energy unless some form of centralized energy-sector planning is in place. Economists and national planners (i.e. governments) attempt to model the interactions between sectors and subsectors so that development plans and financial investments can be undertaken. Each sector has its 'participants' who are usually experts in their own particular field.

Economic theory tries to coordinate the decisions of participants to produce the 'best' outcome. Unfortunately, 'best' in this context is almost impossible to define because participants in the energy game will have different and often conflicting objectives. This section therefore gives, first, a brief outline of what is known as welfare economics and how they can be applied to the electricity industry in order to provide maximum *social welfare* for the various consumers. It is assumed throughout that the well-being of the consumer is paramount and that pricing policies are directed towards producing maximum benefits in consuming electrical energy in efficient and effective ways. However, a good pricing policy should not only provide maximum benefit to consumers, it should also signal any desirable reallocation of resources in society. This

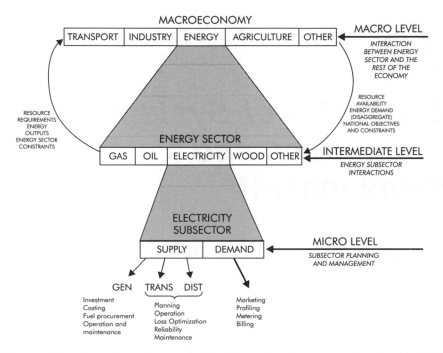

**Figure 12.1** Hierarchy of energy within the national economy showing electricity subsector aspects (*Modified from Monasinghe, Electric Power Economics, Butterworths, 1990. Acknowledgement to Dr Mohan Monasinghe, World Bank*)

latter objective of pricing is extremely difficult to achieve with simple and easy-to-understand tariffs.

As electrical energy cannot be conveniently stored, its production and use are simultaneous. Hence the well-known supply–demand equation must always apply. On the *supply side* are the methods of generation (thermal—coal, oil, gas; nuclear, hydro, wind, solar, etc.), transmission, and distribution of electrical energy; whilst on the *demand side* are the consumers with their large varieties of end-use equipment. Whereas the supply side consists generally of large companies or utilities, the demand side ranges from individual residential consumers to large industries whose existence in a competitive market place depends crucially on stable and economic prices for the energy consumed.

## 12.2 Basic Pricing Principles

Pricing methods (such as a high-day but cheap-night tariffs) can cause consumers, over a period of time, to change their consumption pattern so that their costs are minimized. However, presumably the supplier will still charge sufficient to cover the cost of production and overheads, so it would seem that one group of participants (the consumers) could well benefit without any change to the other group of participants (the suppliers). Such a change, in which one group benefits without affecting any others, is called a *Pareto improvement*, named after the nineteenth-century economist and philosopher, Vilfredo Pareto.

There is obviously a limit to the gains which can be made by the consumer without affecting the supplier, so economic efficiency is said to be maximized where no Pareto improvements are possible, i.e. no alternative pricing policy can make one group of participants better off and no other group worse off. Thus, it is tempting to argue that economic efficiency should enable everyone to benefit in an ideal world. Unfortunately, having everyone on an equal footing does not necessarily mean that everyone is getting good value for money, so a measure of total benefit (supplier plus consumer) is needed.

This produces the idea of measuring *social welfare* as the sum total of the monetary benefits accruing to every individual and group from a given tariff, including many intangible, but nevertheless useful, benefits, reckoned in monetary terms which both suppliers and consumers can experience. A good measure of useful benefit is the willingness to pay for it, leading to the idea of optional alternatives or *choice* for the supplier and/or the consumer.

The willingness to pay is often reflected, in the consumer, by the *sensitivity* to price changes known as *demand elasticity*. For electricity, where the energy consumed is related to income (above a certain minimum level), *income demand elasticity* is given by

$$\eta = \frac{(\Delta Q/Q)}{(\Delta m/m)} \tag{12.1}$$

where $(\Delta Q/Q)$ is the percentage change in consumption as a result of a given percentage change in income $(\Delta m/m)$.

It is useful to have a measure of demand elasticity in deciding upon pricing policies, but, unfortunately, it is extremely difficult to measure because of the many conflicting effects of any income change, e.g. purchase of substitute fuels, changing social climate. These *cross elasticities*, as they are called, also need to be accounted for in any tariff study if the outcome is to be valid. In Figure 12.1, the energy sector would need to be considered for its interactions via cross elasticities.

On the supply side, the generator of electrical energy presumably has a full knowledge of costs of production dependent upon percentage of available capacity that is actually producing electricity and the time schedule. Now, total revenue $(R)$ is the sum of *variable costs* $(VC)$, dependent upon output; *fixed costs* $(FC)$, which are independent of output; and *profit* $(P)$.

Thus

$$P = R - VC - FC \tag{12.2}$$

and if $P$ is to be maximized, then *marginal revenue* ($MR$), defined as increase in revenue per unit increase in quantity sold, should be equal to *marginal cost* ($MC$), the corresponding change in variable costs.

Figure 12.2 shows a *demand curve D* in terms of quantity (kWh) bought by consumers in a given period plotted against *price* that consumers are willing to pay. It is assumed that as the quantity of electrical energy sold increases, its price decreases. The marginal cost of production is shown by line $MC$, which crosses the demand line $D$ at a quantity of electricity (over a period) $Q_1$ at a price $P_1$. The *consumer surplus* is then taken as being proportional to area $CS$, and producer or supplier surplus as proportional to area $PS$. The total surplus is $(CS + PS)$ but this is, of course, divided between a number of consumers and suppliers, any of whom can try to increase their own 'surplus' by varying their demand or their 'price'. It should be noted that by reducing $MC$, the proportion of surplus to suppliers can be increased, thereby producing more income for producers.

For a supplier to maximize profit, then optimal theory suggests that (from equation (12.2))

$$\frac{dP}{dQ} = 0 \text{ for max.} = \frac{dR}{dQ} - \frac{dVC}{dQ} - \frac{dFC}{dQ}$$

where $dQ$ is the change in output.

Now $(dVC/dQ)$ is known as the *short-run marginal cost* (SRMC) and is an important factor in any pricing policy. If $(dFC/dQ)$ is not zero, then

$$\left( \frac{dVC}{dQ} + \frac{dFC}{dQ} \right)$$

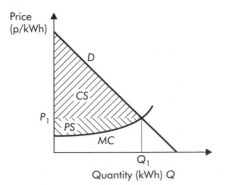

**Figure 12.2**  Price/quantity relationship. $D$ = Demand by consumers; $MC$ = marginal cost of generation; $CS$ = consumer surplus; $PS$ = producer surplus

must be equal to $(dR/dQ)$ for profit maximization, and this is the *long-run marginal cost* (LRMC) which takes into account the changes in the fixed costs (e.g. the cost of installing more generators) to make a profit in the future. The nearer utilities can structure themselves to providing output to SRMC, the more likely they are to make an adequate profit as the actual consumer demand changes. *Economic efficiency* is achieved when a supplier's income is equal to the marginal cost of production.

## 12.2.1 Ramsey pricing

In 1928, Ramsey published a famous paper (Ramsey, 1928) which gave a rule for the pricing of a commodity when SRMC pricing was not possible. Sometimes, SRMC is called *first-best* pricing and any other form of pricing is *second-best*, particularly where utilities are unable to make a profit from SRMC and need to price above it to provide for new or replacement plant.

In the same way that we defined income elasticity $\eta$, we can define *demand elasticity* $\varepsilon$ as

$$\varepsilon = \frac{(\Delta Q/Q)}{(\Delta P/P)} \tag{12.3}$$

where $(\Delta Q/Q)$ is change in consumption and $(\Delta P/P)$ is the corresponding change in price (rather than income). To increase revenue, a supplier could increase price, but preferably only to those consumers (or markets) which are least sensitive to price changes, i.e. their demand elasticity is small. The 'mark-up' in price for the least-sensitive consumers should then alter as little as possible from the SRMC pricing policy. So a reasonable formula for a second-best pricing policy would be to set the price change according to some proportion of the demand elasticity $\varepsilon$. This can be expressed as

$$\text{Price increase (called mark-up)} = \frac{P_i - MC_i}{P_i} = \frac{\lambda}{\varepsilon_i}$$

where $P_i$ = price to consumer group $i$;
$\quad MC_i$ = the marginal cost for consumer group $i$;
$\quad \varepsilon_i$ = the demand elasticity;
$\quad \lambda$ = a weighting factor.

If several groups of consumers are involved then we require that

$$\left(\frac{P_i - MC_i}{P_i}\right)\varepsilon_i = \left(\frac{P_j - MC_j}{P_j}\right)\varepsilon_j = \lambda$$

for the second-best pricing policy in the $i$th and $j$th groups. (This can be proved by using Lagrangian multipliers in a mathematical formulation.) This rule is called the *Inverse Elasticity Rule* (IER) and the pricing policy which maximizes total surplus is known as *Ramsey pricing*.

In a *benefit maximization* calculation, taking into account the various types of consumer, the cost–quantity relationships on the supply side are combined with price elasticity on the demand side to maximize total benefit subject to constraints such as limits on supply capacity, cost constraints imposed by government, and maximum price for 'lifeline' supplies, i.e. a cap to the maximum price any consumer can be asked to pay. Obviously, if electricity supply is a *monopoly* (or any part of supplying consumers is a monopoly business, e.g. transmission), then some imposed constraints are necessary to prevent customers being 'ripped off'. Either direct government controls are required or a 'Regulator' is appointed to carry out the price-control function. If market competition can be set up, then no price regulation is necessary, in theory, except to ensure a 'level playing field' for all participants.

## 12.3    Supply-Side and Demand-Side Options

Because electricity demand and consumption have been notoriously difficult to forecast over the past 15–20 years, it is now recognized that *uncertainty* in forecasting must be accepted. This uncertainty arises for the following reasons:

1.  The price of fuel now fluctuates on worldwide markets almost daily.

2.  Costs of construction vary due to property and land price variations.

3.  Environmental constraints are forever changing, dependent upon pressure groups and 'green' policies.

4.  Uses of electricity for comfort and convenience can change with local climatic conditions.

5.  The effects of economic and political changes are increasingly difficult to estimate.

This uncertainty requires a change in approach, particularly as in the U.K., privatization is producing a rethink of supply-side policies. In fact, one possible approach is to consider supply-side and demand-side options on an equal basis. Economic theory would then tend to support this approach as a way of ensuring an *optimal social outcome*.

### 12.3.1   Supply-side options

The most important option available to the supplier is that of price. It can be viewed as the main coordinating signal, not only to the consumer to influence investment decisions, but also to other suppliers in a competitive economy. By setting prices using the principles given in Section 12.2, a number of varying objectives can be achieved. For example:

- Economic efficiency of the suppliers can be guaranteed.

- Government policies, such as safeguarding future supplies from indigenous resources, subsidies to rural areas, and target rates of return or profit on assets, can be achieved.

- Revenue can be raised to cover costs of investment, management, operations, and maintenance.

- The market share of electricity can be influenced compared with other fuels.

Depending upon the supplier's policy, the options open include:

- scheduling generation to maximize income;

- investing in new plant for future returns;

- purchasing energy from neighbouring utilities or local generators;

- diversifying the business by, for example, supplying heat, aggregates from ash, by-products of flue cleaning, etc.

## 12.3.2 Demand-side options

On the demand or consumers' side of the electricity meter, the options available should endeavour to contribute to the economic efficiency of the system as a whole. These endeavours should be guided mainly by the pricing policies set by the suppliers. Opportunities to exploit demand-side options include:

- use of alternative fuels (i.e. fuel substitution) to achieve objectives, e.g. minimum annual cost, better quality product;

- self-generation or private-contract purchase of electricity;

- improvement in end-use efficiency, e.g. more insulation in buildings and appliances;

- lifestyle changes because of conservation issues;

- end-use storage (hot and cold, compressed gas, etc.).

With proper choice of options, the user theoretically can obtain cheaper operating costs and, because of the shorter lead times compared with increasing generator capacity, the user can end up with less risk-taking on capital investments than on the supply side. Thus, one answer to coping with uncertainty is to encourage consumers to exercise their available options to maximize both supply-side and demand-side benefits.

## 12.4   Load Management and Spot Pricing

One way of maximizing benefits is to have some central coordinated control of demand so that it always matches the power being generated by the suppliers, who, in turn, have a means of measuring the benefit of that demand to the economy. It is well known that such a benefit cannot be measured for every consumer because of the large diversity of needs and objectives. As we have seen, the only appropriate and acceptable measure is that of price and willingness to pay. Consequently, according to competition economic theory, a market in electricity is required corresponding to the markets in oil, coal, gas, etc. The difficulty, of course, is that the infrastructure necessary to inform all consumers of the price fluctuations in the purchase of the next kWh of electrical energy is very costly, since the price will vary with each generator on the system, implying a price change every half-hour at least. There are basically two ways of achieving the objective:

1. The consumer can enter into an agreement with the supplier to exercise direct physical control of electrical appliances when it is advantageous to do so, e.g. because no more generation or transmission capacity is available (peak lopping) or because starting up another generator would push up the price of electricity beyond that which the consumer is willing to pay. This is called *load management* and usually implies direct control of consumer load by the supplier. The methods which can be used are:

   • telecommunication networks—two-way is possible, using telephone, video cabling, etc.

   • radio communications—two-way is expensive, but teleswitching over a radio broadcast band by modulating the carrier can give extensive one-way coverage very cheaply.

   • mains-borne signalling—restricted to areas around injection point on network.

2. Alternatively, the price can be transmitted to the consumer and he/she then determines how to react, either by manual or by automatic control of selected loads.

In both cases, a variation of price broadcast by the supplier is an advantage, known as *spot pricing*. In general, the spot price should follow the SRMC since it should depend upon the availability of generators and the state of the end user. However, it can be shown that SRMC pricing is essential if economic efficiency is to be obtained. It thus tracks the supply–demand balance. Schweppe et al. (1988) have written at length on the various methods and technologies now available for spot-pricing policies. Spot pricing cannot be pre-determined because of changing system conditions, thus leading to *prob-abilistic forecasting* of possible price variations.

Flexible consumers or suppliers who can respond to changing prices will be able to improve their profits by reducing consumption or increasing cogeneration when prices are high and by increasing consumption or reducing generation at times of low prices. Inflexible suppliers who cannot adjust their operations activity to suit evolving prices will experience untoward fluctuations in cash flow. For some participants, the advantages of paying low prices most of the time would not be justified by the risk of occasional periods of higher prices. For these participants, financial instruments can be used for risk sharing. For example, under a system of *forward contracts*, a price is offered for electrical energy transactions at each time period in the future. Thus, consumers can avoid the risk of higher prices by purchasing their future requirements at the fixed price. However, that price will be higher than the usual spot price since it contains a *risk premium* to cover the possible system conditions with higher values of SRMC. Consumers thus pay extra for the increased certainty in the price. However, financial instruments, such as *options*, can be used to reduce the investment risk, if desired.

## 12.5  Electricity Pricing and Markets

Until the early 1980s, almost every electricity utility in the world believed that generation, transmission, and, in many cases, distribution of energy should be an integrated business with one price to the consumer. The actual price must depend upon the average costs to the utility of producing and delivering this energy, so differential prices for large, medium, and small customers was the norm. With the breaking up of the large integrated utilities (unbundling) into separate businesses, as in the U.K. and elsewhere, pricing of electricity services, including the cost of energy, has become much more commercial. The following are practical guidelines for pricing objectives, which are generally acceptable to all types of customer.

1. Prices should be based on economic efficiency, costing resources in terms of fuel, conversion costs, and effects on the environment, not just in purely monetary terms.

2. Prices should be firmly set in accordance with costs, i.e. no hidden or cross subsidies.

3. Prices should ensure commercial viability.

4. Equity between different classes of consumers should be maintained.

5. Tariffs should be as simple as possible and transparent to all customers.

As we have seen, fixed capacity (kW) and variable (kWh) costs are incurred by a utility, but, in addition, there are fixed costs, such as billing, meter installation and reading, and customer services (answering queries and dealing with complaints), which depend on the number of customers. Therefore, a three-part tariff for each customer, made up of a fixed service charge, a kW-capacity charge, and an energy kWh charge, would appear to be the most appropriate. Many tariffs will lump the two fixed charges together for simplicity, leading to the familiar domestic two-part tariff based on household size plus the monthly or quarterly metered kWh (unit) consumption. Larger customers (shops, offices, small industries, workshops, etc.) will have a maximum-demand meter as well as an energy meter, so their tariff will include a price for kW of maximum demand over the year. Additionally, large consumers could have their var consumption (kVArh) metered and charged-for separately, leading to the concept of a separate reactive payment.

## 12.5.1  Electricity market models

The electricity industry is, in principle, just like any other producer or manufacturer of goods in that the electrical product (kWh) is made at a particular location and needs to be transported to the customer who has contracted to purchase that product at a certain price. Normally, some form of a market is set up in a democratic society so that purchasers can 'shop around' for the product that best suits them, at the price they consider to give the best value. For efficiency and expendiency, trading markets have a recognized infrastructure whereby intermediaries purchase and resell commodities in varying quantities and arrange for delivery to varying locations. In the case of electrical energy, there is only one product, and at every time instant the amount produced must equal the amount consumed, otherwise the frequency will vary. Consequently, looked at from a market-structure viewpoint, there must be a direct and continuous connection between generators and the consumers, but the various market intermediaries, such as coordinators or controllers and energy brokers or traders, can be considered as facilitators in this process, as illustrated in Figure 12.3.

Hunt and Shuttleworth (1996) have postulated four market models as illustrating the possible ways that an electricity market can be modelled. These models correspond to varying degrees of monopoly and competition in the electricity industry and represent possible structures for a country to organize its industry. The models are shown in Figures 12.4–12.7 and their characteristics are summarized in Table 12.1.

The four models have these characteristics:

- *Model 1*—Monopolistic or vertically integrated. Generation, transmission, distribution, and supply to the consumer (retailing) are all in one company

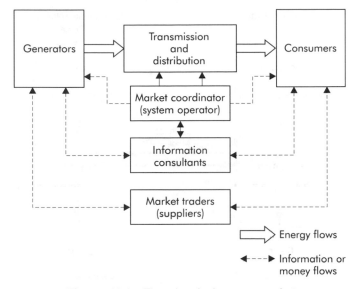

**Figure 12.3**   The electrical energy market

**Table 12.1**   Structural alternatives

| Characteristic | Model 1— Monopoly | Model 2— Purchasing agency | Model 3— Wholesale competition | Model 4— Retail competition |
|---|---|---|---|---|
| Definition | Monopoly at all levels | Competition in generation— single buyer | Competition in generation and choice for Distcos[a] | Competition in generation and choice for final consumers |
| Competing generators | No | Yes | Yes | Yes |
| Choice for retailers? | No | No | Yes | Yes |
| Choice for final customers? | No | No | No | Yes |

[a]Distcos, distribution companies.
(Source: Hunt and Shuttleworth (1996). Reproduced by kind permission of the authors of National Economics Research Association (New York & London))

or utility and there is no choice for anyone. This requires government control or a 'Regulator' to ensure fairness in pricing.

- *Model 2*—Purchasing Agency. This allows a single buyer to purchase energy from competing electricity generators or independent power producers (IPPs). The buyer arranges for delivery over monopoly

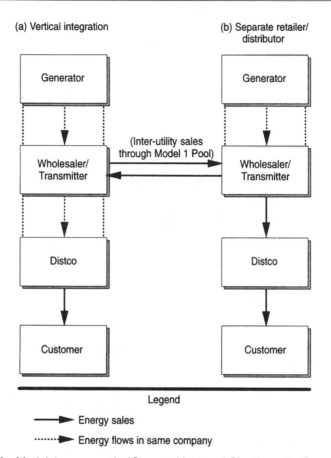

**Figure 12.4**   Model 1—monopoly (*Source: Hunt and Shuttleworth, Competition and Choice in Electricity, Wiley, 1996*)

transmission and distribution systems and sells at advertised tariffs to all consumers (large and small). The Purchasing Agency would purchase by tender or bidding system from the IPPs to ensure the best price is obtained throughout the day and year.

- *Model 3*—Wholesale competition. In this model the distribution companies (Distcos) can buy direct from the producers and deliver to their own supply points over the transmission networks. These networks will allow any Distco to contract for their use, thereby allowing 'open access'. The Distco, however, has a monopoly over supplying the consumers.

- *Model 4*—Retail competition. Here, all consumers can choose their supplier and pay for delivery over the transmission and distribution networks, which are on 'open access'. This is the most that can be obtained by competition,

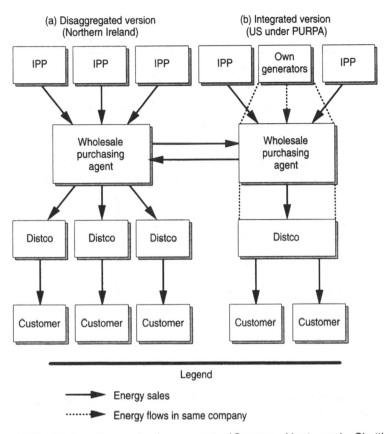

(a) Disaggregated version
(Northern Ireland)

(b) Integrated version
(US under PURPA)

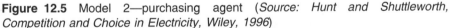

Legend

Energy sales

Energy flows in same company

**Figure 12.5** Model 2—purchasing agent (*Source: Hunt and Shuttleworth, Competition and Choice in Electricity, Wiley, 1996*)

although a Regulator is still required to agree the monopoly transmission and distribution charges, unless alternative delivery networks are built. The Regulator is also required to maintain a fair and equitable market structure for consumers to exercise their choice of supplier.

## 12.5.2 Comments on market models

Most countries are somewhere between Model 1 (monopoly, vertically integrated structure) and Model 4 (completely open, competitive energy trading). There are many problems and difficulties in transferring from one model to another, not least of which is the need to set up an appropriate infrastructure, including the necessary control, metering, and information technology to achieve a free market at reasonable cost. In addition, a considerable re-

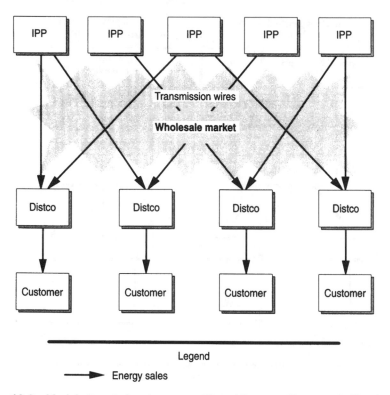

Legend

⟶ Energy sales

**Figure 12.6** Model 3—wholesale competition (*Source: Hunt and Shuttleworth, Competition and Choice in Electricity, Wiley, 1996*)

education and re-training of utility employees is required to enable the 'new' philosophies of marketing and improved customer services to be provided. From the U.K. experience, it is apparent that the technical aspects of providing a reliable electrical energy supply of the specified quality becomes a smaller proportion of the overall activity within the overall supply business, with many of the technical matters being overtaken by energy buying and selling, customer relations, services to the individual consumer (appliance maintenance, billing, data transfer, improved end-use efficiency, etc.), and the need to maintain a healthy share price and public relations image. These aspects require business and management skills combined with a good technical knowledge and understanding.

## 12.5.3 Reactive market

Large consumers are encouraged by their tariff to take supplies as near to unity power factor as possible, thereby helping to reduce the $I^2R$ loss in supply

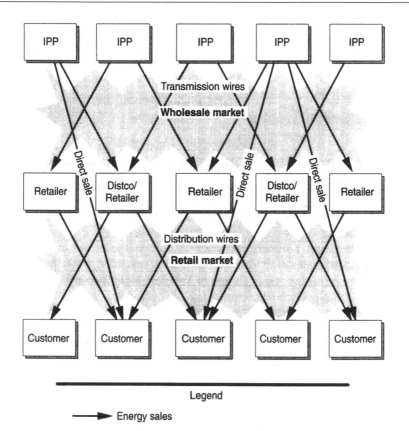

**Figure 12.7** Model 4—retail competition (*Source: Hunt and Shuttleworth, Competition and Choice in Electricity, Wiley, 1996*)

circuits. It is therefore reasonable to expect kVArh to be charged for as a contribution to the increased power loss incurred, particularly in distribution circuits. On the transmission network, reactive power is required to maintain voltage (see Chapter 5) and this has to be provided either from generators by excitation adjustment or by the installation of var compensators (series or shunt) at strategic busbars in the system. If vars are provided by generators through overexcitation (i.e. generated vars), then not only is additional heating of the rotor occurring but also increased losses because of additional stray flux and end-winding eddy currents produced in the stator. To compensate the generator owner for the extra deterioration of the machine, a payment proportional to the MVArh produced is made by the power-system operator to the generator or IPP. Obviously, the transmission network requires to cover investment in reactive compensation and the extra $I^2R$ losses incurred by its operation. Therefore, it can be argued that a reactive market could also be established along similar lines as a real-power pool. Unfortunately, whereas real power can be transmitted over quite long distances without large voltage changes, as Chapter 5 shows, transmission of vars implies a large voltage difference on an uncompensated line, thereby limiting the distance between var sources. This complication makes the setting up of a var market difficult. Negotiated contracts for vars between the power system operator and IPPs could well be preferable.

## 12.6  Demand-Side Management and Least-Cost Planning

Demand-Side Management (DSM) is the process of influencing customer demand throughout 24 hours by encouraging use of electrical energy when prices are low or by assisting customers to employ conservation measures such that their overall energy bill is acceptably low. The main problem in a competitive environment is that both of these desirable goals reduce the income to the supplier. Therefore, regulatory measures must be imposed to achieve set targets, usually associated with environmentally sustainable developments or socially desirable practices. In setting targets, a regulatory body has to be satisfied that all options have been considered, including both supply-side and demand-side possibilities for investment to reduce overall costs. When both supply-side and demand-side options are included, this is known as Least-Cost Planning (LCP). LCP requires customer cooperation for overall benefit maximization.

Figure 12.8 shows the various possibilities for influencing the load-demand shape over the normal day. Most large consumers (especially electrical-energy-intensive industries, such as chemical, steel, cement) have individually negotiated tariffs dependent upon their production needs and demand profile. Metering and billing is a very small part of the overall costs, but cheap energy is often essential for success in their markets. Overall energy/production cost optimization is normally part of their business plan.

### 12.6.1  Benefits of DSM

Considerable benefits can flow from DSM measures, but these need to be realistically costed by surveys or modelling. The following aspects should be considered:

- reduction of peak demand, thereby producing savings in new investment on transmission and distribution capacity;
- avoidance of load shedding where available plant is limited;
- improved load factor on existing plant by making better use of capacity;

| Utility load shape objectives | | Residential | Examples of customer options | |
| --- | --- | --- | --- | --- |
| | | | Commericial | Industrial |
| **Peaking clipping**, or reduction of load during peak periods, is generally achieved by directly controlling customers' appliances. This direct control can be used to reduce capacity requirements, operating costs, and dependence on critical fuels. | 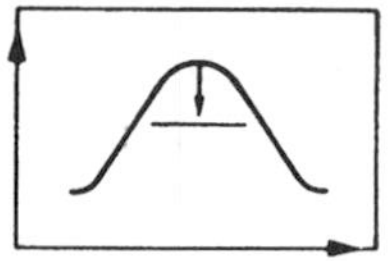 | • Accept direct control of air-conditioners | • Accept direct control of water-heaters | • Subscribe to interruptible rate |
| **Valley filling**, or building load during off-peak periods, is particularly desirable when the long-run incremental cost is less than the average price of electricity. Adding properly priced off-peak load under those circumstances can decrease the average price. | 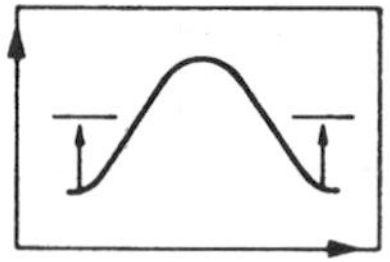 | • Use off-peak water-heating | • Store hot water to augment space-heating | • Add night-time operations |
| **Load shifting**, which accomplishes many of the goals of both peak clipping and valley filling, involves shifting load from on-peak to off-peak periods, allowing the most efficient use of capacity. | 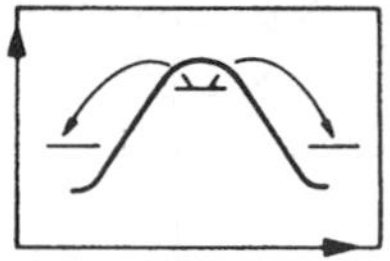 | • Subscribe to time-of-use rates | • Install cool-storage equipment | • Shift operations from daytime to night-time |

**Figure 12.8** Load shape and demand-side alternatives (*Source: Cory, B. J. 'Load management by direct control', internal paper, Department of Electrical Engineering, Imperial College, 1985, Figure 1. Reproduced by permission of EPRI Journal, December 1994, pp 6–15*)

| Utility load shape objectives | Residential | Examples of customer options | |
| --- | --- | --- | --- |
| | | Commerical | Industrial |
| **Strategic conservation** involves a reduction in sales, often including a change in the pattern of use. The utility planner must consider what conservation actions would occur naturally and then evaluate the cost-effectiveness of utility programs intended to accelerate or stimulate conservation actions. |  • Supplement home-insulation | • Reduce lighting use | • Install more efficient processes |
| **Strategic load growth**, a targeted increase in sales, may involve increased market share of loads that are or can be served by competing fuels, as well as developments of new markets. In the future, load growth will include greater electrification — electric vehicles, automation, and industrial process heating. | 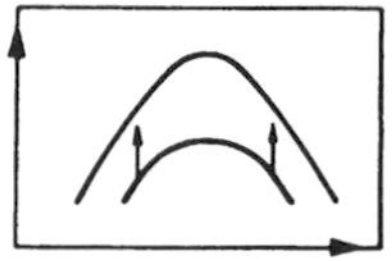 • Switch from gas to electric water-heating | • Install heat pumps | • Convert from gas to electric process heating |

**Figure 12.8**  (*continued*)

- better load forecasting, if elasticity of demand is monitored;

- lower supply costs to customers, including improved cash flow with improved billing and metering facilities.

The largest savings are often due to reduced generation costs as the most commercial and cost-effective generators can be scheduled. The avoidance of investment in new plant has considerable impact, particularly when interest rates are high. If demand elasticities are being continuously monitored for billing purposes, load-forecasting techniques can be further improved with this data. Improved load factors avoid extra start-up, shut-down, and part-load costs on generators. Also, better use of the network reduces use-of-system costs (see later).

Typical DSM methods for residential and small commercial consumers include the following:

- Improved thermal insulation of building structures (loft and wall insulation, double glazing, draught proofing) in cold/temperate climates. In tropical zones, better solar protection and natural cooling is required.

- Use of electrically efficient equipment (heat pumps, microwave ovens, improved refrigerators and freezers with energy labelling to guide purchase, washing machines and dishwashers, low-energy and long-life lamps, etc).

- Better control of electrical appliances, either by local optimal controllers, proximity switches, temperature controllers, etc., or by load management agreements for direct control by the electricity supplier (see Section 12.4).

In evaluating DSM measures, the key factors are as follows:

- load profile and coincidence of peaks, hence the need to disaggregate customers into different groups geographically;

- need to differentiate between groups by consumer type (commercial, domestic, etc.);

- ability to reconfigure supply system to avoid overloads;

- measures are often required to maintain customer participation, e.g. by loyalty cards, bonuses, etc.;

- strategies are required to assist customers to change habits of energy use, e.g. discounts for energy-saving devices.

To become really effective, DSM requires good and continuous contact between suppliers and customers. Ideally, a two-way communication link between the customer's meter, with a wall display, and the supplier enables consumption to be monitored and, if desired, an update on electricity prices and accounting details can be downloaded to the customer. This link can be used for load management control of water-heaters, space-heating, air-conditioning, etc., or the customers can install their own controllers responding to price messages from the supplier. Various communication methods have already been listed but, unfortunately, none are particularly inexpensive relative to an average domestic annual bill of £400 (US$650) or so.

## 12.6.2   Least-cost planning (LCP)

By definition, LCP is a maximization of supplier plus consumer welfare at least cost (see Stoll *et al.*, 1989). Hence there is a need for *integration* of investments by both supplier and consumer to include:

- load management;
- energy conservation;
- generation, including CHP and renewables;
- transmission and distribution capacity;
- interutility energy transfers.

The planning processes should consider *all* the options available to meet the desired objectives, namely:

- minimum cost investment over a given period;
- robustness against changes in market rates, social climate, environmental constraints, etc.;
- maximum cooperation between utility and customer;
- financial viability.

A typical procedure for LCP is illustrated in Figure 12.9, in which various DSM measures are placed in order of ascending cost. Figure 12.9 is sometimes called a *scoping* or *screening* diagram and should include any marginal costs of new generation plus any reinforcements required to the transmission and distribution systems to supply the DSM equipment.

In theory, investment in new plant should not occur until the cost of alternative measures is equivalent to new plant costs. In practice, new plant may be required because of ageing, security of supply, new development, economic boom, etc., as determined by a risk-and-contingency analysis, including uncertainties in forecasts.

## 12.6.3   Comments on DSM

DSM, which may, or may not, include direct load-management control, requires considerable supplier–customer interaction. New metering technolo-

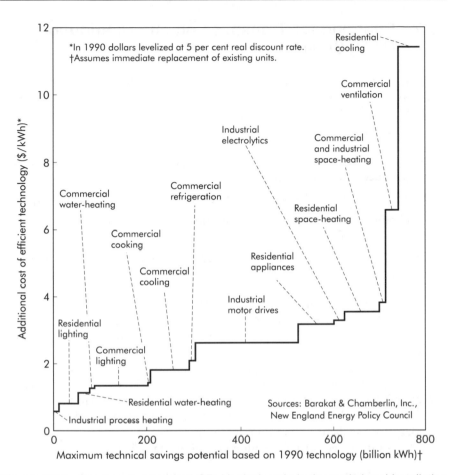

**Figure 12.9** Scoping diagram for LCP. Vertical scale is the capital and installation costs amortized over the life of the replacement unit (*Permission of E.P.R.I., U.S.A*)

gies and communications will facilitate this process, but is it important to keep good relations with the customer through efficient billing and the easy availability of discounts, rebates, etc. Utilities need to set up information technology (IT) infrastructures and to maintain an up-to-date customer data base, including comprehensive profiling of customer demand throughout the day, season, and year. With continuously remotely read meters, the profiling data is readily collected, perhaps by a separate data collection organization, but if regular (monthly or quarterly) meter reading is employed, comprehensive sampling techniques and surveys must be conducted to enable appropriate charges to be levied. These charges should include not only fuel costs related to generation but also delivery charges over the transmission and distribution system.

## 12.7  Charging for Transmission and Distribution (T&D) Services

Chapter 1 introduced the role of the transmission and distribution networks in power systems, but now we must consider the benefits that they provide so that equitable charges for their use can be levied. For this purpose, the services they provide can be summarized as follows:

- to provide energy transport from sources (generators) to consumers (loads) with an acceptable reliability;

- to pool resources and reserves so that security of supply is achieved;

- to obtain benefits of economic operation such that cost of energy to all consumers is minimum at all times;

- to make good use of flexibility of operation for new generators and load connections.

In practice, transmission systems are mostly interconnected in a grid or mesh form so that alternative paths exist between sources and loads. Many distribution networks operate as radial feeders to simplify protection and minimize fault levels.

For charging purposes, it is the practice to monitor all power flows (kW) on these services from major substations and to meter (kWh over specified period) energy injected into the system from generating sources and energy delivered to all customers. In addition, metering of energy over major circuits is also necessary for detailed costing and charging purposes.

A T&D utility needs to charge for the following:

- losses incurred in transporting energy (in kWh);

- operation and maintenance of the system (probably a fixed cost dependent upon capacity or rating);

- reinforcement and extension to maintain customer service and quality (planned by extensive studies and future plausible scenarios);

- new connections and services (related mainly to individual customers when connection and service is negotiated and contract agreed).

The organization required to provide these functions and to allocate the charges is very complex and varies from utility to utility. The 'technical' requirements of monitoring and metering are considered as the job of the engineering staff and associated planners; the collection and collation of data from the system is part of the operation of the system, while the billing and collection of monies falls on the accounting and business division of the utility. Obviously, of vital importance is the data communication between monitoring/

metering points in the network and the operation/control centre, and the information technology infrastructure available between the data centre and the settlement/accounting groups within the organization.

### 12.7.1 Costing transmission and distribution access

Several approaches to costing transmission and distribution (T&D) services and access can be identified. The principal ones are given here.

### (i) Contract path approach

The contract path approach originated in the early days of the industry when systems were interconnected with few tie lines. The approach specifies an electrically continuous path from a generator (or a point of receipt if more than one utility involved) to a point of delivery (i.e. the buyer or an adjacent utility system). The path chosen must have sufficient unused capability to carry the amount of power to be transported. The price is determined to recover the necessary rate of return from the T&D assets assigned for the transport and any other additional costs incurred by the T&D utility due to the transaction.

For transport between closely meshed utilities with strong tie lines, this method is not suitable mainly because of the inability to ensure that the power flows follow the contracted paths. Another serious deficiency with this approach is that it entirely ignores outages of elements in the path, imposing power flows through neighbouring parallel systems which are not parties to the contract. This method is still widely used, particularly in the U.S.A., for pricing wheeling transactions between utilities based on payment per MW or MWkm.

### (ii) Postage stamp method

The postage stamp method differs from the contract path method in that it considers system-wide average T&D costs rather than the costs of specially selected facilities. This method is often referred to as the postage stamp method because it results in the same transport charge irrespective of distance or location. The main argument for this method is that an integrated T&D system is designed to achieve maximum efficiency and reliability at minimum cost on a system-wide basis, and that all users should receive the benefits in such an integrated system.

## (iii) Method based on usage

This method basically concentrates on two measurements: the amount of T&D capacity used, and the per unit cost of that capacity. Measurement of capacity used can be as simple as total megawatts of the transaction (such as the postage stamp method) or involve calculation of estimated or actual power flows and related transmission data. Cost of T&D capacity can be calculated on an embedded, incremental, or marginal basis. It may also be either short-run cost (when there is insufficient time to change capacity) or long-run cost. This method is suitable for transaction pricing between utilities and for the user-specific cost streams in a fully deregulated electricity market.

Cost of the facility is defined as the revenue requirement for the facility on an embedded, incremental, or marginal basis, as defined below.

***Embedded cost***   This is the revenue required to pay for all existing T&D facilities or any added to the system during the life of the transmission contract. It can be calculated from

$$\text{ECC}_i = \sum_{f \in F} \frac{|\Delta\text{MW}_{f,i}| \times \text{EC}_f}{\sum_{s \in S} |\Delta\text{MW}_{f,s}|}$$

where

$\text{ECC}_i$ = annual embedded capacity cost of transaction $i$;

$\Delta\text{MW}_{f,i}$ = change in megawatt flow due to contracted transaction $i$ on facility or circuit $f$;

$\Delta\text{MW}_{f,s}$ = change in megawatt flow due to all contracted transactions on facility or circuit $f$;

$\text{EC}_f$ = annual embedded cost of facility $f$ consisting of depreciation, annuitized cost of capital, taxes and operating/maintenance costs;

$S, F$ = sets of all the transactions $S$ and facilities or circuits $F$ considered in any year.

***Incremental cost***   This is the revenue requirement that is specifically attributed to the T&D service customer. These facilities must be identified for all years across the life of the contract for service. The T&D services customer pays the full cost for any new facilities that the transaction requires; if a new facility would have been built for other reasons at a later time, then the T&D services customer pays the cost to advance the facility's in-service date. If a facility is needed by more than one T&D service customer, then the cost of the facility can be allocated to the incremental customers by usage methods.

*Marginal cost*   This can be defined as the revenue requirement needed to pay for any new capacity on the T&D system. The T&D service customer pays an allocated share of the cost for any new facility that the system requires.

The difference between the three types of cost focuses on the usage allocation function of the facility. Embedded capacity cost includes existing and new customers and charges the embedded cost of all existing and new facilities, thereby providing an average rate for all customers. Incremental capacity cost applies only to new customers who forced the construction of new facilities, and charges the full cost of these facilities over the life time of the facilities. 'New facilities' includes only facilities that are specifically identified as being needed to serve the new customers. Marginal capacity cost applies to all customers requiring new T&D facilities and charges the marginal cost of new facilities anywhere in the system, thereby providing the same marginal rate in £/MW for all marginal customers.

## 12.7.2   Transmission facilities including security

While most distribution facilities can be identified for existing or new customers, since mainly radial circuits and associated components are involved, on the multimesh transmission network this distinction is no longer feasible because of the many paths available between the generators and the loads.

The cost of service provided by a transmission system is spatially and temporally differentiated to take into account the configuration of the network and its inherent capability to operate at times of peak flow. Marginal-cost-based service pricing provides the incentives needed for the network utility to operate and to invest for increased system efficiency, and for generating units and end users to locate efficiently within the network. Unfortunately, it may not necessarily guarantee a sufficiently large return to provide for a financially survivable company.

The transmission network can be seen as one of two businesses:

1.  It is a fixed-asset-based business put in place to guarantee sufficient transmission capability to maintain a reliable and secure supply. Defined in this manner, the business is defined in kW capacity, and is billed to the user (generator or distributor/end user) as a function of their location and their peak demand.

2.  It is a transportation business whose objective is to transport energy at profit. Defined in this manner, the business is denominated in kWh, and is billed to the users as a function of the amount of energy, the location, and the timing of energy input to, or extracted from, the network.

Regardless of the manner in which the transmission business is defined, operating uncertainties exist. These are increased under privatization or unbundling because the network utility no longer has direct control over the decisions of either the supply side or the demand side. Nowadays, in England and Wales, for example, network expansion planning has to be carried out in an environment in which the City financial market information about future generator locations plays an important role. The incentive is for the network utility to provide for the most efficiently constructed and flexibly controlled system.

### 12.7.3  System losses

With a given system and patterns of generation and loads, the estimated losses in kWh per km of system can be calculated. These losses can be costed at the marginal price/kWh for each node or zone of the system and the costs apportioned to users on a kWh basis.

In estimating losses in use-of-system pricing there are a number of possible ways this can be done. These are:

- *Empty system*—the losses are estimated by calculating the flows in all the circuits from a single generator input to a single consumer load as if no other flows are present. Any contract between the generator and load includes a charge for the expected 'copper' losses because of the transport between the two.

- *Vector difference*—contract transport is added to the existing system flows and the difference in losses is included in the cost of that contract. Note that flows opposite to existing flows actually produces a negative incremental loss.

- *Critical path*—the path that the contract transport is expected to take through the system is costed for losses by assuming that the flow is added on top of existing normal flow in the system.

Most of these methods relate to specified flows when a contract in terms of kWh for a specified period up to a given kW maximum is being fulfilled. When many different contracts over many different paths are being accommodated, the complexity of system operation is apparent. For an integrated system, such as NGC, a posted tariff for use-of-system based on capacity is considered to be preferable.

For planning and pricing purposes, network studies based on economic pricing principles are required. Transportation algorithms can provide an insight into the major power flows but to account for security, voltage control, and the necessary stability margins, network studies over a range of loading and generation scenarios are required. These studies can be done through fast d.c. load flows with added group constraints, or through OPFs using full a.c.

equations. From the studies, revenue recovery can be calculated with various pricing policies applied, such as a fixed charge per MW km of system or using a maximization of net benefit formula including security. Whatever basis of charging is employed, for a monopoly utility the total cost recovery per year must be agreed with a regulating authority. It is becoming established that this cost recovery is set at a percentage return on assets to provide a profit margin dependent upon a measure of retail prices (such as the Retail Price Index (RPI) in the U.K.) minus a given percentage factor $X$ per year to provide an incentive towards achieving greater efficiency. This is known as the RPI-$X$ formula and, if set for a fixed number of years, it provides an incentive for the utility to invest such that costs are reduced and profits increase, despite the $X$ penalty.

## 12.7.4  Ancillary services

To maintain reliability and quality of supply on an interconnected transmission system, reserves of real power and reactive power are required, strategically located around the network. In addition, in the event of a dire emergency, such as a system shut-down, a self-start (called 'black-start') capability is required. These reserves, including reactive power provision for voltage control, if not bought by a 'reactive market' process, and plant held in readiness for a 'black-start', need to be purchased by the system operator and charged to system users on some equitable basis.

The ancillary services required are defined in more detail as:

- *Spinning reserves*—fast-response capability held on part-loaded synchronized generators to check any sudden frequency change due to loss of an infeed. Generator capability requires monitoring and periodic assessment to ensure contracted response is available. Response is classified into primary (available within 10 s) and secondary (available within 30 s). Note that if a frequency fall cannot be contained by scheduled part-loaded generators, then automatic load shedding by under-frequency relays is a further safeguard.

- *Standing reserve*—unsynchronized, fast start-up plant (hydro, gas turbine) available within 5 min or less. Demand reduction by operator instruction can also be used in this category.

- *Replacement reserve*—plant that can be brought into service with longer notice in the event of a shortfall in standing reserve. This may be hot-standby plant or deferred-start plant, designated by the system operator.

- *Reactive power (voltage control)*—the requirement of generators to operate at an instructed power factor within registered capabilities and to have a prescribed excitation response on change of voltage. The cost of MVArh is negotiated with individual suppliers, dependent upon type and design of plant. Note that the power-system operators can provide their own reactive power by installing compensators at strategic nodes in the system.

- *'Black-start' capability*—a requirement that a generator can start one or more units without external electrical supply within a given period. Requires provision of self-contained power supply from batteries, diesel generators, gas turbines, etc., to start up power-station auxiliaries.

Most of the ancillary services that are contracted for can be separately and continuously metered or monitored by inspection, thereby the payments for the services can be made through the pool or settlement system. The recovery of these costs from customers is dependent upon the methods adopted for sharing the charges for use-of-system. Hence they can be allocated according to the monitored kW demand or according to the kWh consumed in each billing period.

Finally, it is worth noting that in a competitive environment, contracted services, such as detailed here, can be bid for, using an auction process to ensure that the total ancillary service provision is obtained at minimum cost. Such auctions can include consumers as well as generators, particularly where demand reductions or power factor correction can be beneficial to the overall operation of the power network. In principle, therefore, bidding could be on a daily basis, together with the submission of marginal prices for energy under the pool system. In the U.S.A., electronic auction systems are being established for daily bidding, including MW capacity rights on transmission corridors (known as 'congestion contracts') so that energy trading can flourish, thus leading to minimum prices for all customers.

# Appendix I
## Synchronous Machine Reactances

**Table AI.1**  Typical percentage reactances of synchronous machines at 50 Hz—British

| Type and rating of machine | Positive sequence | | | Negative sequence $X_2$ | Zero sequence $X_0$ | Short-circuit ratio |
|---|---|---|---|---|---|---|
| | $X''$ | $X'$ | $X_s$ | | | |
| 11 kV Salient-pole alternator without dampers | 22.0 | 33.0 | 110 | 2.0 | 6.0 | — |
| 11.8 kV, 60 MV, 75 MVA Turboalternator | 12.5 | 17.5 | 201 | 13.5 | 6.7 | 0.55 |
| 11.8 kV, 56 MW, 70 MVA Gas-turbine turboalternator | 10.0 | 14.0 | 175 | 13.0 | 5.0 | 0.68 |
| 11.8 kV, 70 MW, 87.5 MVA Gas-turbine turboalternator | 14.0 | 19.0 | 195 | 16.0 | 7.5 | 0.55 |
| 13.8 kV, 100 MW, 125 MVA Turboalternator | 20.0 | 28.0 | 206 | 22.4 | 9.4 | 0.58 |
| 16.0 kV, 275 MW, 324 MVA Turboalternator | 16.0 | 21.5 | 260 | 18.0 | 6.0 | 0.40 |
| 18.5 kV, 300 MW, 353 MVa Turboalternator | 19.0 | 25.5 | 265 | 19.0 | 11.0 | 0.40 |
| 22 kV, 500 MW, 588 MVA Turboalternator | 20.5 | 28.0 | 255 | 20.0 | 6.0–12.0 | 0.40 |
| 23 kV, 600 MW, 776 MVA Turboalternator | 23.0 | 28.0 | 207 | 26.0 | 15.0 | 0.50 |

**Table AI.2** Approximate reactance values of three-phase 60-HZ generating equipment—United States (Values in per unit on rated kVA base)

| Apparatus | Positive sequence | | | | | | Negative sequence $X_2$ | | Zero sequence $X_0$ | |
|---|---|---|---|---|---|---|---|---|---|---|
| | Synchronous $X_s$ | | Transient $X'$ | | Subtransient $X''$ | | | | | |
| | Average | Range | Average | Range | Average | Range | Average | Range | Average | Range |
| 2-pole turbine generator (45 psig inner-cooled $H_2$) | 1.65 | 1.22–1.91 | 0.27 | 0.20–0.35 | 0.21 | 0.17–0.25 | 0.21 | 0.17–0.25 | 0.093 | 0.04–0.14 |
| 2-pole turbine generator (30 psig $H_2$ cooled) | 1.72 | 1.61–1.86 | 0.23 | 0.188–0.303 | 0.14 | 0.116–0.17 | 0.14 | 0.116–0.17 | 0.042 | 0.03–0.073 |
| 4-pole turbine generator (30 psig $H_2$ cooled) | 1.49 | 1.36–1.67 | 0.281 | 0.265–0.30 | 0.19 | 0.169–0.208 | 0.19 | 0.169–0.208 | 0.106 | 0.041–0.1825 |
| Salient-pole generator and motors (with dampers) | 1.25 | 0.6–1.5 | 0.3 | 0.2–0.5 | 0.2 | 0.13–0.32 | 0.2 | 0.13–0.32 | 0.18 | 0.03–0.23 |
| Salient-pole generator (without dampers) | 1.25 | 0.6–1.5 | 0.3 | 0.2–0.5 | 0.3 | 0.2–0.5 | 0.48 | 0.35–0.65 | 0.19 | 0.03–0.24 |
| Synchronous condensers (air-cooled) | 1.85 | 1.25–2.20 | 0.4 | 0.3–0.5 | 0.27 | 0.19–0.3 | 0.26 | 0.18–0.4 | 0.12 | 0.025–0.15 |
| Synchronous condensers ($H_2$ cooled at $\frac{1}{2}$ psig rating) | 2.2 | 1.5–2.65 | 0.48 | 0.36–0.6 | 0.32 | 0.23–0.36 | 0.31 | 0.22–0.48 | 0.14 | 0.03–0.18 |

*(Permission of Westinghouse Corp.)*

**Table AI.3** Principal data of 200–500 MW turbogenerators—Russian

| | Units | Values of parameters of turbogenerators of various types | | | | | |
| --- | --- | --- | --- | --- | --- | --- | --- |
| | | 1 | 2 | 3 | 4 | 5 | 6 |
| Power | MW/MVA | 200/235 | 200/235 | 300/353 | 300/353 | 500/588 | 500/588 |
| Cooling of winding | | | | | | | |
| (stator) | | Water | Hydrogen | Water | Hydrogen | Water | Water |
| (rotor) | | Hydrogen[c] | Hydrogen | Hydrogen | Hydrogen | Hydrogen | Water |
| Rotor diameter | m | 1.075 | 1.075 | 1.075 | 1.120 | 1.125 | 1.120 |
| Rotor length | m | 4.35 | 5.10 | 6.1 | 5.80 | 6.35 | 6.20 |
| Total weight | kg/VA (N/kVA) | 0.93 (9.1) | 1.3 (12.7) | 0.98 (9.6) | 1.05 (10.3) | 0.64 (6.26) | 0.63 (6.17) |
| Rotor weight | kg/kVA (N/kVA) | 0.18 (1.76) | 0.205 (2.01) | 0.16 (1.57) | 0.158 (1.55) | 0.11 (1.08) | 0.1045 (1.025) |
| $X_s^a$ | % | 188.0 | 184.0 | 169.8 | 219.5 | 248.8 | 241.3 |
| $X'$ | % | 27.5 | 29.5 | 25.8 | 30.0 | 36.8 | 37.3 |
| $X''$ | % | 19.1 | 19.0 | 17.3 | 19.5 | 24.3 | 24.3 |
| $X_q$ | % | 188.0 | 184.0 | 169.8 | 219.5 | 248.8 | 241.3 |
| $X_0$ | % | 8.5 | 8.37 | 8.8 | 9.63 | 15.0 | 14.6 |
| $\tau_1^b$ | s | 2.3 | 3.09 | 2.1 | 2.55 | 1.7 | 1.63 |

[a]For all reactances, unsaturated values are given.
[b]Without the turbine.
[c]Hydrogen pressure for columns 1–4 is equal to 3 atm (304,000 N/m²) and for column 5 4 atm (405,000 N/m²).
(*Source: Glebov, I. A, C.I.G.R.E., 1968, Paper 11-07.*)

# Appendix II
## Typical Transformer Impedances

**Table A.2.1**  Standard impedance limits for power transformers above 10 000 kVA (60 Hz)

| Highest voltage winding (BIL kV) | Low-voltage winding (BIL kV) For intermediate BIL use value for next higher BIL listed | At kVA base equal to 55°C, rating of largest capacity winding | | | | | | | |
|---|---|---|---|---|---|---|---|---|---|
| | | Self-cooled (OA), self-cooled/forced-air cooled (OA/FA) self-cooled/forced-air, forced-oil cooled (OA, FOA) Standard impedance (%) | | | | Forced-oil cooled (FOA and FOW) Standard impedance (%) | | | |
| | | Ungrounded neutral operation | | Grounded neutral operation | | Ungrounded neutral operation | | Grounded neutral operation | |
| | | Min. | Max. | Min. | Max. | Min. | Max. | Min. | Max. |
| 110 and below | 110 and below | 5.0 | 6.25 | | | 8.25 | 10.5 | | |
| 150 | 110 | 5.0 | 6.25 | | | 8.25 | 10.5 | | |
| 200 | 110 | 5.5 | 7.0 | | | 9.0 | 12.0 | | |
| | 150 | 5.75 | 7.5 | | | 9.75 | 12.75 | | |
| 250 | 150 | 5.75 | 7.5 | | | 9.5 | 12.75 | | |
| | 200 | 6.25 | 8.5 | | | 10.5 | 14.25 | | |
| 350 | 200 | 6.25 | 8.5 | | | 10.25 | 14.25 | | |
| | 250 | 6.75 | 9.5 | | | 11.25 | 15.75 | | |
| 450 | 200 | 6.75 | 9.5 | 6.0 | 8.75 | 11.25 | 15.75 | 10.5 | 14.5 |
| | 250 | 7.25 | 10.75 | 6.75 | 9.5 | 12.0 | 17.25 | 11.25 | 16.0 |
| | 350 | 7.75 | 11.75 | 7.0 | 10.25 | 12.75 | 18.0 | 12.0 | 17.25 |
| 550 | 200 | 7.25 | 10.75 | 6.5 | 9.75 | 12.0 | 18.0 | 10.75 | 16.5 |
| | 350 | 8.25 | 13.0 | 7.25 | 10.75 | 13.25 | 21.0 | 12.0 | 18.0 |
| | 450 | 8.5 | 13.5 | 7.75 | 11.75 | 14.0 | 22.5 | 12.75 | 19.5 |
| 650 | 200 | 7.75 | 11.75 | 7.0 | 10.75 | 12.75 | 19.5 | 11.75 | 18.0 |
| | 350 | 8.5 | 13.5 | 7.75 | 12.0 | 14.0 | 22.5 | 12.75 | 19.5 |
| | 450 | 9.25 | 14.0 | 8.5 | 13.5 | 15.25 | 24.5 | 14.0 | 22.5 |

| 750 | 250 | 8.0 | 12.75 | 7.5 | 11.5 | 13.5 | 21.25 | 12.5 | 19.25 |
|---|---|---|---|---|---|---|---|---|---|
| | 450 | 9.0 | 13.75 | 8.25 | 13.0 | 15.0 | 24.0 | 13.75 | 21.5 |
| | 650 | 10.25 | 15.0 | 9.25 | 14.0 | 16.5 | 25.0 | 15.0 | 24.0 |
| 825 | 250 | 8.5 | 13.5 | 7.75 | 12.0 | 14.25 | 22.5 | 13.0 | 20.0 |
| | 450 | 9.5 | 14.25 | 8.75 | 13.5 | 15.75 | 24.0 | 14.5 | 22.25 |
| | 650 | 10.75 | 15.75 | 9.75 | 15.0 | 17.25 | 26.25 | 15.75 | 24.0 |
| 900 | 250 | | | 8.25 | 12.5 | | | 13.75 | 21.0 |
| | 450 | | | 9.25 | 14.0 | | | 15.25 | 23.5 |
| | 750 | | | 10.25 | 15.0 | | | 16.5 | 25.5 |
| 1050 | 250 | | | 8.75 | 13.5 | | | 14.75 | 22.0 |
| | 550 | | | 10.0 | 15.0 | | | 16.75 | 25.0 |
| | 825 | | | 11.0 | 16.5 | | | 18.25 | 27.5 |
| 1175 | 250 | | | 9.25 | 14.0 | | | 15.5 | 23.0 |
| | 550 | | | 10.5 | 15.75 | | | 17.5 | 25.5 |
| | 900 | | | 12.0 | 17.5 | | | 19.5 | 29.0 |
| 1300 | 250 | | | 9.75 | 14.5 | | | 16.25 | 24.0 |
| | 550 | | | 11.25 | 17.0 | | | 18.75 | 27.0 |
| | 1050 | | | 12 | 18.25 | | | 20.75 | 30.5 |

(*Permission of Westinghouse Corp.*)

# Appendix III

## Typical Overhead Line Parameters

**Table A.3.1** Overhead-line parameters—50 Hz (British)

| Parameter | 275 kV | | 400 kV | |
|---|---|---|---|---|
| | $2 \times 113\,\text{mm}^2$ | $2 \times 258\,\text{mm}^2$ | $2 \times 258\,\text{mm}^2$ | $4 \times 258\,\text{mm}^2$ |
| $Z_1\,\Omega/\text{km}$ | $0.09 + j0.317$ | $0.04 + j0.319$ | $0.04 + j0.33$ | $0.02 + j0.28$ |
| $Z_0\,\Omega/\text{km}$ | $0.2 + j0.87$ | $0.14 + j0.862$ | $0.146 + j0.862$ | $0.104 + j0.793$ |
| $Z_{mo}\,\Omega/\text{km}$ | $0.114 + j0.487$ | $0.108 + j0.462$ | $0.108 + j0.45$ | $0.085 + j0.425$ |
| $Z_P\,\Omega/\text{km}$ | $0.127 + j0.5$ | $0.072 + j0.5$ | $0.075 + j0.507$ | $0.048 + j0.45$ |
| $Z_{pp}\,\Omega/\text{km}$ | $0.038 + j0.183$ | $0.033 + j0.182$ | $0.035 + j0.177$ | $0.028 + j0.172$ |
| $B_1\,\mu\text{mho}/\text{km}$ | 3.60 | 3.65 | 3.53 | 4.10 |
| $B_0\,\mu\text{mho}/\text{km}$ | 2.00 | 2.00 | 2.00 | 2.32 |
| $B_{m0}\,\mu\text{mho}/\text{km}$ | 5.94 | 7.00 | 7.75 | 8.50 |

$Z_1$ Positive-sequence impedance.
$Z_0$ Zero-sequence impedance of a d.c. 1 line.
$Z_{m0}$ Zero-sequence mutual impedance between circuits.
$Z_p$ Self-impedance of one phase with earth return.
$Z_{pp}$ Mutual impedance between phases with earth return.
$B_1$ Positive-sequence susceptance.
$B_0$ Zero-sequence susceptance of a d.c. 1 line.
$B_{m0}$ Zero-sequence mutual susceptance between circuits.
Note: A d.c. 1 line refers to one circuit of a double-circuit line in which the other circuit is open at both ends.
The areas quoted are copper equivalent values based on $0.4\,\text{in}^2$ and $0.175\,\text{in}^2$.

**Table A.3.2** Overhead line data—a.c. (60 Hz) and d.c. lines

| | 500–550 kV—a.c. | | 700–750 kV | h.v.d.c. | |
| --- | --- | --- | --- | --- | --- |
| Region: | Pacific | Canada | Canada | Mountain | Pacific |
| Utility: | So. California Edison Co. | Ontario Hydro | Quebec Hydroelectric Co. | U.S. Bureau of Reclamation | Los Angeles Dept. of W. & P. |
| Line name or number | Lugo–Eldorado | Pinard–Hanmer | Manics.–Levis | Oregon–Mead | Dalles–Sylmar |
| Voltage (nominal), kV; a.c. or d.c. | 500; a.c. | 500; a.c. | 735; a.c. | 750 (±375); d.c. | 750 (±375); d.c. |
| Length of line, miles | 176 | 228 | 235.92 | 560 | 560 |
| Originates at | Lugo sub | Pinard TS | Manicouagan | Oregon border | Oregon border |
| Terminates at | Eldorado sub | Hanmer TS | Levis sub | Mead sub | near, Los Angeles |
| Year of construction | 1967–69 | 1961–63 | 1964–65 | 1966–70 | 1969 |
| Normal rating/cct, MVA | 1000 | — | 1700 | 1350 mW | 1350 mW |
| STRUCTURES | | | | | |
| (S)teel, (W)ood, (A)lum | S | S, A | S | S | S or A |
| Average number/mile | 4.21 | 3.7 | 3.8 | 4.5 | 4.5 |
| Type: (S)q, (H), guy (V) or (Y) | S (S-56) | V (A-51, S-51) | –; (S-71) | S, T (S-72, 73) | S, T (S-72, 73) |
| Min 60-Hz flashover, kV | 850 | 1110 | — | — | 915 |
| Number of circuits: Initial; Ultimate | 1; 1 | 1; 1 | 1; 1 | 1; 1 | 1; 1 |
| Crossarms (S)teel, (W)ood, (A)lum | — | S, A | S | S | S or A |
| Bracing (S)teel, (W)ood, (A)lum | — | — | S | S | S or A |
| Average weight/structure, lb | 19 515 | 10 800; 4730 | 67 3000 | 11 000 | — |
| Insulation in guys (W)ood, (P)orc, kV | No guys | Nil | — | None | — |
| CONDUCTORS | | | | | |
| Al, ACSR, ACAR, AAAC, 5005 | ACSR 84/19 | ACSR 18/7 | ACSR 42/7 | ACSR 96/19 | ACSR 84/19 |
| Diameter, in.; MCM | 1.762; 2156 | 0.9; 583 | –; 1361 | –; 1857 | 1.82, 2300 |
| Weight 1 cond./ft, lb | 2.511 | 0.615 | 1.468 | 2.957 | 2.68 |
| Number/phase; Bundle spacing, in | 2; 18 | 4; 18 | 4; 18 | 2; 16 | 2; 18 |

**Table A.3.2**  (Continued)

| Parameter | Col 1 | Col 2 | Col 3 | Col 4 | Col 5 |
|---|---|---|---|---|---|
| Spacer (R)igid, (S)ring, (P)reform | S | R | R | S or P | — |
| Designed for—A/phase | 2400 | 2500 | — | 1800 | 3380 |
| Ph. config. (V)ertical, (H)orizontal, (T)riangular | H | H | H | H | H |
| Cond. offset from vertical, ft | — | — | — | — | — |
| Phase separation, ft | 32 | 40 | 50 | 38–41 | 39–40 |
| Span: Normal, ft; Maximum, ft | 1500; 2764 | 1400; 2550 | 1400; 2800 | 1150; 1800 | 1175; 1800 |
| Final sag, ft; at °F; Tension, 10³ lb | 46 at 130F; 19.1 | 45.0 at 60F; 3.4 | 226 at 120F; 6.2 | 40 at 120F; 12.4 | 35 at 60F; 13.0 |
| Minimum clearance: Ground, ft; Structure, ft | 40; 12.5 | 33; 10.5 | 45; 18.3 | 35; 7.75 | 35; 7.8 |
| Type armour at clamps | Performed | Nil | None | — | None |
| Type vibration dampers | Stockbridge | Spacer damper | Tuned spacer | — | Stockbridge |
| Line altitude range, ft | 1105–4896 | 800–1300 | 50–2500 | 2000–7000 | 400–8000 |
| Max corona loss, kW/three-ph. mile | 5 | — | 74 ($\frac{1}{2}$ in. snow/h) | — | — |
| **INSULATION** | | | | | |
| Basic impulse level, kV | 2080 | 1800 | — | — | — |
| Tangent: (D)isk or (L)ine-post | D | D | D | D | D |
| Number strings in (V) or (P)arallel | 2V | 1 or 2P | 4V | Single | Single |
| Number units; Size; Strength, 10³ lb | 27; $5\frac{3}{4} \times 10$; 30 | 23; $5\frac{3}{4} \times 10$; 25 | 35; $5 \times 10$; 15 | —; —; — | —; —; — |
| Angle: (D)isk or (L)ine-post | D | D | D | D | D |
| Number strings in (V) or (P)arallel | 2V | 3P | 4P | Single | — |
| Number units; Size; Strength, 10³ lb | 25; $6\frac{1}{4} \times 10\frac{3}{8}$; 40 | 26; $5\frac{3}{4} \times 10$; 25 | 35; $6\frac{1}{4} \times 10$; 36 | —; —; — | —; —; — |
| Insulators for struts | None | None | — | — | None |
| Terminations: (D)isk or (L)ine-post | D | D | — | D | D |
| Number strings in (P)arallel | 2P | 3P | — | — | 4P |
| Number units; Size; Strength, 10³ lb | 25; $6\frac{1}{4} \times 10\frac{3}{8}$; 40 | 26; $5\frac{3}{4} \times 10$; 25 | —; —; — | —; —; — | —; —; — |
| BIL reduction, kV or steps | 3.5 steps | — | — | — | — |
| BIL of terminal apparatus, kV | 1150–1425 | 1675–1900 | 2050–2200 | — | 1300 |
| **PROTECTION** | | | | | |
| Number shield wires; Metal and Size | 2; 7 No. 6 AW | 2; $\frac{5}{16}$ in. galv. | 2; $\frac{7}{16}$ in. galv. | 1; — | 2; — |

**Table A.3.2**   (Continued)

| | 500–550 kV—a.c. | | 700–750 kV | h.v.d.c. | |
| Region: | Pacific | Canada | Canada | Mountain | Pacific |
| Utility: | So. California Edison Co. | Ontario Hydro | Quebec Hydroelectric Co. | U.S. Bureau of Reclamation | Los Angeles Dept. of W. & P. |
|---|---|---|---|---|---|
| Shield angle, deg; Span clear, ft | 24.5; 28 | 20; 48 | 20; 50 | 30;– | 20;– |
| Ground resistance range, $\Omega$ | – | 15 | – | – | 30 max. |
| Counterpose: (L)inear, (C)rowfoot | C | L | L | – | – |
| Neutral gdg: (S)ol, (T)sf; | S; 0 | S;– | S;– | S;– | S;– |
| Arrester rating, kV; Horn gap, in | 420; None | 480/432; None | 636; – | –; – | –; – |
| Arc ring diameter, in.: Top; Bottom | None | None | –; – | –; – | –; – |
| Expected outages/100 miles/year | 0.5 | 1 | – | – | 0.2 |
| Type relaying | Ph. compar., carr. | Dir'l compar. | – | – | – |
| Breaker time, cyc; Reclos, cycles | 2; None | 3; None | 6; 24 | –; – | –; – |
| VOLTAGE REGULATION | | | | | |
| Nominal, % | ±10 | – | – | Variable | Grid control |
| Synch. cond. MVA; Spacing, miles | None | None | –; – | –; – | None |
| Shunt capac. MVA; Spacing, miles | None | None | 330; 236 | –; – | On a.c. terminals |
| Series capac. MVA; Compens., % | None | None | None | –; – | –; – |
| Shunt reactor, MVA; Spacing, miles | –; – | – | – | –; – | – |
| COMMUNICATION | | | | | |
| (T)elephone circuits on structures | None | None | None | None | None |
| (C)arrier; Frequency, kHz | C; – | C; 50–98 | C; 42.0 and 50.0 | None | None |
| (P)ilot wire | None | – | P | None | None |
| (M)icrowave; Frequency, MHz | M; 6700 | –; – | M; 777.35 | M; – | – |

*(Permission of Edison Electric Institue)*

# Bibliography and Further Reading

These references are provided for the advanced study of electric power systems or for further in-depth study as required for planning, operational, and costing purposes. A key is provided to indicate the most important books or articles that should be looked at first if seeking more information or historic developments on topics in the denoted chapters. All books and papers will have more references to consult.

| Books | Chapter |
|---|---|

Adkins, B. and Harley, R., *General Theory of Alternating Current Machines*, Chapman and Hall, London, 1978.  —  3 †

Anderson, P. M. and Fouad, A. A., *Power System Control and Stability*, Iowa State University Press, Ames, Iowa, 1977.  —  8 †

Arrillaga, J., *High Voltage Direct Current Transmission*, Peter Peregrinus on behalf of the Institution of Electrical Engineers, UK, 1983.  —  9 †

Arrillaga, J. and Arnold, C. P., *Computer Analysis of Power Systems*, Wiley, Chichester, 1990.  —  6 †

Berrie, T. W., *Power System Economics*, Peter Peregrinus on behalf of the Institution of Electrical Engineers, UK, 1990.  —  12 *

Berrie, T. W., *Electricity Economics and Planning*, Peter Peregrinus, on behalf of the Institution of Electrical Engineers, UK, 1992.  —  12 *

Bewley, L. W., *Travelling Waves on Transmission Systems*, Dover Books, New York, 1961  —  10 †

---

Key: † worth consulting for advanced study; * read for background knowledge; ** classic book or article

# Books                                                      Chapter

Bickford, J. P., Mullineux, N. and Reed, J. R., *Computation of Power System Transients*, Peter Peregrinus on behalf of the Institution of Electrical Engineers, UK, 1976.    10 †

*British Electricity International: Modern Power Station Practice*, Vols L and K, Pergamon, Oxford, 1991.    1,10 †

Debs, A. S., *Modern Power System Control and Operation*, Kluwer Academic, Boston, 1988.    4,5 *

*E.H.V. Transmission Line Reference Book*, Edison Electric Institute, New York, 1968.    1,10 †

El Abiad, A. H. and Stagg, G. W., *Computer Methods in Power System Analysis*, McGraw-Hill, New York, 1968.    6,7 †**

*Electrical Transmission and Distribution Reference Book*, Westinghouse Electric Corporation, East Pittsburgh, Pennsylvannia, 1964.    1,10 †

Electricity Council, *Power System Protection*, 3 vols, Peter Peregrinus on behalf of the Institution of Electrical Engineers, UK, 1991.    11 †

*Electricity Supply Handbook*, published by *Electrical Review*, yearly (March).    1 *

Elgerd, O. I., *Electric-Energy Systems Theory—An Introduction*, 2nd edn, McGraw-Hill, 1983.    3,4,6,7,8 *

Fitzgerald, A. E., Kingsley, C. and Umans, S., *Electric Machinery*, 5th edn, McGraw-Hill, 1990.    3 *

Flurscheim, C. H. (Ed.), *Power Circuit Breaker Theory and Design*, Peter Peregrinus on behalf of the Institution of Electrical Engineers, UK, 1987.    10 †

Fouad, A. A. and Vittal, V., *Power System Transient Stability Analysis Using the Transient Energy Function Method*, Prentice Hall, Englewood Cliffs, New Jersey, 1992.    8 †

Giles, R. L., *Layout of E.H.V. Substations*, Cambridge University Press, Cambridge, 1970.    11 †

Glover, J. D. and Sarma, M., *Power System Analysis and Design*, 2nd edn, PWS, Boston, Massachusetts, 1994.    6,7 *

Gonen, T., *Electric Power Distribution System Engineering*, McGraw-Hill, 1986.    1,3 *

Grainger, J. J. and Stevenson, W. D., Jr, *Power System Analysis*, McGraw-Hill International, 1994.    3,4,6,7,8 *

Greenwood, A., *Electrical Transients in Power Systems*, Wiley-Interscience, 1971.    10 †

Gross, C. A., *Power System Analysis*, 2nd edn, Wiley, New York, 1986.    2,6 †

Hindmarsh, J., *Electric Machines*, 4th edn, Pergamon, London, 1984.    3 *

Hunt, S. and Shuttleworth, G., *Competition and Choice in Electricity*, Wiley, Chichester, 1996.    12 †

---

Key: † worth consulting for advanced study; * read for background knowledge; ** classic book or article

# Books                                                    Chapter

Johns, A. T. and Salman, S. K., *Digital Protection for Power Systems*, Peter Peregrinus on behalf of the Institution of Electrical Engineers, UK, 1996.                                                                 11 *

Kirchmayer, L. K., *Economic Operation of Power Systems*, Wiley, New York, 1958.                                                                   4 **

Kundur, P., *Power System Stability and Control*, McGraw-Hill, New York, 1994.                                                               4,5,8 †

Lakervi, E.and Holmes, E. J., *Electricity Distribution Network Design*, Peter Peregrinus on behalf of the Institution of Electrical Engineers, UK, 1995.                                                                   1 †

Lander, C. W., *Power Electronics*, 3rd edn, McGraw-Hill, 1996.        9 †

Looms, J. S. T., *Insulators for High Voltages*, Peter Peregrinus on behalf of the Institution of Electrical Engineers, UK, 1990.                    10 *

Machowski, J., Bialek, J. W., and Bumby, J. R., *Power System Dynamics and Stability*, Wiley, Chichester, 1997.                            8 †

Munasinghe, M., *Electric Power Economics*, Butterworths, Oxford, 1990.                                                                          12 †

Pai, M. A., *Power System Stability. Analysis by the Direct Method of Lyapunov*, North Holland Publishing, Amsterdam, 1981.                     8 †

Pai, M. A., *Energy Function Analysis for Power System Stability*, Kluwer Academic, Boston, Massachusetts, 1989.                                   8 †

Pavella, M. and Murthy, P. J., *Transient Stability of Power Systems, Theory and Practice*, Wiley, Chichester, 1994.                              8 †

Phadke, A. G. and Thorpe, J. S., *Computer Relaying for Power Systems*, Wiley, Chichester, 1988.                                                11 †

Rudenberg, R. *Transient Performance of Electrical Power Systems*, MIT Press, Boston, 1969.                                                     10 †

Say, M. G., *Alternating Current Machines*, 4th edn, Pitman, London, 1976.                                                                       3 **

Schweppe, F., Caramanis, M. C., Tabors, R. D. and Bohn, R. E., *Spot Pricing of Electricity*, Kluwer Academic, Boston, Massachusetts, 1988.                                                                       12 †**

Sterling, M. H., *Power System Control*, Peter Peregrinus on behalf of the Institution of Electrical Engineers, UK, 1978.                        4,5 *

Stoll, H. E. (Ed.), *Least-Cost Electric Utility Planning*, Wiley, New York, 1989.                                                              12 †

Sullivan, R. L., *Power System Planning, McGraw-Hill*, 1977.           1 *

Turvey, R. H. and Anderson, D., *Electricity Economics*, Johns Hopkins University Press, Baltimore, Maryland, 1977.                             12 **

Weedy, B. M. *Underground Transmission of Electric Power*, Wiley, Chichester, 1980.                                                              3 *

Key: † worth consulting for advanced study; * read for background knowledge; ** classic book or article

## Books                                                           Chapter

Wood, A. J. and Wollenberg, B. F., *Power Generation, Operation and Control*, 2nd edn, Wiley, New York, 1996.                          4,5 *
Yao-nan, Yu, *Electric Power System Dynamics*, Academic Press, New York, 1983.                                                        4 †


## Articles                                                        Chapter

Adair, E. R., 'Electrophobia', *I.E.E.E. Power Engineering Review*, **December** (1996), 3–8.                                           1 *
Annestrand, S., "Bonneville Power Administration—Prototype 1100/ 1200 kV line', *Trans. I.E.E.E.*, **PAS-96** (1977), 357–366.         1,4 **
Barthold, L. O. and Carter, G. K., 'Digital travelling wave solutions, 1—single phase equivalents', *Trans. A.I.E.E.*, **80** (1961).  10 †
Bickford, J. P. and Doepel, P. S., 'Calculation of switching transients with particular reference to line energisation', *Proc. I.E.E.*, **111** (1967).                                                           10 †
Booth, E. S., *et al.*, 'The 400 kV grid system for England and Wales', *Proc. I.E.E.*, **9A** (1962).                                  1 *
Brown, H. E., Person, C. E., Kirchmayer, L. E. and Stagg, G. W., 'Digital calculation of three-phase short circuits by matrix method', *Trans. A.I.E.E.*, **79**, Pt III (1960), 1277.                         6 **
Brown, H. E., Carter, G. K., Happ, H. H. and Person, C. E., 'Power flow solution by impedance matrix iterative method', *Trans. I.E.E.E.*, **PAS-82** (1963), 1.                                             6 *
Carpentier, J., Contribution à l'étude du dispatching economique', *Bulletin de la Société Française des Electriciens*, Ser. 8, **3** (1963).                                                             4 †**
Carson, J. R., 'Wave propagation of overhead wires with ground returns', *Bell System Tech. J.*, **5** (1926), 539.                    10 **
Casson, W. and Sheppard, H. J., 'Technical and economic aspects of the supply of reactive power in England and Wales', *Proc. I.E.E.*, **108A** (1961).                                                         3 **
Cory, B. J. and Lewis, P. (Eds), 'The Reorganisation of the Electricity Supply Industry—A Critical Review, *Power Engineering Journal I.E.E.*, **April** (1997).                                             12 *
Dopazo, J. F., Klitin, O. A., Stagg, G. W. and Van Slyck, L. S., 'State calculation of power systems from line flow measurement', *Trans. I.E.E.E.*, **PAS-89** (1970), 1698–1708.                          11 *
Dunnett, R. M., Calviou, M. C. and Plumptre, P., 'Charging for the use of transmission by marginal cost methods', Presented at 14th Power Systems Computation Conference (P.S.C.C.), Avignon, France, August 1993.                                                      12 †

Key: † worth consulting for advanced study; * read for background knowledge; ** classic book or article

## Articles                                                                Chapter

Farmer, E. D., Cory, B. J. and Perera, B. L. P., Optimal pricing of
transmission and distribution services in electricity supply', *Proc.
I.E.E.: Generation, Transmission and Distribution*, **142**, Pt 1 (1996),
1–8.                                                                        12 *

Fortescue, C. L., 'Method of symmetrical co-ordinates applied to the
solution of polyphase networks', *Trans. A.I.E.E.*, **37**, Pt II (1918).   7 **

Friedlander, E., 'Transient reactance effects in static shunt reactive compensators for long a.c. lines', *I.E.E.E. Trans.*, **PAS-95** (1976), 1669.   5 *

Glimn, A. F. and Stagg, G. W., 'Automatic calculation of load flows',
*Trans. A.I.E.E.*, **76**, Pt III (1957), 817.                              6 **

Gyugyi, L., *et al.*, 'Principles and application of static thyristor-controlled shunt compensation, *I.E.E.E. Trans.*, **PAS-97** (1978) 1935.      5,9 *

Ham, P. A. L., 'Electronics in the control of turbine generators', *I.E.E.
Electronics & Power*, **May** (1978), 365–369.                              4 †

Handschin, E., Schweppe, F. C., Kohlar, J. and Fiechter, A., 'Bad data
analysis for power system state estimation', *Trans. I.E.E.E.*, **PAS-94**
(1975), 329–337.                                                            11 †

Happ, H. H., 'Optimal power dispatch—a comprehensive survey',
*I.E. Trans.*, **PAS-96** (1977), 841–853.                                  4 *

I.E.E.E. Committee Report, 'Computer representation of excitation systems', *Trans. I.E.E.E.*, **PAS-87** (1968) 1460–1464.                      5 †

I.E.E.E. Committee Report, 'Dynamic models for steam and hydrotur-
bine in power system studies', *Trans. I.E.E.E.*, **PAS-92** (1973), 1904–
1915.                                                                       4 †

I.E.E.E. Committee Report, 'Excitation system models for power system stability studies', *Trans. I.E.E.E.*, **PAS-100**, No. 2 (1981), 494–
509.                                                                        5 †

I.E.E.E. Task Force on Load Representation for Dynamic
Performance: Standard load models for power flow and dynamic
performance simulation', *Trans. I.E.E.E.*, **PWRS-10**, No. 3 (1995),
1302–1312.                                                                  3 †

Kidd, W. L., 'Load and frequency control', *Elec. J.*, London, **1961**.    4 **

Laughton, M. A. and Davies, M. W. H., 'Numerical techniques in
solution of power system load-flow problems', *Proc. I.E.E.*, **111**,
No. 9 (1964).                                                               6 **

Mann, B. J. and Morrison, I. F., 'Digital calculation of impedance for
transmission line protection', *I.E.E.E. Trans.*, **PAS-90** (1971), 270–
279.                                                                        11 **

Miles, J. G., 'Analysis of overall stability of multimachine power systems', *Proc. I.E.E.*, **109A** (1961), 203.                                8 **

Milne, A. G. and Maltby, J. H., 'An integrated system of metropolitan
electricity supply', *Proc. I.E.E.*, **114** (1967), 745–755.               1 *

---

Key: † worth consulting for advanced study; * read for background knowledge; ** classic book or
article

## Articles                                                                  Chapter

Moran, F., 'Power system automatic frequency control techniques',
   *Proc. I.E.E.*, **106A** (1959).                                              4 *
Paris, L., 'Influence of air-gap characteristics on line-to-ground switch-
   ing surge strength', *Trans. I.E.E.E.*, **PAS-86** (1967), 936.             10 †
Paris, L., 'The future of U.H.V. transmission lines', *I.E.E.E. Spectrum*,
   **44** (1969).                                                              10 **
Park, R. H. and Robinson, P. H., 'The reactance of synchronous
   machines', *Trans A.I.E.E.*, **47** (1928).                                  3 **
Ramsey, F. P., 'A mathematical theory of saving', *Economic Journal*,
   **38**, (1928), 543–559.                                                    12 **
Robert, R., Micromachines and microreseaux', *C.I.G.R.E.*, Paris, **1950**,
   paper 338.                                                                   8 **
Rockefeller, G. D., 'Fault protection with a digital computer', *I.E.E.E.
   Trans.* **PAS-88** (1969) 438–461.                                          11 **
Sasson, A. M., 'Non-linear programming solutions for load-flow, mini-
   mum loss and economic dispatching problems', *I.E.E.E. Trans.*,
   **PAS-88** (1969), 399.                                                      6 *
Sato, N. and Tinney, W. F., 'Techniques for exploiting the sparcity of
   the network admittance matrix', *Trans. I.E.E.E.*, **PAS-82** (1963), 944.   6 †**
Schmill, J. V., 'Optimum size and location of shunt capacitors on dis-
   tribution feeders', *I.E.E.E. Trans.*, **PAS-84** (1965), 825.              3 *
Schweppe, F. C. and Handschin, E., 'Static state estimation in power
   systems', *Proc. I.E.E.E.*, **62** (1974).                                  11 **
Shakeshaft, G., 'General purpose turbo-alternator model', *Proc. I.E.E.*,
   **110** (1963), 703–713.                                                     3 **
Stott, B., 'Review of load flow calculation methods', *Proc. I.E.E.E.*, **62**
   (1974), 916–929.                                                            6 **
Tinney, W. F. and Walker, J. W., 'Direct solutions of sparse network
   equations by optimally ordered triangular factorisation', *Proc.
   I.E.E.E.*, **55**, No. 11 (1967), 1801–1809.                                6 †**
Ward, J. B. and Hale, H. W., 'Digital computer solution of power flow
   problems', *Trans. A.I.E.E.*, **75** (1956), 111.                           6 **
Ward, J. B., 'Equivalent circuits for power flow studies', *A.I.E.E. Trans.*,
   **PAS-68** (1949), 373–382.                                                 6 **
Wells, D. W., 'Method for economic secure loading of a power system',
   *Proc. I.E.E.*, **115**, No. 8 (1968) 1190.                                 4 **
Wu, F. F., 'Fast decoupled load flow analysis', *Trans. I.E.E.E.*,
   **PAS-96** (1977), 268–275.                                                  6 †

Key: † worth consulting for advanced study; * read for background knowledge; ** classic book or article

# Index